Ecology of a Changed World

Front cover: Between 1970 and 2021, the number of people in the world doubled to nearly 8 billion and the number of chickens increased fivefold to 26 billion. The number of wild birds in North America declined by 30% (the illustration runs to 2017, when bird abundances were estimated). There may now be fewer birds in North America than people in the world. See Figures 11.2, 12.3, and 24.5. Illustration by Allison Johnson.

Ecology of a Changed World

TREVOR PRICE

University of Chicago, Chicago

Illustrated by

AVA RAINE

UNIVERSITY PRESS

Oxford University Press is a department of the University of Oxford. It furthers
the University's objective of excellence in research, scholarship, and education
by publishing worldwide. Oxford is a registered trade mark of Oxford University
Press in the UK and certain other countries.

Published in the United States of America by Oxford University Press
198 Madison Avenue, New York, NY 10016, United States of America.

© Oxford University Press 2022

All rights reserved. No part of this publication may be reproduced, stored in
a retrieval system, or transmitted, in any form or by any means, without the
prior permission in writing of Oxford University Press, or as expressly permitted
by law, by license, or under terms agreed with the appropriate reproduction
rights organization. Inquiries concerning reproduction outside the scope of the
above should be sent to the Rights Department, Oxford University Press, at the
address above.

You must not circulate this work in any other form
and you must impose this same condition on any acquirer.

Library of Congress Cataloging-in-Publication Data
Names: Price, Trevor, 1953– author.
Title: Ecology of a changed world / Trevor Price ; illustrated by Ava Raine.
Description: New York : Oxford University Press, [2022] |
Includes bibliographical references and index.
Identifiers: LCCN 2021029388 (print) | LCCN 2021029389 (ebook) |
ISBN 9780197564172 (hardback) | ISBN 9780197564196 (epub)
Subjects: LCSH: Ecology. | Global environmental change. |
Biodiversity—Environmental aspects.
Classification: LCC QH541.P735 2021 (print) |
LCC QH541 (ebook) | DDC 577—dc23
LC record available at https://lccn.loc.gov/2021029388
LC ebook record available at https://lccn.loc.gov/2021029389

DOI: 10.1093/oso/9780197564172.001.0001

Royalties for this book will be donated to savingnature.org

Contents

Preface	vii
Acknowledgments	ix
About the Companion Website	xi

1. The Changed World	1

PART 1. THE RISE AND FALL OF POPULATIONS

2. Population Growth	9
3. Population Regulation	21
4. Interactions between Species: Mutualisms and Competition	31
5. Predation and Food Webs	43
6. Parasites and Pathogens	52
7. Evolution and Disease	64
8. The Human Food Supply: Competition, Predation, and Parasitism	75
9. Food Security	83

PART 2. THE THREATS TO BIODIVERSITY

10. Prediction	97
11. Human Population Growth	110
12. Growth of Wealth and Urbanization	121
13. Habitat Conversion	132
14. Economics of Habitat Conversion	143
15. Climate Crisis: History	155
16. Predictions of Future Climate and its Effects	166
17. Pollution	176
18. Invasive Species	184

vi CONTENTS

19. Introduced Disease	194
20. Harvesting on Land	205
21. Harvesting in the Ocean	215
22. Harvesting: Prospects	226

PART 3. AVERTING EXTINCTIONS

23. Species	239
24. Population Declines	248
25. Extinction	261
26. Species across Space	273
27. Island Biogeography and Reserve Design	283
28. The Value of Species	294

Notes and References	303
Subject Index	331
Species Index	337

Preface

This book has been updated over 15 years in order to keep up with the rapid changes of the twenty-first century, and it will continue to be updated through an associated website, which also contains appendices to chapters, and worked examples. The coronavirus pandemic, the Black Lives Matter movement, exceptional fires and heatwaves exmplify the changes and the challenges of our times. A book on the science behind the biodiversity crisis has much to say about the pandemic and climate change. The SARS CoV-2 virus crossed into humans most likely from the hunting of bats for food. So, we might ask, what would happen if there were no bats in the world? The science is clear. Every time a species is lost, others become more common, making new disease transmission and virulence even more likely. We can see this in the rapid spread of coronavirus through the dense human population. We can also observe more directly effects of bat loss in North America, where white nose disease inadvertently introduced from Europe in 2006 has killed millions of bats, which in turn has been linked to increased insecticide use by farmers attempting to combat the insect pests the bats would otherwise have eaten. This book moves beyond specific examples and beyond disease. We want to quantify biodiversity loss and the consequences for human well-being as we go forward.

What about Black Lives Matter? As sports writer Barney Ronay put it: "so much unhappiness is created, so much talent is lost, so many people who should be doing things and have opportunities to do those things, don't receive those opportunities." (Complete citations are in the references section at the end of the book.) What can we say about connections between these injustices and conservation? Asymmetries and inequalities lie beyond race. They include gender, sexual orientation, disability, caste, religion, nationality, and wealth. Such inequalities are not only morally indefensible, but contribute to the crisis of nature. As a middle-class white male, my concern with the natural world comes from privilege and past experience. Others have not had this same fortune, and one feels that nothing but good could come out of more such opportunity.

In this book I focus on one important inequality: that of wealth. Aside from the other benefits of having money, many are not able to buy enough food, despite there being more than enough food now produced to feed everyone. Like other forms of inequality, poverty is not something I have personally had to deal with, but it is something I have witnessed firsthand, working in India. Wealth disparities impact conservation greatly, from the direct effects of being poor

viii PREFACE

(e.g., it leads people to hunt bats) to the more general lack of opportunity that is associated with all types of discrimination. Within countries, wealth inequality continues to increase, but the economic growth of Asian and South American countries has meant that across the world inequality has been decreasing, at least until the recent economic downturn. On average, people have been becoming richer and healthier, and we hope this trend will pick up again soon. Such welcome changes have huge implications for the conservation of biodiversity. These changes are covered in the book, and many of the consequences surely apply more generally to the mitigation of all social injustices.

Acknowledgments

Chris Andrews, Bettina Harr, Julia Weiss, and several anonymous reviewers read the whole book. Various pieces have been read by Erin Adams, Sarah Cobey, Ben Freeman, Peter Grant, Sean Gross (who made the compelling suggestion to delete a chapter), Rebia Khan, Kevin Lafferty, Karen Marchetti, Robert Martin, Natalia Piland, Yuvraj Pathak, Uma Ramakrishnan, Mark Ravinet, Matthew Schumm, David Wheatcroft, and anonymous reviewers. Many people have responded to requests for information, including David Archer, Sherri Dressel, Clinton Jenkins, David Gaveau, Kyle Hebert, David McGee, Loren McClenachan, Nate Mueller, Natalia Ocampo-Peñuela and Stuart Sandin. The copy editor, Betty Pesagno, made an excellent and thorough review. Kaustuv Roy has always been supportive and a great friend. I appreciated the R programming environment (citations are in the Notes and References section at the end of the book). I particularly wish to thank Angela Marroquin and Bettina Harr for much help with the figures and Ava Raine for her outstanding drawings.

The book is dedicated to local activists, who are at the frontline of conserving the planet's biodiversity, sometimes at considerable risk to themselves. They include Homero Gómez González and Raúl Hernández Romero, who were murdered in 2020, apparently by illegal loggers. Their work involved conserving the winter habitat of the emblematic monarch butterfly which migrates from the eastern United States to a few mountain tops in central Mexico. Not so long ago the monarch population numbered in the many hundreds of millions, but in the last 10 years, never more than 100 million (Chapter 24).

About the Companion Website

http://global.oup.com/us/companion.websites/9780197564172/ins_res/l0k-res/.

Oxford has created a website to accompany *Ecology of a Changed World*.

As the world continues to change, the site will be used to post regular updates to various chapters and figures.

In addition, the website carries study questions and their solutions, plus an appendix that covers the measurement of uncertainty and some mathematical derivations.

The reader is encouraged to consult this resource as a complement to the book.

1

The Changed World

About 66 million years ago, an asteroid hit what is now the Yucatan region of Mexico, setting off a major extinction of species on Earth. Remarkably, over the present century, the environment will change more rapidly than it has over any corresponding timespan since that devastating impact. Sixty-six million years is a long time, and the magnitude of the changes we will face in the near future is especially remarkable because these changes have been brought about entirely by humans. Some changes are for the better, at least for humanity: more people than ever are living longer, healthier lives, and standards of living in Asia and Latin America are improving at an impressive rate. Yet, the negative impacts of population growth and consumption on the environment are huge and may catch up with us sooner rather than later, making it crucial that we appreciate the basics of ecology in the context of what is happening. That is the purpose of this volume.

Mass extinctions in the past have been defined as periods of exceptionally high species loss (as indicated by the fossil record of mollusks on the continental shelf, which is more complete than that on land.) Scientists have traditionally recognized five such events as having happened in the past 600 million years, each characterized by the loss of at least 75% of all species. A mass extinction comparable to those of the big five could happen as soon as the next 100 years (see Chapter 25). Indeed, many important species have already been lost. The geological record indicates that recovery and diversification of species to pre–mass extinction levels can take as long as 5–10 million years. Further, the loss of a species means loss of its scientific, cultural, aesthetic, recreational, health, and economic value. But it is also likely to have far-reaching consequences for other species, which are connected to one another through various paths, including predation, parasitism, mutualism, and competition. Consequently, when a few species are lost, many others may follow or become dramatically reduced in numbers, and some may become common pests. Any mass extinction will undoubtedly have negative consequences for human quality of life and even life itself. We remain highly dependent on nature and its resources.

In a world out of balance, it is impossible to predict with much certainty what will happen. Scenarios range from small impacts that deprive people from visiting wild places, with associated economic impacts for tour operators, up to the spread of devastating diseases that eliminate us and/or our food. The recent arrival of the coronavirus in our midst is one case in point. It is unfortunate that so much work and

Ecology of a Changed World. Trevor Price, Oxford University Press. © Oxford University Press 2022.
DOI: 10.1093/oso/9780197564172.003.0001

2 ECOLOGY OF A CHANGED WORLD

research are reactionary, staving off pandemics and other crises as they appear rather than proactively focusing on preventing the imbalances that cause such problems.

The last mass extinction marked the beginning of the Cenozoic era. The Cenozoic era is divided into seven different epochs, each characterized by distinct kinds of animals and plants, as preserved in the fossil record. Transitions between epochs are driven by climate change and extinctions. The Pleistocene epoch, which began about 2.5 million years ago, saw the initiation of periods of cooling and glaciation interrupted by interglacial periods that experienced temperatures occasionally up to about 1°C higher than today (2°C higher than 100 years ago). The end of the last glacial period about 12,000 years ago was followed by the rise of agriculture and consequent increased human population density. This transformative change in earth's ecology initiated the seventh epoch of the Cenozoic, the Holocene.

The consensus is that we are entering an eighth epoch, again marked by rapid and large-scale environmental changes, termed the Anthropocene. People debate when we should date the start of this epoch. In keeping with the way past epochs are defined, the main criterion is that a characteristic feature in the fossil record should be widespread across the world, so hypothetical future intelligent life could map its beginning. One research group has placed the date precisely at 1610, which is associated with a rapid cooling of the earth, as recorded by oxygen isotopes (Chapter 15). This group somewhat controversially attributes that cooling to the death of 50 million humans in the Americas due to disease, followed by subsequent growth of forests, thereby removing the greenhouse gas CO_2 from of the atmosphere. Whether that is a plausible scenario is considered several times in this book. Others prefer 1950 as the date, which marked the start of what has become known as the "Great Acceleration" in development. For example, we manufacture more than 350 million tonnes (see Table 1.1 for conversion of units) of plastic and 4 billion tonnes of concrete a year, and these products were just becoming available in 1950. Because epochs are defined by a feature preserved in the rocks, the signature development defining the start of the Anthropocene in this case is the nuclear fallout associated with the atom bomb, its testing, and, in two cases, its detonation in population centers.

Today's loss of species differs from previous extinctions in at least four key ways. First, humans both intentionally and unintentionally transport organisms all over the world, mixing up communities and affecting the way species interact with each other (Europeans imported diseases that killed an estimated 50 million North and South Americans in the 1500s; see Chapter 19). Second, we have removed large animals, often predators, which has ramifying effects on other species. Third, changes are happening very quickly, and the threats are multiple, compounding each other. Fourth, humans usurp a large fraction of the earth's resources (see Fig. 1.1); left unchecked, few resources will be left for other species for the indefinite future. These topics will be revisited in detail in this book.

Table 1.1. Weights and Measures*

	Metric	Equivalents
Weights	Kilogram (kg)	2.2 pounds
	Tonne = 1000 kg	1.1 US tons (0.98 UK tonnes)
Distances	Kilometer (km)	0.62 miles (approximately 5/8 mile)
Area	Hectare (ha) = 10,000 m^2	2.2 acres
Volumes	Liter	2.11 US pints (1.76 UK pints)
Greenhouse gas	Carbon (tonnes)	Carbon dioxide/3.667 (tonnes)

*Metric is used throughout this text. Carbon emissions are sometimes reported in terms of CO$_2$ and sometimes in terms of carbon. Carbon is used in this book.

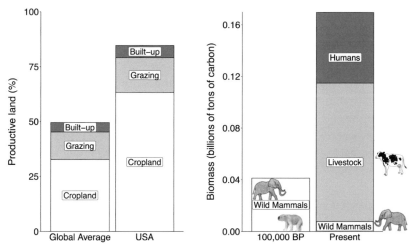

Figure 1.1 *Left:* An estimate of the proportion of the world's annual historical plant growth that is on land now used by humans (either to grow crops, graze domestic animals, or be used for cities and towns) and what the proportion would be if everyone had the same consumption pattern as the United States does at present, with no change in agricultural practices (see Chapter 13). *Right:* Estimated biomass of the world's mammals 100,000 years ago and at the present day. (Y. Bar-On, R. Phillips, and R. Milo. [2018]. The biomass distribution on earth. *Proceedings of the National Academy of Sciences* 115: 6506–6511). Images are those of the largest Australian mammal, *Diprotodon octatum*, which went extinct 44,000 years ago, and the African elephant.

4 ECOLOGY OF A CHANGED WORLD

In 2002, the naturalist E. O. Wilson published an important book, *The Future of Life*, in which he summarized the major threats to nature using the acronym HIPPO: Habitat loss, Invasive species, Pollution, Population sizes of humans affecting everything else, and Overharvesting. Presently, habitat loss is still thought to be the largest contributor to the endangerment of species, and it is clear that many species introduced from one place in the world to another have caused considerable harm in their new location. Harvesting has led to dramatic species declines and continues to do so, most notably in fisheries, but rather surprisingly in many tropical forests as well. As noted above and repeatedly emphasized throughout this book, the decline of species that humans hunt affects population sizes of many other species too.

Wilson briefly discussed but did not elaborate on three other ingredients that contribute to threats to nature. First, the problem is not just too many people. The human population is presently increasing at the rate of about 200,000 people per day and population size is the ultimate stressor, but rapid increases in consumption are at least as great a threat to the environment. Real income per person has increased by 50% since the year 2000, to an average of almost $11,000 per year in 2020 (this is an average, of course, and is skewed somewhat by the steep increases in income of the very wealthiest; see Chapter 12). Second, climate change has developed over the past 15 years into a full-blown emergency, aptly named the Climate Crisis. With a 1°C global rise over the past 100 years, heatwaves, droughts, floods, and unusual storms are already affecting humans directly, as well as indirectly through impacts on other species. Given that carbon emissions continue to grow, albeit with a slight downturn during the pandemic, and the amount of CO_2 in the atmosphere is higher than it has been for at least 2.5 million years and perhaps much longer, it is unclear how and if we can mitigate these effects. Third, disease outbreaks are having devastating impacts on many species, including our own. While many of these outbreaks are a result of transport of, for example, a virus, from one location where its effects are mild to one where they are debilitating (hence part of the general problem of invasive species), they also involve jumps between species.

To fully summarize the most pressing threats to our planet, we update Wilson's HIPPO acronym by adding a C for climate change and a D for disease: COPHID. COPHID is the organizing framework for this book, further modifying the earlier acronym by removing the P for population. This is done to emphasize that population and wealth are not the direct causes of environmental deterioration, but have instead created those direct threats. The organizing framework used here was originally introduced in a paper published in 2003, which asked why amphibians (frogs, toads, salamanders) had been declining steeply at least since the early 1980s.

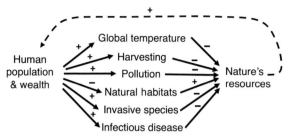

Figure 1.2 Paths of influence from human populations to the natural world. The six threats in the center correspond to Climate Change, Overharvesting, Pollution, Habitat loss, Invasive species, and Disease (COPHID). A positive sign (+) indicates that an increase in the factor at the base of the arrow causes an increase in the factor that the arrow points to (and also that a decrease causes a decrease). A minus sign means that an increase causes a decrease and a decrease causes an increase. To get the overall effect of human influences on natural resources through, for example, harvesting, one multiplies the values along the connecting paths, which in all cases is one "+" and one "–" so the product is negative (population increase causes harvesting to go up, which causes natural resources to go down). To calculate the total effect of human population size on species, one would then sum across all six connecting paths ([human → pollution → nature] + [human → habitat → nature], etc.). All paths are negative, so the sum must be negative. One can extend this figure by considering how each factor affects humans. The path from natural resources to humans is clearly positive. The summed path from humans to pollution (+) to nature (–) to and back to humans (+) is the product of +, –, +, which is negative. Such a negative feedback loop may be one ultimate control on human population size. In principle, one can place values on the arrows (e.g., a value of 0.1 on the arrow leading to harvesting would imply that a tenfold increase in human influence would cause harvesting to increase by a factor of one), but these values are typically uncertain and not constant.

Figure 1.2 depicts these various factors and their impacts as a path diagram in which the factors are connected by arrows. The sign of each arrow indicates how the factor at the arrowhead would change given a change in the magnitude of the factor at the base. A plus sign means the factor at the arrowhead changes in the same way as the one at the base (i.e., if the base factor increases, so does the head factor, and if the base factor decreases, so does the head factor). A minus sign means the factor at the arrowhead changes in the opposite direction (i.e., if the base increases, the head declines, and if the base decreases, the tip increases). Such diagrams are a useful way to model the world and are considered further in Chapter 4.

Rapid changes in a complex world make it very difficult to predict what will happen in the future. One possible and favorable outcome is that as standards

6 ECOLOGY OF A CHANGED WORLD

of living and educational opportunities improve, the trend for people to have smaller families will continue. The human population may then stop growing as early as midcentury and start to decline. It is also plausible that as standards of living rise and natural areas become scarcer, nature itself will become more highly valued, with a greater clamor for, and investment in, its preservation. In this view, the time we are living through now is a bottleneck for life. This reasoning led Wilson to declare in the *Future of Life* (p.189): "The central problem of the new century, I have argued, is how to raise the poor to a decent standard of living worldwide while preserving as much of the rest of life as possible." This statement summarizes the organization of the book. The first section (Chapters 2–9) considers ecological and evolutionary principles underlying the factors that cause populations to increase or decrease, whether they be humans or other species; the second section (Chapters 10–22) describes COPHID in more detail; and the third section (Chapters 23–28) explores the history of extinctions, and underlying principles of conservation biology that inform the goal of reducing future extinctions.

PART 1
THE RISE AND FALL OF POPULATIONS

2

Population Growth

A population is a collection of individuals belonging to one species, present in a specified geographical area. Sometimes the population refers to the whole species and sometimes to a subset. Population size is the number of individuals in the population. As of October 2021, the human population size was approaching 7.9 billion, and the population size of London, for example, was more than 9.3 million. Predicting how the numbers of both humans and other species will change in the future is essential to understanding environmental impacts. This chapter:

(1) Investigates simple models of population growth, which inevitably lead to either impossibly large population sizes or extinction. The word "model" is used to mean a mathematical description of an ecological process.
(2) Considers the value of a model in understanding the past and predicting the future.

2.1 Models of Population Growth

Fruit flies are used widely in labs across the world because they reproduce quickly, so it is possible to investigate genetic and evolutionary change in a few months or years. They are also useful for ethical reasons because no one is concerned about the scientist's daily slaughter of flies. A female fruit fly in a test tube with plenty of food lays about 100 eggs, which within 20 days themselves become 100 reproducing fruit flies. The 50 or so females among those 100 flies are themselves able to lay 100 eggs, if each is placed in its own test tube. I estimated that about 250 billion flies are needed to tightly pack into the buildings of a typical university (based on the volume of buildings and the number that can fit into a test tube). To cover the earth (land and sea) a mile (1.6 km) deep in flies requires perhaps 7,500,000,000,000,000,000,000,000 (7.5×10^{24}) flies, which is more than 1,000 times the number of stars in the universe. How long would it take our reproducing fruit flies to achieve these population sizes? The answer is that it would take about 4 months to fill the university and another 6 months to fill the world. A shortage of food means we have no need to shut down the fruit

Ecology of a Changed World. Trevor Price, Oxford University Press. © Oxford University Press 2022.
DOI: 10.1093/oso/9780197564172.003.0002

10 ECOLOGY OF A CHANGED WORLD

Table 2.1. Symbols Used.

N	Size of a population (number of individuals).
t	Time, often in years.
N_0	The initial size of a population, at time $t = 0$.
N_t	The size of a population t units later.
b	Birth rate, babies/individual/time interval.
d	Death rate, deaths/individual/time interval.
r	$= b-d$, *per capita* growth rate, for geometric or exponential growth. More generally, the intrinsic rate of increase (Chapter 3).
K	The number of individuals at carrying capacity (Chapter 3)

fly labs. Environmental resources, such as food and habitat, limit the growth of populations.

One might think that it would take much, much, longer to fill the world with fruit flies if food is unlimited. But consider the math. After one generation (20 days), 50 females are breeding. After two generations (40 days), 50×50 females are breeding. After three generations (60 days), 50^3 females are breeding. And after t generations 50^t females are breeding. Ten months is 15 generations. Enter 50^{15} into a spreadsheet and multiple by two to account for males, and you will get close to the result for filling the world.

We have built a model that describes the growth of the fly population:

$$N_t = N_0 \times 50^t$$

N_t is the population size (number of females) at time t in generations after the present day ($t = 0$) (see Table 2.1). N_0 is the number of females present at day 0, which in the above example was 1. It is immediately obvious that a model is often needed because in the fruit fly example, intuition can be misleading. A model is useful for at least two other important reasons:

(1) A model clearly states the assumptions. In the case of the flies, the assumptions were that a female consistently gives rise to 50 daughters and then dies, and that all these daughters survive to reproduce.

(2) If past changes do not match predictions, a model can be used to identify how the assumptions are violated. For example, the fact that fruit flies are not taking over the world must be either because females produce fewer

POPULATION GROWTH   11

daughters or because fewer daughters survive. In the case of the fruit flies, it is because few daughters survive.

To predict the number of females we would observe at some time in the future, given the present number, we only need know the number of daughters a female has. The number of surviving daughters is known as a parameter of the model, and we will give it the symbol B. Thus, we could write the equation above more generally:

$$N_t = N_0 \times B^t$$

This equation describes geometric growth. In the fruit fly example, the parameter, $B = 50$ females, but the equation would be true for all values of B. (N_0, the starting number of females, is also a parameter, but it is more often called the initial condition.) To make predictions, we need values for all parameters in the model. For example, if each female had $B = 6$ surviving daughters and we started with $N_0 = 2$ females, then after $t = 3$ generations, $N_3 = 2 \times 6^3 = 432$ females would be present, which again seems quite a large number for such innocuous production.

Birth and Death Rates

The fruit fly example illustrates the usefulness of a simple mathematical model. To make the model generally applicable, we separate change in the numbers of individuals over a certain time interval into those added (births) and those subtracted (deaths). The decomposition into change in the numbers of births and deaths enables us to evaluate much more easily why populations are increasing or decreasing. One parameter, the birth rate, b, is the average number of surviving babies produced by all individuals over a given time interval. (Table 2.1 lists all the symbols.) The birth rate for humans from July 1, 1992 to June 30, 1993 was $b = 0.0245$ persons/person/year (Table 2.2) and is often written as a percentage (i.e., 2.45%). While twins and early mortality affect the actual number of people who gave birth, it appears that about one in forty of all people had a baby in that year. This is surprising, given that about half of all people are male and many females were too young or too old to reproduce.

The second parameter, the death rate, d, is the number of individuals dying per individual present at the beginning of the time interval (equivalently, the proportion of all individuals that died over the time interval, or the probability of an individual dying). Over that same 1992–1993 interval, the death rate was $d = 0.009$ persons/person/year. That is, just under 1% of all people alive at the beginning of the year died (Table 2.2).

12 ECOLOGY OF A CHANGED WORLD

Table 2.2. World Human Population, 1992–1993.*

Population size	5.5 billion
Births	135 million
Birth rate, b	0.0245 babies/person/year
Deaths	50 million
Death rate, d	0.009 people/person/year

*Data are from July 1 to June 30 (United Nations, https://population.un.org/wpp). Note that these are estimates, and other databases differ slightly.

The number of babies present at the end of the year is $b \times N$, where N is the number of people present at the beginning. From mid-1992 to mid-1993, the number of babies born was 0.0245 x 5.5 billion = ~135 million. Over this same period, the number of individuals dying, $d \times N$, was ~50 million.

The change in population size (ΔN) per time interval (ΔT) is then.

$$\frac{\Delta N}{\Delta T} = bN - dN = (b - d) \times N$$

Between mid-1992 and mid-1993, the world added $\dfrac{\Delta N}{\Delta T}$ = individuals, more than twice the current population size of California.

For many applications, birth and death rates are commonly collapsed into a single parameter, $r = b - d$. Here, we define r as the per capita growth rate—the number of individuals added over the time interval as a fraction of the number present at the beginning. If $r > 0$, the birth rate exceeds the death rate, and the population increases. If $r < 0$, the death rate exceeds the birth rate, and the population decreases. The per capita growth rate is an important quantity, cropping up in statistical reports all the time, not only for population increases, but also for such things as increases in prices over years, growth in productivity, and declines in populations. In humans from 1992 to 1993, r is estimated as $b - d = 0.0245 - 0.009 = 0.0155$ individuals/individual/per year. As in the case of birth and death rates, in most reports, one sees r written as a percentage (i.e., the human population increased between 1992 and 1993 by $r = 1.55\%$).

The critical assumption of the model we now develop is that the birth and death rates do not change over time. If these rates do not change, we can project future human population size, just as we did in the fruit fly example, by multiplying the population size measured on July 1 each year by 1.55% (Fig. 2.1). As in that example, the plot shows geometric increase. The number of people added to

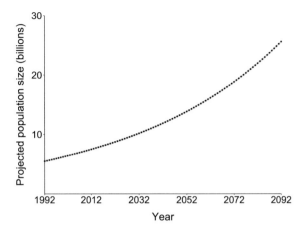

Figure 2.1 In the middle of 1992, the human population was estimated to be 5.5 billion, and it grew by 1.55% that year (Table 2.2). If the growth rate were to stay constant, future population sizes can be predicted by iteratively multiplying the population size by 1.55% (illustrated by the points), demonstrating a geometric increase. The smooth gray curve drawn through the points follows the exponential equation, $N_t = N_0 e^{rt}$ (Appendix). N_0 and N_t are two population sizes, separated by t years, and r is the *per capita* growth rate (individuals added/individual/year; see Figure 2.2).

the population is higher every year than it was in the preceding year. This is because at the beginning of every year there are more people reproducing than in the previous one. Given that the increase from 1992 to 1993 was 85 million, 5.585 billion people were present at the end of the year. To predict the increase over the next year, we assume that birth and death rates do not change and we multiply this number by 1.55%, giving a total of 86.6 million people added the following year. That process can be repeated iteratively to give projected population size through time, as in Figure 2.1.

Model of Exponential Growth

The census intervals in Figure 2.1 were taken every year on July 1, but of course many populations, including that of humans, change continuously. We can represent such continuous growth as a smooth curve drawn through the points (Fig. 2.1), which is then termed the exponential curve. The two types of curves describe exactly the same relationship, but whereas geometric growth measures population at equally spaced time intervals, exponential growth records

14   ECOLOGY OF A CHANGED WORLD

change continuously. To derive a formula for the exponential curve, bring the time points close together by writing:

$$\frac{\Delta N}{\Delta T} = rN$$

as

$$\frac{dN}{dT} = rN$$

The only difference is that the time interval is small. The formula for the exponential is then developed through integration (see the online Appendices for this chapter). It is as follows:

$$N_t = N_o e^{rt}$$

This formula is exceptionally useful because it can be used to predict population size at any time into the future. Here, e^{rt} means the number e (the irrational number 2.71828 . . .) raised to the power rt (the e part is what gives the curve the name "exponential"). If we have values for the present population size N_o and the per capita growth rate, r, we can plug these into the equation to predict population sizes in the future. For example, if the human population continued to grow at the 1992 rate of 1.55% per year, we would predict that population on July 1, 2020 would be $5.5 \times e^{0.0155*28} = 8.49$ billion, but we could also predict the population size on any other day. In this book, we will frequently refer to and employ the exponential curve. It is one of only two equations we consider (the other one, the logistic one, appears in Chapter 3). Neither equation is particularly intuitive, and both are worth spending some time thinking about (see the online questions).

The use of the exponential equation requires that we know the value of r. We could obtain r by measuring the population at two time points (e.g., the beginning and end of the year). Then we could rearrange the equation $N_t = N_o e^{rt}$, as follows:

$$r = \ln\left(\frac{N_t}{N_0}\right) / t$$

where ln symbolizes natural logarithm. Assuming that the population is truly growing exponentially (i.e., that birth and death rates do not change), all we would have to do is measure the population size at two points, take the log of

POPULATION GROWTH   15

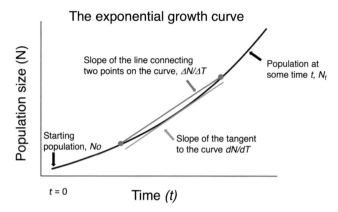

Figure 2.2 To model continuous growth, replace $\frac{\Delta N}{\Delta T} = rN$ in the text with $\frac{dN}{dT} = rN$. Here, $\frac{dN}{dT}$ is the slope of the tangent to the curve at a given population size (N) and is termed the instantaneous rate of population increase. It is the increase we would see if the rate at that particular point was fixed, which will underestimate the true increase because at every time step, more individuals are reproducing. We may approximate the tangent by measuring the difference in population sizes across a certain time interval $\frac{\Delta N}{\Delta T}$. The shorter the timescale over which we measure $\frac{\Delta N}{\Delta T}$, the more similar the slope is to the tangent. Provided r is measured over a timescale that the increase is less than 10%, the use of $\frac{\Delta N}{\Delta T}$ is a good approximation, but beyond that one should use the exact formula for r.

their ratio, and divide by the time separating the two points. This is what the United Nations does in its reports, despite presenting the data as if it had been calculated as the proportionate increase over a year, which is the difference between the population sizes divided by the starting population:

$$r \cong \left(\frac{N_t - N_0}{N_0} \right) / t$$

where the $\cong$ symbol means that the two sides are approximately equal. The right side of the equation is how we presented r in the section on geometric growth above, but the geometric growth estimate of r and the exponential growth estimate only hold true when the estimate is made over a short period of time. It is for this reason that when geometric growth is modeled one often sees the birth rate, death rate, and growth rate written, respectively, with the capital letters B, D, and R.

16 ECOLOGY OF A CHANGED WORLD

Figure 2.2 illustrates the difference between the two methods: the first formula is based on the tangent to the curve of population growth (that is the instantaneous rate of change), whereas the second is derived from the difference between two points on the curve. In practice, provided that the change is measured over a short enough time interval that the population increase is less than about 10%, the two methods yield very similar figures. In other words, $\frac{\Delta N}{\Delta T}$ is a good estimate of $\frac{dN}{dT}$ when measured over a sufficiently short time. In our earlier discussion of geometric growth in humans, we equated the percentage increase across a year with r, which was acceptable because that increase was small (1.545%). The exact method gives a correct value of 1.533%, which is scarcely different (see the Appendices). However, later in this chapter we consider the spread of a disease, where the percentage increase was measured over a year as 170%. In this case, r must be estimated using the exact method, $r = \ln(N_t/N_o)/t$. Once we have an accurate estimate of r, we can easily predict population size at any time point, so the remainder of the book only uses the continuous (exponential), rather than the discrete (geometric) model.

Doubling Times

An important feature of geometric and exponential growth is that the doubling time—the time it takes for a population to double in size—is constant. Thus, a population grows from ten to twenty in the same time as it grows from one million to two million. The doubling time can be derived from the equation, $N_t = N_o e^{rt}$, by setting $N_t = 2N_o$. Rearranging, we get

$$t_{double} = \ln(2)/r$$

Here $\ln(2)$ is the natural logarithm of 2, or approximately 0.7. Thus, if $r = 0.01$ individuals/individual/year, the doubling time would be about 70 years. We can also turn this around and estimate r from the time it takes a population to double in size.

2.2 The Value of a Model

Predicting the future is fraught with difficulties (Chapter 10), but with a model in hand, we can state with confidence what would happen if the assumptions of the model were to be met. To reiterate, the value of the model is not only that intuition often fails us but that the assumptions that underlie predictions are explicit.

This brings us to the second use of a model, which is to help understand what happened in the past. What are the most important factors we need to invoke to explain a particular change? Human population size has not increased as fast as expected from our 1992 projection, which predicted a population size in 2016 of about 1 billion more people than there actually were. The assumption of constant birth and death rates must be violated, and in this case birth rates have been declining (Chapter 11).

The most useful models are those that can sufficiently account for trends in the data but make as few assumptions as possible. As we add more assumptions to a model, we introduce more parameters to explain the data. This will always help improve the fit of the model to observed patterns but also increase the number of plausible explanations, making them difficult to separate. For example, in the growth model for fruit fly populations, we assumed females always lay 50 eggs and then die after reproducing. If we find fewer flies than the model predicted and are prepared to assume that females continue to lay 50 eggs, then the only possible explanation is that more young are dying. Alternatively, if we include in the model both the possibility that some young do not survive or that fewer eggs are laid, then myriad possible combinations of number of eggs, hatching eggs, and survivors can explain the same pattern. Trying to work out how each contributes becomes increasingly difficult. It is not helpful to add more parameters if one factor explains most of the observations anyway, because we can use that parameter to well predict the future.

Use of a Model: Early Spread of HIV in the United States

An excellent example of the value of a model comes from an application of exponential growth to predict the transmission of the human immunodeficiency virus (HIV) in the United States. HIV is the cause of AIDS, which attacks the cells involved in the immune response, making people especially vulnerable to other diseases, such as tuberculosis. HIV is thought to have initially arrived in the United States in the late 1960s, probably from Haiti, to which it had been brought from West Africa, where it had been contracted from a chimpanzee. The virus was first identified in 1981. Worldwide, about 37 million people may be currently infected, of which perhaps two-thirds are in Africa and 1 million in the United States.

Cases of AIDS reported in the United States increased approximately exponentially over the first few years that records were kept (Fig. 2.3, left side). This created a great deal of worry because the steep trajectory shown in Figure 2.3 suggested that HIV might rapidly infect a large fraction of the population. At the time, few studies had been done, and little was known about how

18 ECOLOGY OF A CHANGED WORLD

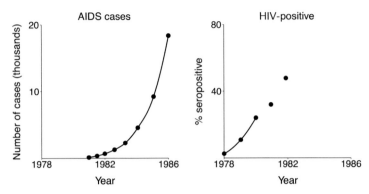

Figure 2.3 *Left:* Increase in the number of AIDS cases reported in the United States during the early years of infection. *Right:* Increase in HIV presence in blood samples from the San Francisco homosexual community. Lines are fit by eye and appear approximately exponential. On the right graph, the curve fit indicates a rise from 3.5 to 27% over the first two years. Using the formula $r = \ln(N_t/N_o)/t$, this gives a growth rate of $r = 1$ new seropositive sample/seropositive sample/year. (The U.S. Center for Disease Control; R. M Anderson and R. M. May. [1987]. Transmission dynamics of HIV infection. *Nature* 326: 137–141)

easily it was transmitted. Models of population growth were used to estimate transmission rates. One analysis addressed the early spread of the virus in San Francisco.

As part of a study on the presence of hepatitis B in the gay community in San Francisco, blood samples from 785 individuals were stored from 1978 onward. Fortunately, the samples were preserved in such a way that they could subsequently be screened for HIV and used to estimate the rate of infection. The fraction of seropositive samples (Fig. 2.3, right) shows the rapid rise in infections. This and other studies suggest a doubling time of 8–10 months during the early stages of infection (i.e., t_{double} is approximately 0.7 years). As we showed above, the per capita rate of increase, r, can be determined from the doubling time:

$$r = \ln(2) / t_{\text{double}}$$

From this we get $r = 1$ infection per infected person per year. The death rate over the first few years of infection was negligible ($d = 0$), so r is equivalent to the birth rate, b. In other words, one infected individual was estimated to, on average, infect one more individual each year during these early stages.

Additional data indicated that gay and bisexual men in this population averaged about twenty sexual partners per year. Given the estimate of 1 infection/infected person/year, this value implied a transmission rate of about 0.05 per partner, which seemed low to investigators at the time. However, despite the relatively small sample size and simplifying assumptions of the model, this early estimate of the transmission rate is in accord with the present-day understanding that HIV is not easy to contract.

We have an excellent illustration of how models can be used to inform. First, one could use the model to ask if HIV is likely to be sustainable. That is, what factors are required for $r = b - d$ to become negative? The model was made more complex by including additional parameters. Assuming that the most sexually active individuals become infected quickly, adding an extra term to the model to account for this leads to lower estimated transmission rates among the remaining population. The revised model thus predicts that individuals with few partners per year may be quite unlikely to transmit the virus. This in itself may be sufficient for the virus to become extinct. Similarly, the spread of coronavirus in 2020 initially followed an exponential curve but soon increased more slowly, as chances of transmission were reduced by social distancing and quarantine (Chapter 19).

Indeed, the model can be used to ask how changing habits might affect the chances of transmission and persistence of AIDS. In San Francisco, the average number of partners per month declined twofold from 1982 to 1984, and the use of condoms rapidly increased. If we assume that infected individuals who practice safe sex or develop AIDS are no longer sources of new infections, according to the models the level of infection may decline, perhaps below sustainable limits. In the case of AIDS in the United States at least, these practices led to a decline in new infections up to about 2012. The number has since leveled off at about 39,000 per year, apparently because some sections of the population do not practice safe sex and have limited access to health care.

2.3 Conclusions

The simplest model of population growth leads, often quite rapidly, to impossibly large population sizes. In 1798, Malthus, aware of the power of geometric growth, recognized that something must act to keep growing populations in check. In his book, he argued that food was the final limit on human population growth, but the desperate conditions so produced will lead to other factors that raise the death rate. First, he observed, war might help slow population growth. But if that fails, pestilence will "sweep off their thousands and tens of thousands." However, if "success" is incomplete, "gigantic inevitable famine stalks in the rear, and with one mighty blow levels the population with the food of the world." The growth rate of any population must on average be close to zero. But the growth rate of a population can be slowed not only by increasing mortality (d going up) but by a decreasing birth rate (*b* going down). Malthus was correct when he noted that populations must reach a limit, but it turns out he was wrong that this inevitably leads to war, disease, and famine. For humans, population growth is slowing, but remarkably that seems to be happening through the more benign process of a declining birth rate rather than an increasing death rate. We explore these demographic shifts in Chapter 11.

3

Population Regulation

In Chapter 2 we introduced a simple model of population growth that assumed unchanging birth and death rates over time. If that assumption is met and the birth rate exceeds the death rate, a population increases in size forever. On the other hand, if the death rate exceeds the birth rate, the population decreases continuously and eventually becomes extinct. In nature, all populations vary in size over time, but they generally fluctuate within certain bounds and neither get very small nor grow excessively large. In this chapter, we investigate limitations on a species population size. The chapter covers:

(1) An introduction to competition for resources as one means of regulating population size.
(2) Derivation of a simple model of competition by adding an additional parameter to the model of exponential growth. This model predicts a stable long-term population size and thus, in principle could apply in nature.
(3) Consideration of real examples of population fluctuations to ask what causes deviations from the predictions of this model.
(4) Application of the model to the problem of sustainable fisheries.

3.1 Competition for Resources

The gray heron is a large bird that eats fish, frogs and other animals, and typically breeds in small colonies. Bird watchers in the United Kingdom have surveyed gray heron populations over the past 80 years (Fig. 3.1). A particularly harsh winter in 1963 caused a major drop in their numbers. Prior to that year, populations seemed to fluctuate around an average of about 7,000 pairs, and afterward around an average of about 8,000 pairs, as suggested by the horizontal lines shown in Figure 3.1. These numbers might approximate the carrying capacity (K) of the environment, which is the number of individuals the environment can support as a result of available resources. Apparently, the carrying capacity increased after 1963 (i.e., K = about 7,000 pairs in the early years, and K = about 8,000 pairs in the later years). The higher value in later years reflects improvements in water quality—the provision of new habitat as artificial ponds created through gravel extraction filled with water—and increased feeding

Ecology of a Changed World. Trevor Price, Oxford University Press. © Oxford University Press 2022.
DOI: 10.1093/oso/9780197564172.003.0003

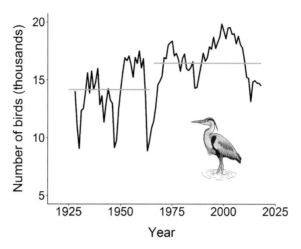

Figure 3.1 Estimated number of herons breeding in England and Wales combined. The gray horizontal lines are drawn by eye and are considered here to approximate carrying capacity during earlier and later years. (D. Massimino, I. D. Woodward, M. J. Hammond, et al. [2019] BirdTrends 2019: trends in numbers, breeding success and survival for UK breeding birds. BTO Research Report 722. BTO, Thetford. www.bto.org/birdtrends)

opportunities at freshwater fisheries. The number of individuals present may also be affected by predation, notably by humans, but the carrying capacity concept assumes that food and other resources, such as nest sites, are the main limit. Fluctuations around the carrying capacity baseline occur for many reasons that affect birth and death rates. For example, major drops in population size associated with cold weather probably act largely through reduced food supply (e.g., ponds freeze over).

Density Dependence and Density Independence

It should be apparent from Figure 3.1 that something keeps heron populations within certain bounds: they neither grow to be very large nor decline to be very small. Consider the possibility that the same amount of food is available every year to a population. When the population size is low, each female garners a large share, and she raises many offspring (birth rates increase). When populations are large, however, each female has less food and she raises fewer offspring (birth rates decrease). In this case, food supply is a density-dependent factor, ultimately responsible for keeping the population size regulated between certain limits. The test for density dependence is to plot birth rates or death rates against population

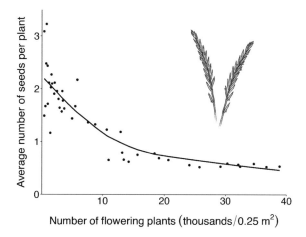

Figure 3.2 Seed counts on a grass (the dune fescue, *Vulpia fasciculata*) from Wales, UK, after an experiment in which plants were grown at different densities. A decline in birth rate with increasing density implies density dependence. (A. R. Watkinson and A. J. Davy. [1985]. Population biology of salt-marsh and sand dune annuals. *Vegetatio* 62: 487–497)

size. Figure 3.2 shows the number of seeds per grass following an experiment where grasses were planted on sand dunes at different densities. At higher densities, each plant produced fewer seeds.

Competition for food, water, and other resources such as nest sites is a major cause of density-dependent population regulation. As the population increases, more individuals compete for the same quantity of a resource. Competition may take two forms: interference and exploitative. In interference competition, individuals fight with each other to gain access to resources. In exploitative competition, individuals use up available resources—that is, they compete without directly interacting with one another. Interference and exploitative competition are often both present in the same system. For example, grasses grown at high densities on sand dunes have low seed set, which is mostly attributed to competition for water (exploitative competition). However, they may also exhibit aggression in the form of chemicals sent out by one plant that inhibit the root growth of others. In another example, foxes fight with each other to maintain territories (interference) but will often sneak into a neighbor's territory for food (exploitation).

Various density-independent factors are superimposed on density-dependent factors. Density independence implies that the birth rates or death rates are unaffected by the number of individuals present. If the government issues hunting permits based on population size so that roughly the same proportion of animals is killed each year, this would be a density-independent factor (death rate, d, = constant). Density-independent factors cannot in themselves regulate populations. If only density independence operates, the population will eventually grow to a very large size or become extinct. While we focus on density-dependent factors for this reason, most episodes of mortality and reproduction involve both density dependence and density independence. Herons may die because of extreme cold even if they are at very low densities (density independence), but the mortality rate would certainly be higher if the population was large, leading herons to compete for open water sites in which to fish (density dependence).

3.2 Model of Logistic Growth

In the exponential model we considered in Chapter 2, the change in population size was given by the equation:

$$\frac{dN}{dT} = rN,$$

where N is the population size at the beginning of the time interval and r is equal to the per capita growth rate (individuals added/individual/unit time). Exponential growth may be approximated in nature over short intervals, but it cannot lead to a stable population. The simplest model that does lead to a stable population is the logistic growth curve, which introduces an extra parameter, K, to the exponential model. To obtain the logistic growth equation, we multiply the exponential equation by $1 - \frac{N}{K}$:

$$\frac{dN}{dT} = rN\left(1 - \frac{N}{K}\right)$$

The left side of the equation is the change in population size over a certain time interval. From the right side, one can see that when N is very small, that is, $\frac{N}{K}$ is close to 0, the population growth rate is close to rN (i.e., exponential). We might expect this to be so because competition is low. That explains why we need to define r as the intrinsic rate of increase and not the per capita growth rate as it was presented in Chapter 1; it is the per capita growth rate expected when no density-dependent factors are operating, and the population is growing exponentially.

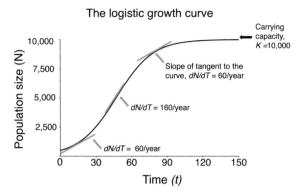

Figure 3.3 The logistic growth curve. The smooth curve plots the logistic equation, $\frac{dN}{dT} = rN\left(\frac{K-N}{K}\right)$. Population size grows from a small number to K, the carrying capacity (10,000 individuals); r, the intrinsic growth rate, is 0.064 individuals/individual/year. Tangents to the curve at three population sizes (1,000, 5,000, 9,000) give the number of individuals added over a year. Note that population growth is highest at intermediate densities. One heuristic way to see why is to consider that at low population sizes, one individual might produce three offspring (three total); at intermediate population sizes, two might produce two (four in total); and at high population sizes three individuals produce one each (again three in total).

In the logistic equation, when $N = K$, $1 - \frac{N}{K} = 0$ so the population growth rate is zero, it is neither increasing nor decreasing. A plot of logistic population growth starting from a small number, with $K = 10,000$ individuals and $r = 0.064$ individuals/individual/year is shown in Figure 3.3. This is the second (and last) mathematical model in the book, and we will use it in several chapters as the simplest model of population growth with density-dependent regulation.

3.3 Fluctuations of Population Size in Nature

The logistic curve is a simple model for evaluating predictions of density dependence. Some population increases seem to fit the model quite well. For example, the wildebeest population in the African Serengeti declined to low levels in the early 1960s due to rinderpest, a viral disease related to measles that spilled over from domestic cattle (the word "rinderpest" comes from the German meaning "cattle plague"). Once a vaccine was developed for cattle, rinderpest was no longer a threat to the wildebeest population. The population initially increased and then leveled off in a way that roughly matches logistic growth

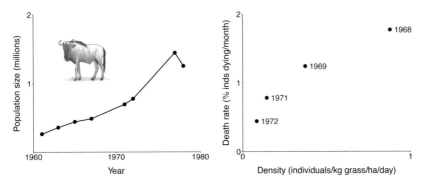

Figure 3.4 *Left;* Population size of wildebeest in the Serengeti, Tanzania, between 1960 and 1980, based on aerial censuses. *Right:* Death rate of wildebeest in the dry season plotted against density with respect to measured food resources. (A. R. E. Sinclair. [1979]. The eruption of the ruminants. In: A. Sinclair and M. Norton-Griffiths, M. [eds.] *Serengeti: dynamics of an ecosystem.* Chicago: University of Chicago Press; A. R. E. Sinclair and M. Norton-Griffiths. (1982). Does competition or facilitation regulate migrant ungulate populations in the Serengeti? A test of hypotheses. *Oecologia* 53: 364–369)

(Fig. 3.4). Figure 3.4 shows how food availability correlates with number of individuals surviving and implies density-dependent growth. The death rate, *d,* was four times higher in 1968 than in 1970, associated with just one-tenth the amount of food available to an individual.

Most populations do not follow such a simple pattern. Instead, fluctuations in population size are common, attributed to the many factors that affect death and reproduction (e.g., Fig. 3.1). The population sizes of some species vary enormously from one year to the next. In November 2004, a birdwatcher on a 230 km drive in Morocco estimated that he passed 69 billion desert locusts on the ground and "the air was full of them." In other years, locusts are scarcely seen. Large fluctuations in locust population size in China have been monitored over a 2,000-year period. Locust outbreaks most commonly occurred in cool and dry periods, likely because such periods were associated with both floods and droughts, which generate receding waters, leaving wet soil favorable for locust development. Between 2019 and 2020, east Africa suffered from a devastating locust outbreak, the worst, in fact, for more than 25 years. This event was linked to the increasing frequency of tropical storms, generated by a warming ocean, which created lakes in the deserts of Saudi Arabia and receding waters. That event was followed by another storm that prolonged the wet conditions and hence food, and yet another one whose winds helped the locusts spread south.

Besides climate, populations of locusts fluctuate dramatically because adults produce offspring in such large numbers that they overshoot the carrying capacity of the environment. Under these conditions, the large number of individuals present eat all the food, causing the population to collapse to small numbers. This phenomenon is called delayed-density dependence because the effects on reproduction or survival do not operate instantly, as they do in the logistic growth model. We can observe effects of delayed-density dependence on both predators and prey. When predators become scarce, the prey can recover and grow to a large population size, which in turn increases the predator's food availability. Such predator–prey interactions may lead populations of predators and prey to cycle over time.

3.4 Maximum Sustainable Yields

In the logistic model, when populations are at low density individuals have access to plentiful resources. Therefore, each individual has a high chance of survival and produces many offspring—the per capita growth rate is high. Consequently, the population starts to grow, competition for resources increases, and the per capita growth rate steadily declines, until, at carrying capacity, it is equal to zero. At this point, each individual exactly replaces herself: one mother has one surviving daughter.

Unlike the per capita growth rate, which steadily declines as populations grow, the number of individuals added to the population per unit time, $\frac{dN}{dT}$, is highest at intermediate population sizes (Fig. 3.3). When populations are small, few reproducing individuals are present in the population, even if they are producing many offspring. With a moderate number of reproducing individuals population growth is higher even though each of these individuals has fewer surviving offspring than an individual would in a very small population. When there are many individuals in the population, they have few surviving offspring, and population growth rate is again low.

In the logistic model, the point of maximum population growth is reached when $N = K/2$ (see the discussion in the Appendices for this chapter, in the section "Maximum Sustainable Yield in the Logistic Equation")—that is, when the population size is at half the carrying capacity. This is known as the point of maximum sustainable yield because it is the population size at which one can sustainably remove the most individuals. For the parameters in Figure 3.3, one could continue to remove 160 individuals if the population was maintained at half the carrying capacity, but one would need to remove fewer individuals to maintain populations at either higher or lower levels. On the other hand, if one

28 ECOLOGY OF A CHANGED WORLD

always removed more than 160, the population would decline to zero. In the rest of this chapter, we develop the principle of maximum sustainable yields further, drawing on fisheries.

Some early attempts to manage fisheries simply used the model of logistic growth to guide the harvest. The pre-fishing population size was assumed to be the carrying capacity. The intrinsic rate of increase is estimated (e.g., from growth rates in a depleted population) and the quota accordingly set. Chapter 20 describes this approach for estimating sustainable yields from hunting of antelope in West Africa. However, the logistic equation is a simple model of density dependence, and all populations deviate to some extent, as illustrated by the examples of herons (Fig. 3.1), locusts, and wildebeest (Fig. 3.4). Many complexities, both of the density-dependent and density-independent type, come into play. More sophisticated models of population growth have tried to take some of these complexities into account. One prominent goal is to estimate the maximum economic yield, which depends not only on the number of fish but also on their size, as well as the number of boats required to catch the fish. We consider a model developed for George's Bank, an area larger than the state of Massachusetts and of relatively shallow water (several to tens of meters) off the northeast coast of the United States.

George's Bank used to be home to millions of cod. Yields declined throughout the twentieth century because of harvesting beyond the maximum sustainable yield. Presently, total fish biomass is less than 10% of estimated historical levels (see Chapter 21). In 1993, Canada declared a moratorium on the fishing of cod in the northern part of George's Bank, which it controls, and placed limits on the catch of other species. Between 1990 and 1994, cod on the southern part, which the United States controls, declined by 40%. The United States also closed some areas but allowed limited fishing in others.

The model evaluates the effect of fishing on total fish biomass summed across all twenty-one commercial fish species, including cod, by adding more parameters to the simple logistic growth equation (which as we have seen has two parameters, r and K, plus the initial condition, N_0). Among these parameters are those for how quickly an individual grows in the presence of competitors (fish grow more quickly to a large size when they have abundant food) and the effects of predation (e.g., cod eat herring). Values for these parameters were estimated from field studies.

The exploitation rate is defined as the proportion of the mass of all fish that is removed each year. A 50% exploitation rate implies that when fish are rare, few fish are caught, but when they are abundant, many are caught. The modelers found that the maximum sustainable biomass that can be harvested, termed the multispecies maximum sustainable yield, is achieved with about a 50%

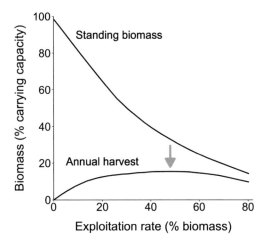

Figure 3.5 A model for fisheries exploitation on George's bank off Massachusetts. Mass is the total weight of fish in the sea, as a proportion of what it would be without any harvesting, considered to be the carrying capacity. Exploitation, on the X axis, is the proportion of the mass available that is removed in a year. The arrow indicates the point of maximum sustainable yield, where the exploitation rate = 0.48. Note that this is a model and that no data are plotted. However, data are used to estimate the various parameters, such as r, K, and individual growth. (B. Wörm, R. Hilborn, J. K. Baum, et al. [2009]. Rebuilding global fisheries. *Science* 325: 578–585)

exploitation rate (Fig. 3.5). Unlike the simple logistic growth model, however, this occurs when the total biomass is at about 35% of the maximum possible, partly because low densities increase not only per capita growth rates (as in the logistic model), but also the rate at which individuals grow.

According to the model, the mass of fish caught would be about the same with a harvesting rate of 20% or 80% of the standing biomass (Fig. 3.5). In the case of a 20% harvest, many fish remain, only a few fishermen are needed to haul in the fish, and the environment is much closer to its natural state. In the case of 80% only a few fish are present, and many species would be considered collapsed, defined as a population <10% of its historical size, as cod is at present. Further, many fishermen are needed to catch the fish because the fish are rare. George's Bank remains highly regulated, yet total biomass levels remain a fraction of what would be optimal for sustainable yields. Chapters 21 and 22 further consider the state of fisheries, including cod on George's Bank, and describe in more detail a model that is currently being used for fishery management in Alaska.

3.5 Conclusions

Density dependence is critical to the persistence of populations. When population numbers are low, birth rates go up or death rates go down, so numbers increase. Under conditions of density dependence, the maximum rate of growth of a population (i.e., individuals added per unit time) should be at some intermediate density. This occurs when sufficient numbers of individuals are around to reproduce, but not so many that competition between them is severe. If the goal of fisheries is to maximize yield over the long term, it is important to estimate this number. However, even if we fit complex models (as in the George's Bank example), many other factors not in the model come into play (e.g., disease, climate change), so one needs to apply model predictions very cautiously. In some well-regulated fisheries, a variable quota is set to ensure that in years of low numbers not too many fish are harvested (see Chapter 22).

4

Interactions between Species

Mutualisms and Competition

Chapter 3 developed a model of population regulation in which the carrying capacity of the environment sets the population size of a species. Carrying capacity depends on the availability of limited resources, such as food. Individuals of the same species compete with each other for these resources, thereby regulating the population, but individuals from different species also compete for these resources, thereby affecting each other's population size. In addition to competition, other species interactions, such as predation and disease, affect mortality and reproduction. More gray herons means fewer fish. More humans means more cows and chickens because we provide food for them, but fewer fish in the sea because we eat them.

The next three chapters ask how interactions among species affect their abundance and distribution. This chapter focuses on competition, with the following goals:

(1) To introduce the main kinds of interactions between species.
(2) To ask how competition between species limits the population sizes and geographic ranges of these species.
(3) To consider how competition sets an upper limit on the number of species found in one place.
(4) To briefly consider how competition affects evolution.

The following two chapters describe predation, parasitism, and disease, which are additional species interactions that affect population numbers.

4.1 Interactions between Species

We use pairs of arrows between species to illustrate how a change in the population size of one species affects the population size of another species (Fig. 4.1). Consider a single arrow in the diagram. The sign on the arrow tells how a change in abundance of the species at the base would affect the abundance of the species at the arrowhead. A positive sign means that when the base species goes up or down in abundance, the arrowhead species goes up or down in the same direction. A negative sign means that when the base species goes up or down in

Ecology of a Changed World. Trevor Price, Oxford University Press. © Oxford University Press 2022.
DOI: 10.1093/oso/9780197564172.003.0004

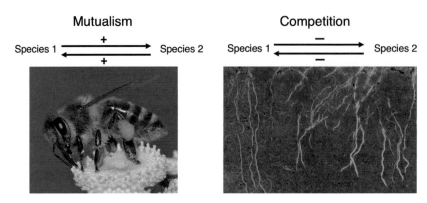

Figure 4.1 (Color plate 1) The lines indicate the effect of a change in the population size of the species at the base on the population size of the species at the arrowhead. *Left:* Bees and flowers form a mutualism, where an increase in the population size of one causes an increase of the other, and a decrease in one causes a decrease of the other. *Right:* Roots of ground ivy (left thin roots, green in the color version) and wild strawberry compete for nutrients and water. An increase in one species causes a decrease in the other, and a decrease in one species leads to an increase in the other. (H. de Kroon. [2007]. How do roots interact? *Science* 318:1562–1563) Photos (*Left*: Andreas Trepte, www.photo-natur.de, *Right*: Marina Semchenko).

abundance, the arrowhead species abundance changes in the opposite direction. For example, a decrease in foxes would cause rabbits to go up and an increase in foxes would cause rabbits to go down; thus, the arrow from fox to rabbit would be negative.

Figure 4.1 diagrams two common interactions. In a mutualism, both arrows carry positive signs. Bees gain nectar from plants and pollinate them in the process. Fewer bees means fewer plants, and fewer plants means fewer bees. A second classic mutualism is between nitrogen-fixing soil bacteria and some plants such as peas and soybeans. The bacteria draw nitrogen from the atmosphere and convert it into nitrates and ammonia that can be used by the plant (termed reactive nitrogen). The plants support the bacteria in special root nodules. Because plants cannot fix atmospheric nitrogen themselves, those that support nitrogen-fixing bacteria have an advantage in locations where nitrogen is a limiting nutrient. Mutualisms imply that the decline of one species will cause the other to decline as well, a topic we consider in Chapter 16 in our discussion of how corals are suffering from the loss of their associated algae in a warming ocean. Mutualisms are not only important in nature but also in the services that nature provides for us (for example, the decline in bees as pollinators has become a serious problem; see Chapter 28). However, the positive association between a pair of mutualists implies that the interaction between them cannot regulate population numbers. Instead something external to the system (e.g., for corals, ocean warmth; for bees, disease) affects one of the species, with ramifications for the other. In the absence of any external checks, a positive feedback between mutualists would result in either increase without limit or decrease to extinction. In this way, mutualisms differ from competition, predation, and parasitism, where one species limits the other directly and stable coexistence is theoretically possible.

4.2 Competition

The rightside of Figure 4.1 depicts interspecific competition, whereby an increase in either species causes the other to go down. Interspecific competition is similar to competition between members of the same species, but in this case individuals of one species use resources that would otherwise be available to individuals of the other species. As with competition between individuals belonging to the same species, interspecific competition may operate through interference and/or exploitation. Wild strawberry roots grow toward and "attack" ground ivy roots, which retreat (Fig. 4.1). The mechanisms likely include both exploitative competition (e.g., for water) and interference competition through sending out inhibitory chemicals.

In any one location in the world, some species are more common than others. Furthermore, all species are restricted to certain habitats and specific regions. Competition between species is one important factor setting both abundances and distributions.

Niches

The term "ecological niche" is used to describe the range of resources utilized, and conditions occupied, by a species. The dimensions of a species' niche include abiotic factors, such as the range of temperatures that can be tolerated, and biotic factors, such as the size of prey that can be consumed. J. Grinnell, who introduced the term in 1924, stated that no two species can occupy for long the same identical niche, because one will surely be slightly more efficient than the other when competing for the same resources. In the 1930s, W. Gause experimentally tested this idea with single-celled organisms belonging to the genus *Paramecium*. *Paramecium* are common in freshwater and typically feed on bacteria. When Gause grew two species of *Paramecium* in separate test tubes along with a constant supply of bacteria, each population grew on a trajectory resembling the logistic growth curve (Fig. 4.2). When pairs of species

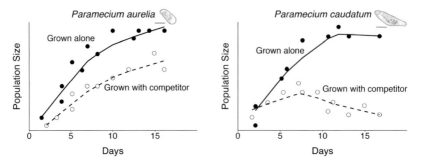

Figure 4.2 When grown alone in culture, each of two species of *Paramecium* roughly follow a logistic growth curve. When grown together, *Paramecium aurelia* has a lower initial growth rate and lower population asymptote, indicating competition from *Paramecium caudatum*. *Paramecium caudatum* suffers even more. Population size here is a measure of biomass, not numbers. By the time the experiment was discontinued, *Paramecium caudatum* had declined to low numbers and would be eventually lost. The scale bars below the drawings are 0.1 mm. (G. F. Gause [1934]. *The struggle for existence*. Baltimore: Williams and Wilkins), pp. 101–102)

were grown together, each depressed the other's growth rate. In the example shown in Figure 4.2, *Paramecium aurelia* outcompeted *Paramecium caudatum* in the test tube environment and *Paramecium caudatum* declined toward elimination.

According to Grinnell's reasoning, the coexistence of species in nature derives from the wide range of available resources, such as different types of prey. For a pair of species living under natural conditions, one species is superior at exploiting one resource and can reduce it to a level at which the other cannot maintain itself on that resource, but the other species is better at using a second resource. The two species can then coexist. Figure 4.3 illustrates one resource axis, prey size, for two species of Darwin's finches on the Galápagos Islands. The large ground finch occupies what may be termed the large seed niche, and the medium ground finch the medium-sized seed niche. First, the large ground finch cracks seeds that the medium ground finch cannot. Second, the medium ground finch can harvest small seeds more efficiently, and because it is smaller, it requires fewer seeds each day to maintain itself. The result is that when medium seeds become scarce, the medium ground finch can persist, even though the large ground finch would not be able to support itself. In consequence, in the presence of both medium-sized and large seeds, both finch species coexist: each is most efficient at consuming one or the other seed size.

Figure 4.3 Seeds and two species of Darwin's finches on Isla Daphne Major, Galápagos. The large ground finch, *Geospiza magnirostris,* can crack the large items (cactus seeds and fruits from a plant *Tribulus cistoides,* which contains a few seeds in a woody case), whereas the medium ground finch finds these tasks difficult. The medium ground finch, *Geospiza fortis,* can eat medium-sized seeds more efficiently than the large ground finch to the point that they become so scarce that the large ground finch could not persist. We can speak of large seed and medium seed niches. (Redrawn from P. R. Grant and B. R. Grant. [2006]. Evolution of character displacement in Darwin's finches. *Science* 313: 224–226)

Competition between Species Occupying Environments of Different Quality

It is possible for two species to consume similar resources but live in different locations, which may sometimes be quite close to each other. Consider a scenario in which a large-bodied species and a small-bodied species exploit the same food. The larger species aggressively excludes the smaller one from locations with high food abundance, but, being larger, each individual will require more food. If the region in which these species live varies in food abundance, we expect to find the large species in high-food areas (because it can get enough food and physically drive out the small species) and the small species in low-food areas (because the large species cannot find enough food to support in these habitats). Such gradients can therefore determine patterns of species distribution and abundance via their effects on competitive interactions.

A classic example concerns two species of barnacles that coexist in Scotland (Fig. 4.4) and engage in interference competition for space on rocks. After a free-living larval stage, barnacles settle down on a rock, grow a shell, and lead a sedentary life sifting the water for food particles. One species of barnacle lives in the upper zone, where it is rarely covered by water and therefore feeds relatively infrequently. It is resistant to drying out and over periods of lengthy low tides, at the location where the two species co-occur, the species that is found lower down has much poorer survival (top graph in Fig. 4.4). This second species is regularly under water and can feed frequently. When the species from the splash zone is moved down, by simply moving a rock with young barnacles, it flourishes if the lower species is consistently removed, but the lower species can squeeze it off the rock because it is larger, with a heavier shell (lower two panels in Fig. 4.4). We conclude that the physical gradient permits these two barnacles to coexist and

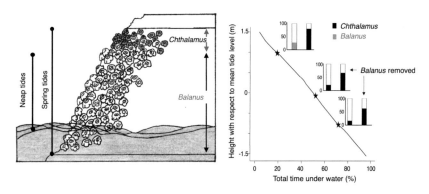

Figure 4.4 *Left*: Distribution of two species of barnacle, *Chthamalus stellatus* (upper species) and *Balanus balanoides*, on a rocky shore in the Firth of Clyde, Scotland. Top of bars indicate average high tide during spring tides (maximum amplitude) and neap tides (7 days after spring tides, minimum amplitude). *Right top*: Survival of *Balanus* and *Chthamalus* at mean neap high tide level, monitored between February and May 1955 (most mortality appears to have been caused by six consecutive warm days over the March neap tide, during which the barnacles were continuously exposed). *Lower histograms*: Survival of *Chthamalus* transported to two locations lower down the shore. In half the treatments *Balanus* growing near *Chthamalus* were regularly removed with a pen knife. Survival was monitored over one year (above) or 6 months (below). (J. H. Connell. [1961]. The influence of interspecific competition and other factors on the distribution of the barnacle *Chthamalus stellatus*. *Ecology*, 42:710–723; J. H. Connell. [1961]. Effects of competition, predation by *Thais lapillus*, and other factors on natural populations of the barnacle *Balanus balanoides*. *Ecological Monographs*, 31:61–104)

thus promotes greater species diversity than we would see in a more homogeneous environment. It is interesting that the main identified cause of mortality for the second species at its upper range limit was a consequence of the abiotic environment (desiccation), whereas the main cause of mortality at lower range limit of the first species was competition. But one must note that usually both the abiotic and biotic environments work synergistically. For example, at its upper range limit, the lower species may eventually evolve to exploit food even in drier conditions, if none of the space was continually preempted by the upper species.

The barnacles compete for space. Space may be partitioned over short distances as well as over very large areas, such as continents. In this way, competition along resource gradients can influence the geographical range of a species. Here is a possible example from India: The yellow-browed warbler and the greenish warbler are two similar and common Asian birds, but the greenish warbler is about 30% larger (8.5 grams cf. 6 grams). During the spring, the two species breed in the forests of the Himalaya and much of temperate Asia. They migrate to India for the winter, where they feed on insects and spiders in the tree crowns (Fig. 4.5). At this time of year, insects and spiders are more abundant in the south of India than in the north of the country, corresponding to the warmer and wetter conditions found there. The larger warbler is common in the south and rare in the north, likely because food supplies are insufficient for it in the north. However, the smaller warbler is confined to the north (Fig. 4.5). Its restriction to the south cannot be because of food and instead appears to be through aggression: the larger species can displace the smaller species. In the overlap zone, the two species fight with each other.

Competition from Introduced Species

Plants and animals introduced from one region of the world to another are a major threat to native species (Chapter 18). About half of the 2,200 flowering plant species present on the Hawaiian Islands have been introduced from elsewhere. Here we consider an example of the competitive effects of a species introduced to Hawaii on the native flora. The fayatree is a small tree/shrub native to islands in the eastern Atlantic. The tree was brought to Hawaii at the end of the nineteenth century, probably as an ornamental plant. It now forms dense forests in areas that have recently lost their plant cover, especially in locations of recent volcanic eruptions. Lava flows contain little reactive nitrogen and the fayatree is successful because it carries nitrogen-fixing bacteria in its root nodules. No native Hawaiian plant on these soils carries bacteria capable of fixing nitrogen. Consequently, the fayatree outcompetes the native tree when both start off as young seedlings by growing faster and shading out light. The fayatree also

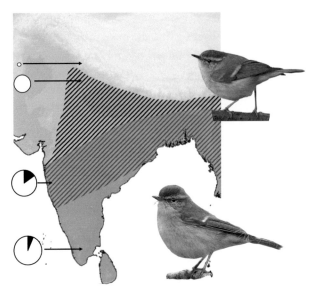

Figure 4.5 Approximate geographical ranges of two common warblers that overwinter in south Asia. The yellow-browed leaf warbler rarely occurs further south, but the greenish warbler regularly occurs in locally productive habitats further north of the mapped limit. Circles are measures of relative numbers of arthropods (insects and spiders) in late January, based on 5 years of sampling at the southern point, 1–2 years for the others. The dark portion of each circle is the fraction of arthropods larger than 4 mm, which formed 30% of the greenish warbler diet and 8% of the yellow-browed leaf warbler diet at the site of range overlap. (S. Gross and T. Price. [2000]. Determinants of the northern and southern range limits of a warbler. *Journal of Biogeography* 27:869–878, M. Katti and T. Price. [2003]. Latitudinal trends in body size among overwintering leaf warblers, genus *Phylloscopus. Ecography* 26: 69–79. (Photos: Natthaphat Chotjuckdiku (yellow-browed warbler) and Dibyendu Ash)

completely changes the nutrients in the soil, adding almost four times as much nitrogen to the soil each year than is found in rainfall and animal feces. These high-nitrogen levels set the stage for other species that are more efficient growers than the fayatree, which includes both native and other introduced species.

4.3 Competition and Coexistence

Examples in the previous section highlight two questions that are foundational to the science of ecology and critical to the conservation of biodiversity: (1) Why

INTERACTIONS BETWEEN SPECIES   39

are more species found in one place than another? (2) Why, even in one place, are some species rarer than others? These are big questions. Many factors are involved, and their importance varies from place to place, but competition clearly provides at least part of the explanation. Following Grinnell's principle that no two species can occupy the same niche, we might expect that the number of niches in a habitat determines the number of species that live there. Sometimes this appears to be the case. For example, the large ground finch and the medium ground finch coexist on Isla Daphne Major in the Galápagos Islands by consuming different sizes of seeds, therefore occupying different niches (Fig. 4.3). If only one class of seed were present, we would expect only one of these species to be present, and on islands without a supply of large seeds, the large ground finch is indeed absent.

A difficulty with more generally applying the number of niches argument comes from ascertaining the number of niches available. In fact, on Isla Daphne the two large food items illustrated in Figure 4.3 require different handling skills, and a third finch species, the cactus finch, is present. The cactus finch specializes on cactus seeds, which it can extract from the cactus fruit by using its long beak. The other two finch species, which have stubbier beaks, have to wait until the seeds are exposed. If the cactus finch did not exist, would we know there was a niche available for it? Continuing with Isla Daphne, a few intermediate-sized seeds are present on the island, which presumably could be best consumed by a finch species of intermediate beak size. This suggests that these intermediate size seeds might be an unoccupied niche and that four species could coexist on Daphne. The seeds are usually not sufficiently common for enough individuals to persist, especially given that they can be eaten by both the large ground finch and the medium ground finch. However, in 1981, a moderate-sized male belonging to another species immigrated from a distant island, hybridized with a medium ground finch, and generated a new population on the island that was intermediate in size between the medium and large ground finch. While this population may not persist over the long term and the role of intermediate sized seeds in enabling its establishment is not clear, the example does show how resources, such as food, may be more or less finely partitioned, with the result that more or fewer species can coexist.

Models have been used to predict the number of species that can coexist when they are competing for resources. The simplest ones assume that each species grows according to the logistic growth curve, but still several extra parameters need to be added. Two of the extra parameters are the extent to which one species depresses the growth rate of the other, and the abundance of the different resources. Figure 4.6 illustrates the outcome of one such model. In this example, individuals of different body sizes compete with each other for food that varies in size, with their body sizes correlating with the ability to efficiently consume a

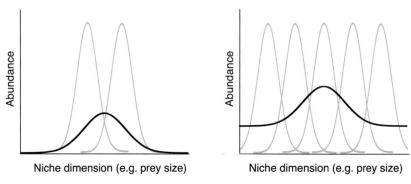

Figure 4.6 One model to explain why more species are found in one place than another. The thick black line diagrams the abundance of prey of different sizes. The thin lines draw the relative abundance of individual body sizes for particular species, with each size matched to the prey of a certain size. For example, the right curve in the left figure shows that in this species a few individuals have an intermediate body size, and a few have a large body size, with most in between. When prey abundance is low (left), only two species are able to persist. But when prey abundance is higher (right), many more species can persist, each using a different portion of the prey size spectrum.

particular size of prey (as in Darwin's finches, but prey sizes vary continuously, rather than falling into discrete classes). A key result is that when food is scarce, only two species can persist, but when it is common, many more can persist, with each one being a resource specialist (i.e., occupying a smaller niche).

Alternative Theories of Coexistence

In addition to the niche-based theories of coexistence described in the previous section, other mechanisms of coexistence are important. For example, in the case of the *Paramecium* experiments (Fig. 4.2), if the population sizes of *Paramecium aurelia* are repeatedly depressed for some reason, they may never exclude *Paramecium caudatum*. Predators that focus on the most common species, on the best competitor, or consume a large number of individuals of all species, could cause species to coexist. In the absence of the predator, however, the best competitor may outcompete the others. An example of how predation results in the coexistence of multiple prey species is described in Chapter 5.

Disturbances such as fires, hurricanes, and natural tree falls in forests are common. As disturbed areas open up, the first plants to colonize tend to be good dispersers (e.g., grasses), moving from patch to patch as open habitat becomes available, only to be replaced by more slowly growing, superior competitors (e.g., trees). Each habitat has an associated community of animals. The intermediate disturbance hypothesis states that the largest number of species will be maintained in a small region if there is always some disturbance. Without any disturbance, species that occupy ephemeral habitats would disappear. But with too much, those slow-growing, slow-arriving species could not persist.

Another suggestion for how species can coexist is that it may take a long time for one species to outcompete another if they occupy very similar niches. The neutral theory of species coexistence holds that some species are so similar that in the extreme there is no tendency for one to replace the other. In other words, species can coexist for a long time precisely because they are so similar in their resource requirements. These ideas may especially help to explain the great diversity of tropical trees (Chapter 26).

Finally, species can be found in some locations where the death rate is higher than the birth rate ($b - d < 0$), but the population is maintained by immigration from elsewhere. Experimental transplants of plants have shown that this is particularly common along elevational gradients, where dispersal distances are short and environmental conditions change quickly. By contrast, over large geographical distances, it appears that more species could persist in a site but have been unable to reach it. Much controversy surrounds the role of each mechanism discussed, which we revisit when we consider why more species occur in tropical rather than temperate regions (Chapter 26).

4.4 Competition and Evolution

The rise of large mammals (including, eventually, us) was associated with the removal of competition from dinosaurs following the asteroid impact 66 million years ago. One can consider that large mammal niches were previously occupied by dinosaurs. Once dinosaurs went extinct, small-sized mammals, which had already persisted for more than 100 million years, were able to evolve to fill the vacated niches. The rapid formation of many species from a single ancestor, triggered by the presence of empty niches, is termed adaptive radiation. One can study adaptive radiation most easily in recently formed, simple, environments. The classic example is Darwin's finches in the Galápagos Islands. Perhaps two million years ago a flock of finches arrived from South America. Subsequently, fourteen species evolved from one ancestral species, radiating to occupy a variety of niches. Two of these species are the large ground finch and the medium ground finch (Fig. 4.3). They also include two species of warbler finches, which have thin, sharp beaks used to eat insects and nectar, and the woodpecker finch that uses sticks to extract grubs. Both the warbler finches and the woodpecker finch are presumed to have evolved in the Galápagos because no warblers or woodpeckers were present when they arrived.

4.5 Conclusions

Competition between species clearly affects population sizes. Many examples exist where the increase in the population size of one species causes another to decrease, attributable to competition. For example, along mountainsides as the climate warms, some species at the base of the mountain extend their range up, competitively forcing other species to move even further up and perhaps eventually off the top (Chapter 15). Two factors are involved: the direct effect of heat on individuals ability to survive and reproduce, and the competitive edge to the species adapted to warmer conditions. In many cases, it is not simply two factors, but many, including abiotic (e.g., climate, pollutants) and biotic (e.g., competition, predation) factors that interact. This is what makes the future so difficult to predict (Chapter 10) and potentially so dangerous. Next, we investigate predation and parasitism as two biotic forces that have demonstrably caused more devastation than competition but also exacerbate any effects from competition.

5

Predation and Food Webs

Predation is the species interaction in which an individual of one species, the predator, kills and eats an individual of another, the prey. In a path diagram, the arrow from predator to prey carries a minus sign (more predators means fewer prey), whereas the arrow from prey to predator carries a plus sign (more prey means more predators). To emphasize this asymmetry, the predator is placed above the prey (Fig. 5.1; compare with Figure 4.1, where competitors are placed side by side).

Competition for food, the topic of Chapter 4, is a consequence of two species consuming the same resource. Therefore, the single path between two competitors (Fig. 4.1) is properly written as the product of the paths connecting the two species through their prey (Fig. 5.1). Consider one of the predator species. This predator eats the prey species, so the path from predator to prey is negative (−). The prey is also eaten by the other species, so the path from the prey to the other species is positive (+). When we multiply the signs of the two paths, one positive and one negative, it becomes clear that the effect of one predator on the other predator is negative; the two predators compete with each other.

This chapter expands on this theme to consider the interactions of multiple species in a community. We consider:

(1) The concept and consequence of food chains, in which some animals consume other animals that consume yet other animals, and so on.
(2) The concept of a food web, which depicts arrows between not only predators and prey, but also competitors.
(3) How predation and competition together affect the number of species that can coexist in a community.

5.1 Food Chains

A food chain stretches from producers (plants, algae) to primary consumers, which eat producers; to secondary consumers, which eat animals that eat plants; to tertiary consumers, which eat animals that eat animals that eat plants. One such food chain is shown in Figure 5.2.

Ecology of a Changed World. Trevor Price, Oxford University Press. © Oxford University Press 2022.
DOI: 10.1093/oso/9780197564172.003.0005

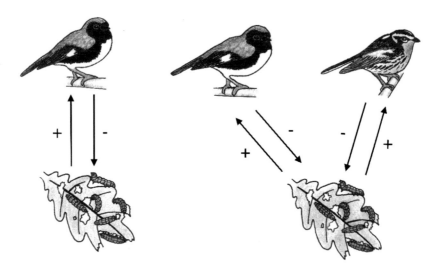

Figure 5.1 *Left:* Path diagram of predation (bird eats insect). *Right:* Path diagram of competition, drawn as two bird species feeding on the same insect prey (compare with the right part of Figure 4.1, which can be derived from this figure).

It is rare to see more than four levels in a food chain, in part because more than 90% of all the energy at one level is lost as heat and waste. Consequently, tertiary consumers (e.g., eagles) should be rare, implying that a quaternary consumer (e.g., an eagle that eats eagles) would be so rare that it would be unable to persist. However, many tertiary consumers do have mechanisms to survive when food is scarce, allowing them to have longer lives than those further down the food chain. For example, snakes have such a low metabolic rate that they can survive for months without eating. In this group, the king cobra is a quaternary consumer, largely eating other snakes. Sharks are both secondary and tertiary consumers and are common, but no animal seems to regularly prey upon them. These examples imply that the reasons for so few levels in the food chain are more complex than just energy loss.

If one traces the path up a food chain, one can see that an increase in primary production should cause an increase in all of the consumers in the chain, as can also be calculated by multiplying the signs along the arrows pointing up in Figure 5.2 (i.e., $+ \times + = +$). On the other hand, as one follows the food chain down, an increase in the top predator should cause a decrease in its prey, which causes an increase in the next level down, and so on (i.e., $- \times - = +$). In consequence, an increase in a predator causes an increase in species two levels below that predator in the food chain. For example, if birds are prevented from eating the caterpillars because a cage has been placed around young oak trees, more oak

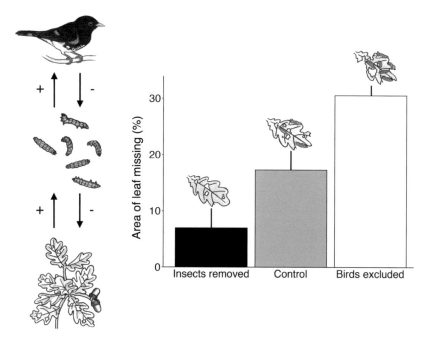

Figure 5.2 *Left:* Short food chain in a Missouri forest. Birds eat caterpillars, which eat oak leaves. *Right:* Spraying insecticide to remove insects from small trees halves the amount of leaf eaten, whereas caging trees to keep out birds doubles the amount eaten. The study is based on 90 trees (30 in each treatment), repeated over 2 years. The bars are standard errors, demonstrating that the treatments significantly affect the area of leaf. (R. J. Marquis and C. J. Whelan. [1994]. Insectivorous birds increase growth of white oak through consumption of leaf-chewing insects. *Ecology* 75: 2007–2014)

leaf is eaten (Fig. 5.2). More birds means more oak growth because they reduce the number of caterpillars.

The population size of species at intermediate levels in the food chain may be affected by both the abundance of species that eat them and the abundance of species they eat. The debate about the relative importance of top-down control by predators or bottom-up control by available food has been particularly applied to insects. Some researchers believe that top-down control predominates. Others note that defenses against herbivory (e.g., sequestration of toxic chemicals in leaves) could cause plants to be less available to consumers than they appear, causing food shortages that translate into regulatory bottom-up control. The experiment described in Figure 5.2 clearly shows that birds limit insects (top-down control), but it also seems likely that an increase in plant growth would increase

46 ECOLOGY OF A CHANGED WORLD

insect abundance (bottom-up control). In general, we expect both bottom-up and top-down control to affect a species' abundance.

The Story of the Sea Otter

Sea otters historically bred along the entire Pacific North American coast, where they inhabit kelp forests. Native Americans used to hunt sea otters and may have extirpated them from a few places, but otter populations appear to have been relatively stable until 1791, when the fur trade started. One hundred years later, sea otter numbers were low, and the species was threatened with extinction. At that time, the population in southern California had been reduced to an estimated 30 individuals. As a result of protection over the last 100 years, the otter has rebounded. At some locations, otters have been deliberately introduced, and at others they found their way to various locations naturally. Introductions are not easy, partly because sea otters have been known to swim more than 100 km across the sea and travel another 300 km along the shore to get back to where they had been before.

Sea otters eat sea urchins, which are a prime grazer of the Pacific's kelp forests. These forests are sometimes over 50 m tall, taller than most forests on land. Kelp forests contain many specialized species, at least when they are not overfished. In Alaska, areas that have not been recolonized by otters typically have no kelp forest, but they have many urchins, which feed on the young kelp.

As they moved into Alaska, otter numbers increased from the late 1960s to the late 1980s from 500 to 5,000. This rapid rate of growth of about 18% per year likely occurred because the otters could eat so many sea urchins. As numbers increased, sea urchin numbers decreased, and so the kelp forests have started to recover. The otter is termed a keystone species (organisms that have outsized ecological impacts relative to their biomass). For a clarifying illustration, in the building trade, the keystone is the wedge-shaped stone at the top of a building's arch, whose removal would cause the whole structure to collapse.

The otter story does not end with its recovery and the increase in kelp. In western Alaska along the Aleutian Islands, the otter went into a major decline in the 1990s, following a decrease in the number of seals and sea lions (Fig. 5.3). Over 6 years, between the islands of Kiska and Seguam at the end of the Aleutian chain, otter numbers declined from about 50,000 to about 10,000, implying that 7,000 additional animals died per year. The decline in the sea otter population coincided with the first sighting of an attack by a killer whale on otters, which was soon followed by several more recorded attacks. A killer whale that eats only otters is estimated to require about 1,800 otters per year, so just four whales could have been responsible for this precipitous decline. Killer whales form small

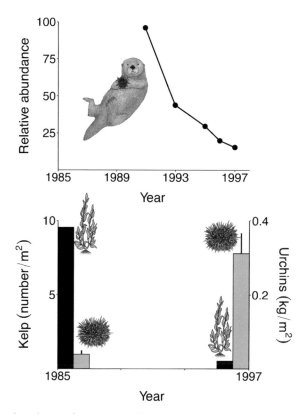

Figure 5.3 Abundance of otters at Adak Island in the Aleutian Island chain off Alaska, as well as measurements of sea urchins (gray bars) and kelp (black bars) before and after the otters declined. (J. Estes, J. M. T. Tinker, T. M. Williams, et al. [2008]. Killer whale predation on sea otters linking oceanic and nearshore ecosystems. *Science* 282:473–475)

groups called pods, which specialize in certain prey. A few whales evidently discovered that otters are good to eat.

Although otter hunting by whales could have arisen through a chance discovery, these whales may have turned to otters because alternative prey had been reduced to low numbers. The large plankton-feeding whales that are an important target for killer whales had been reduced to low numbers by whaling. Seals and sea lions also declined, perhaps because there were fewer fish due to fishing or because killer whales switched to eating more seals and sea lions once the whales were removed. The decline in the otter population has had expected ramifications down the food chain, with an increase in sea urchins and a decrease in kelp (Fig. 5.3).

48 ECOLOGY OF A CHANGED WORLD

Trophic Cascades

One important conclusion from the otter and whale story is that large animals have a major impact on the abundance of other species. As we will document later in this volume, humans have hunted out large animals in many places and are continuing to do so. The loss of these animals generates a trophic cascade down the food chain that produces dramatic effects on species abundances, with some becoming very common once their predator is removed and others being lost from the system. An interesting finding is that the mere presence of predators can keep prey species at low levels by causing changes in behavior. For example, the baboon population in West Africa has increased in recent years; this development is associated with the loss of lions and other large cats (sometimes called "the baboonification of Africa"), even though lions do not usually eat baboons. When lions are present, baboons spend a great deal of time up in trees, neither feeding nor mating, but when lions are absent, the baboons spend much more time on the ground, engaged in both activities. Such behavioral effects of predators on prey are termed trait-mediated effects. The trait in this case would be the baboon's behavior, which is affected by the presence of a lion.

5.2 Food Webs

On the Channel Islands off the Californian coast, a marine reserve established in 1978 contains a large kelp forest. Based on the presence of trophic cascades in similar communities in Alaska, one might expect otters to be present, but they have yet to recolonize this area. What, then, keeps urchin levels low and kelp forests thriving in this reserve? It turns out that the California spiny lobster is an important predator of sea urchins. The lobster ranges from central Baja California to just north of Los Angeles. In the presence of lobsters, urchins are at low levels and the kelp forests are lush (Fig. 5.4). Thus a decline of otters in southern California may have had less of an effect on the kelp forests than it did in Alaska, at least until humans overharvested lobsters. This shows how similar communities in different geographic regions can exhibit varying patterns of trophic interactions.

Note that where lobsters and otters do co-occur, they compete for urchins. Adding lobsters to the food chain generates a food web, with species arranged both horizontally and vertically. Relationships are further complicated because otters eat spiny lobsters (Fig. 5.5). If spiny lobsters increase, will otters increase or decrease? One path, called the direct path because it is a result of an interaction between the two species, predicts an increase. The other path, called the indirect path because it is mediated by another species, predicts a decrease. What would

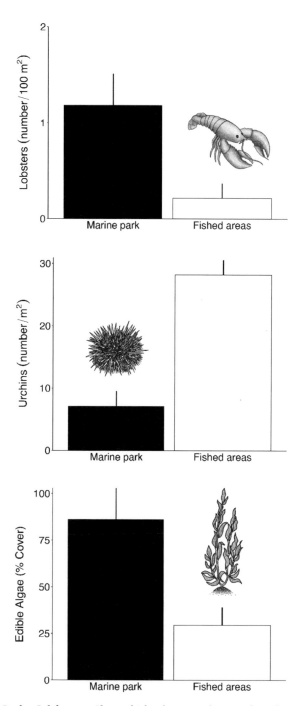

Figure 5.4 In the Californian Channel Islands, sea urchins are heavily impacted by lobsters, as shown by a comparison of a protected area (on the small island of Anacapa) with unprotected areas. Consequently, kelp is more abundant in protected areas. The standard errors are based on surveys at 2 sites in the protected area and 14 sites in the unprotected area. (K. D. Lafferty. [2004]. Fishing for lobsters indirectly increases epidemics in sea urchins. *Ecological Applications* 14: 1566–1573)

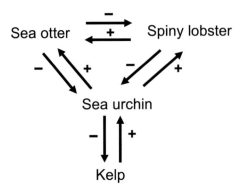

Figure 5.5 A portion of the food web containing some species in southern California.

actually happen depends on the strengths of the two paths. The effect of otters on lobsters is simpler. Both paths from otters to lobsters predict a decrease (one path the consequence of predation and the other the consequence of competition), so lobsters should decrease if otters increase.

5.3 Communities

A community is a group of interacting species that inhabit the same place. A community may contain hundreds of species, whose interactions can be diagrammed as a complex food web. As noted in Chapter 4, if every species has a different niche, multiple species can coexist. Many other factors might affect what species are present and their abundance. Other mechanisms outlined in Chapter 4 include disturbances that create a diversity of different niches over time and space, and also the possibility that species are so similar that they persist together in the same niche for a long time, without displacing one another.

Predation is also an important factor. Darwin was the first to point this out, noting that more grass species coexist in an area grazed by sheep or rabbits than in areas without grazers. In a classic experiment on the Washington state coast, starfish were removed from rock pools over several years. The number of species covering the rocks declined to one, the mussel, which is the dominant competitor. When starfish are present, they specialize in consumption of mussels, freeing up space for other species, notably limpets and chitons (which like the mussel are mollusks), and barnacles, which would otherwise be pushed out by the mussel.

Although predators can maintain diversity in an ecological community, that is not always the case, and it is perhaps more common for predators to depress community diversity. We saw an example of this effect in the case of the sea urchin, which can destroy kelp forests and the species associated with it.

5.4 Conclusions

The way species are connected critically affects the way the world will change in the future. In principle, we can work out the effects of one species on another by tracing all paths in the food web. To do that we would need to know not only the sign of interaction, but also how much a change in the numbers of one species causes a change in the numbers of another to which it is directly linked. We would also need to assume that these effects do not change over time. But even without precise knowledge of magnitudes, it is clear that effects can be very large. Climate change, pollution, invasive species, and predators may directly affect the abundance of individual species. Changes in the abundance of even one species can have huge ramifying effects throughout the food web.

6

Parasites and Pathogens

A parasite is an organism that exploits another living organism, called the host, causing it some degree of harm but rarely killing it. In a path diagram, parasitism is therefore a $+,-$ interaction, just like predation; parasites harm hosts, but hosts benefit parasites. Most parasites live on or in their host, but some do not (Fig. 6.1). A single parasite usually does not usually kill its host, but some parasites multiply within the host to the extent that the more susceptible host individuals do die. The term parasite is most generally applied to animals such as ticks and to plants such as mistletoe, which parasitizes trees. Bacteria, viruses, and fungi act as parasites but are given a different name: pathogens. Parasites and pathogens have major effects on the population size of their hosts, including humans, and we devote this and the next chapter to their effects. This chapter

(1) Describes the basic biology of parasites.
(2) Considers how parasites affect the population sizes of their hosts.
(3) Discusses implications for communities of species.
(4) Introduces pathogens and associated diseases with respect to humans.
(5) Considers the principles of vaccination.
(6) Introduces a model to determine the fraction of people that need to be vaccinated to prevent disease spread.

In the next chapter, we discuss how the evolution of parasites affects the degree to which they harm their hosts.

6.1 Parasites

Ectoparasites, such as ticks and lice, live on the outside of an animal, whereas endoparasites live on the inside. Endoparasites include many kinds of worms and crustaceans. The world contains many more parasites than predators. For example, two deer in a Costa Rican rainforest carried a total of fifteen large parasite species, including five species of ticks and a parasitic fly on the outside (ectoparasites) and a fly larva, seven species of nematodes (roundworms), and a tapeworm on the inside (endoparasites). In that same forest, an iguana had seven species of roundworms living in just one part of its gut.

Ecology of a Changed World. Trevor Price, Oxford University Press. © Oxford University Press 2022.
DOI: 10.1093/oso/9780197564172.003.0006

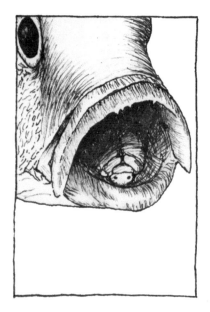 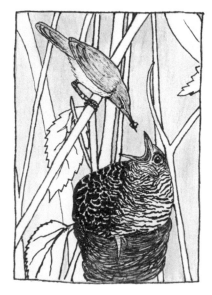

Figure 6.1 Two remarkable parasites. *Left:* The tongue worm eats the tongue of a fish and then acts as the new "tongue." *Right:* The Eurasian cuckoo lays its eggs in other birds' nests. When the young cuckoo hatches, it pushes the host's eggs or young out of the nest, and then the host (here, a reed warbler) raises the cuckoo as if it were its own offspring.

Transmission

A body should be a good place to live. If the parasite successfully establishes itself, it is surrounded by food in a more or less constant environment. The problem is finding and establishing in the host. Many parasites are transmitted when a host individual comes close to another one (e.g., sexually transmitted diseases are passed on during copulation, and mites move from parent birds to their offspring when they feed them). Other parasites, such as nematode worms that live in the intestine, rely on their eggs being eaten after being passed out in feces.

Yet other parasites have a primary host in which the adult form lives, with other host species aiding in transmission. Eggs or larvae are passed from the primary host through one or sometimes more intermediate hosts that facilitate movement from one individual to another. The guinea worm from northern Africa has one intermediate host. Its primary host is a human. Its intermediate host is a water flea (a crustacean, related to barnacles, crabs, etc.). The larval form of the guinea worm passes into humans when they drink water containing infected water fleas. The worms develop inside the human body, and the females

54 ECOLOGY OF A CHANGED WORLD

and males subsequently mate, whereupon the males die, but the females grow for up to a year. They eventually migrate to the legs and feet and exit through the skin when the infected person wades into water (people immerse themselves in water because the worm makes the legs feel very hot). The worm then releases millions of larvae, where they search for the water fleas. The worm can be controlled and even eliminated if people drink boiled water. That strategy has substantially reduced the number of cases of human infection over the past 25 years. According to the World Health Organization, more than 3.5 million cases of infections in people in 1986 declined to about 1,000 in 2011 and just 20 in 2020, albeit with some recent infections of dogs. Its decline is not likely to be lamented.

Many other parasites have more than one intermediate host. Consider the 2- to 3-cm lancet fluke (a kind of flatworm), which has three hosts in total:

(1) Adults live in the liver of many vertebrates, including sheep. These flatworms are hermaphrodites: that is, each individual makes both sperm and eggs. One fluke lays 20,000–50,000 eggs a day, which pass out of the gut with the feces, some to be ingested by a snail.

(2) The eggs hatch in the snail, and the larvae themselves reproduce asexually inside the snail. The snail covers hundreds of larvae at a time in slime and coughs up the slime ball.

(3) An ant eats the slime ball. One or more of the larvae then migrate to the ant's brain, causing the ant to march up a blade of grass, attach to it, and wait to be eaten by a passing sheep.

In these examples, the parasite alters the host's behavior to its own benefit. The guinea worm causes an infected human to wade into water, and the larval fluke causes an ant to walk up a grass blade. In the case of the flatworm larva and the ant, the flatworm actually causes the ant to commit suicide. Manipulation of hosts is a recurring theme in parasite biology.

6.2 Parasite Effects on Host Populations

Parasite and Host Population Cycles

Interactions between parasites and hosts lead to a variety of complicated fluctuations in the numbers of both host and parasite. If a parasite has only one host, the presence of many hosts increases the chances that one host will encounter another, raising the chances of parasite transmission. At very low host population sizes, the parasite may go locally extinct because an infected

Figure 6.2 *Left:* In crowded conditions, parasites are easily transmitted and epidemics may break out. *Right:* In uncrowded conditions, a host may die before the parasite is transmitted, resulting in the parasite disappearing. However, if host individuals continue to actively seek each other out (e.g., as mates), this effect will be lessened, and the parasite may be maintained even in small populations.

individual from the host species dies before it encounters an uninfected host (Fig. 6.2). Consequently, the host population may then recover.

One of the best examples of population fluctuations driven by a parasite is that of the red grouse on Scottish moors. The grouse is shot for sport. While shooting obviously reduces numbers, it is not the main cause of the grouse population decline. Indeed, in some years, grouse are at such low numbers that few are allowed to be shot. Because grouse shooting is an industry worth millions of dollars, the causes of its population crashes have been thoroughly studied. The culprit is a nematode worm, which infects the grouse when the birds ingest the worm's eggs. More than 10,000 worms may be present in a single grouse. Worm infections lower offspring production and affect adult survival, in part because infection makes the grouse smellier, thus making it easier for predators to locate the nest. Red grouse populations cycle across years (Fig. 6.3). In a remarkable experiment, researchers prevented population crashes by catching the grouse and giving them a dose of worm-killing medicine. The effectiveness of this treatment in stabilizing the grouse population definitively shows how a parasite can have a major ongoing impact on population size. In natural conditions, when red grouse are at low numbers, infection drops. The grouse populations increase and the cycle begins anew.

In the grouse example, transmission of the parasite from one host to another depends on host density; this is what allows the host population to recover. Transmission of this kind is termed density-dependent transmission. In other cases, transmission is not affected by the density of a host, but only by

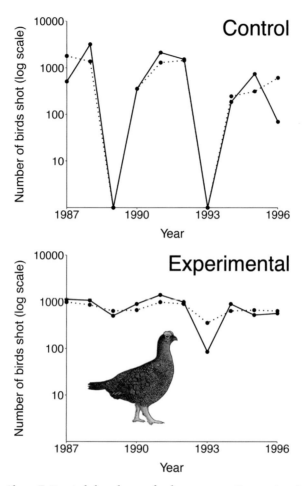

Figure 6.3 *Above:* Estimated abundance of red grouse over time on two Scottish moors (indicated by the dashed and solid lines respectively), based on the number shot by hunters. *Below:* Estimated numbers at two other nearby locations where many birds had been given a deworming treatment. (P. J. Hudson, A. P. Dobson, and D. Newborn. [1998]. Prevention of population cycles by parasite removal. *Science* 282:2256–2258)

the proportion of hosts that are affected. For example, if a female always seeks a partner to mate with, however difficult it is to find him, the probability of her becoming infected from a sexually transmitted disease depends only on the probability of the male being infected. This kind of transmission is termed frequency-dependent transmission and will not result in a rescue effect when host populations become rare.

6.3 Parasites and Communities

In California's salt marshes, such as Carpinteria Reserve just east of Santa Barbara, herons and other birds are parasitized by a flatworm, *Euhaplorchis californiensis,* that has two intermediate hosts, a snail and a fish. The heron sheds the flatworms' eggs on to the mudflats, where a snail may eat them (the native California horned snail is also parasitized by 16 other flatworm species). The larvae hatch inside the snail where they multiply and destroy the snail's gonads, castrating males and females alike. Flatworm larvae (termed the *cercaria* larva) swim out of the snail and penetrate a fish where they encyst in the fish's brain, causing the fish to swim to the surface, flash, contort, and jerk, making it 10 to 30 times more likely to be eaten by herons in experimental studies. In sum, the parasite causes the total numbers of fertile snails and fish to decrease. Herons may not be greatly affected by the parasite, and their numbers may even increase because the parasite alters fish behavior, making it easier to get food. This example illustrates how important it is to consider parasites as contributors to other species' population sizes. A more or less complete food web has been worked out for the Carpinteria Salt Marsh, and when parasites are added, it becomes orders of magnitude more complex (Fig. 6.4).

6.4 Infectious Diseases in Humans

Human infectious diseases result from both parasites such as the guinea worm and malaria (a single-celled organism), and pathogens including bacteria (e.g., cholera) and viruses (e.g., measles). In 2019, about 5 million people died from all infectious diseases combined (Table 6.1). Just under half were in Africa, including one million children under the age of 14. However, in total, mortality from infectious diseases has been reduced by about half between 2000 and 2019, attributed to improved sanitary conditions and vaccinations. COVID-19 reversed this decline, adding more than 5 million deaths by the end of 2021.

The World Health Organization tallies the effect of diseases using a metric termed the DALY (Disability Life Adjusted Years), which quantifies the number of years lost due to preventable causes over a (hypothetical) healthy life. The tally includes a weighting to account for both premature death, and healthy years lost without death (someone who lives to the age of 60 but cannot work from their 50th birthday because they have been debilitated by a disease has a higher DALY than someone who is healthy up to the age of 60 and is then killed in a traffic accident). In 2019, infectious diseases contributed 3% to the DALY in the Americas, but 33% in Africa, given the high infant mortality. Measles causes deaths primarily of infants (more than 100,000 in Africa in 2019) and so contributes a

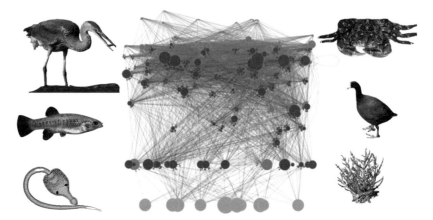

Figure 6.4 (Color plate 2) Food web from Carpinteria Bay. Each circle is a species in the web, with plants along the base (circle size corresponds to abundance). Arrows indicate the flow of energy and include cases where predators eat parasites inside prey. Plants are low in the food web, with predators like herons and their parasites at the top (figure courtesy J. P. McLaughlin and K. D. Lafferty). *Left:* Part of the life cycle of a flatworm. The cercaria larva (below) is the flatworm's free-living stage between California horned snail and the California killifish (illustrated). The fish is consumed by birds such as the great blue heron, where the larva matures into an adult worm, infesting the liver. *Right:* pickleweed, eaten by mice and snails, and the American coot and striped shore crab, both of which eat algae, but also animals. (Photos, on left, from below by Todd Huspeni, Zach Alley, and Alain Carpentier. Photos on the right from below are from Wikipedia (CC A 2.0), M. Baird (CC A 2.0), P. Tillman (CC BY-SA 3.0). (K. D. Lafferty, R. Hechinger, J. Shaw, et al. [2006]. Food webs and parasites in a salt marsh ecosystem, pp. 119–134, in S. K. Collinge and C. Ray, eds., *Disease ecology: Community structure and pathogen dynamics*. Oxford: Oxford University Press)

Table 6.1. Global Deaths Attributed to Major Communicable Diseases*

	Deaths 2000	Deaths 2019
AIDS-related (virus)	1,379	675
Diarrheal disease (bacteria, viruses, protists)	2,647	1,525
Tuberculosis (bacteria)	1,738	1,200
Malaria (protist)	721	410
Measles (virus)	544	165
Total infectious disease	8,613	5,101

*Units are 1000s of people (from the World Health Organization, www.who.int).

great deal to the DALY. In contrast, COVID-19 causes deaths primarily in older people (on average COVID-19 mortality has reduced an individual's lifespan by 16 years). In Africa, restrictions imposed by COVID-19 resulted fewer or delayed measles vaccinations, thereby threatening hundreds of thousands of children. Possibly the result will be an increase in the DALY above what it would be if instead measles vaccinations had gone on as before, despite the expected increase in COVID-19 infections. The impact of COVID-19 is considered in more detail in Chapter 19.

Recent devastating diseases in humans have resulted from contact with other animal species. Of the two forms of human immunodeficiency virus, known as HIV-1 and HIV-2, one is related to HIV from chimpanzees and the other to HIV from a monkey known as the sooty mangabey. The viruses causing ebola and COVID-19 are derived from bats. Influenza was originally a disease of birds. Although influenza now circulates permanently in humans, an outbreak in 1918–1919 described below, was likely a new infection from a bird. Domestication of some animals and the eating of wild ones partially explain the recent appearance of these diseases in human populations. An additional reason for their spread lies in increased human density due to the rise of agriculture and urbanization, increasing contact with infected individuals. As noted, high population densities facilitate disease transmission, whereas low densities may actually cause the disease to die out (Fig. 6.2).

6.5 Immunity and Vaccination

The number of new infections that one infected person causes is an important quantity; it is termed the reproductive value and is symbolized by R. The reproductive value at time $t = 0$, Ro, is the number of new infections coming from an infected person when a disease enters a population of susceptible individuals (although we read this as time $t = $ zero, Ro is pronounced R naught). Measles is particularly infectious over a four-to-nine-day period and $Ro = 15$. The upside of surviving a measles infection is that individuals subsequently become immune. Thus, with time most encounters do not involve a transmission event. Before vaccines were developed, measles occasionally broke out in cycles (Fig. 6.5). In pre-vaccination England, for example, these outbreaks occurred mostly in the spring because people move about more during this time of year.

In England, cases of measles were rare in many years because most people had caught the disease before and thus had developed immunity. Over time, a pool of susceptible individuals builds up. At some point, a threshold is crossed, and measles can once again grow rapidly. In this case, unlike the case for the

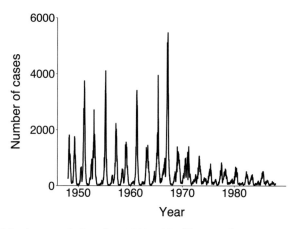

Figure 6.5 Measles cases in London, 1948–1987. The measles vaccine was first developed in 1963, and an improved form came into widespread use in 1968. Data collated by B. Bolker (ms.mcmaster.ca/~bolker/measdata.html)

red grouse (Fig. 6.3), the outbreaks are not driven by changes in the total numbers of the host, but by changes in the numbers of the host that can be infected (Fig. 6.6; compare with Fig. 6.2). Although details depend on how people are distributed and move about, measles may require human populations of about 300,000 to be maintained. With population sizes smaller less than this, at some point in time so few people are susceptible that infected individuals do not encounter a susceptible one, R drops below 1 and the disease dies out. This phenomenon is termed herd immunity because even if you have not contracted the disease, you are not likely to be infected if most individuals already have become immune.

The presence of herd immunity illustrates how vaccines can eliminate a disease, even if everyone in a population is not vaccinated. A vaccinated individual develops immunity without necessarily getting sick from the disease. Vaccination thus reduces the number of susceptible individuals in a population, and if that number goes below a threshold, the disease will die out instead of spreading. Even if a few people in the population are not vaccinated, vaccination of a high percentage of the population means that the remaining susceptible individuals will not encounter each other frequently enough for the disease to persist. In the case of measles, if we assume that fifteen encounters lead to transmission in a susceptible population ($Ro = 15$), then vaccination of fourteen out of fifteen people (93%) will result in one transmission event per infection ($R = 1$), provided the vaccine is 100% efficient. More generally, the fraction of people one needs to vaccinate for the disease to be eliminated should be greater than ($Ro - 1$)/Ro. Vaccination of about 95% of the Romanian population in the mid-1980s

PARASITES AND PATHOGENS 61

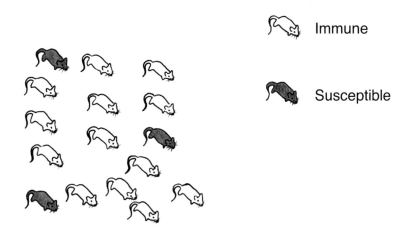

Figure 6.6 Once infected, individuals may become resistant to subsequent infection, which from the point of view of a parasite is equivalent to death of the host (compare with Fig. 6.2). In this case, unlike death, if individuals do continue to come into contact with others, e.g. to mate, they will often not carry the parasite, so the parasite may not be able to maintain itself.

led to virtual elimination of measles, with occasional small outbreaks thought to have resulted from introduction by visitors.

6.6 Model of Disease Dynamics

In Chapter 2 we derived a model of the early stages of the spread of HIV based on the exponential equation. We now consider the essentials of a model that traces the complete progress of a disease through the population, accounting for a slowdown in transmission rates due to death and the development of immunity, and resulting in the reproductive value dropping below one. The model expands on the one introduced in Chapter 2, to examine the spread of influenza in Chicago during the 1918–1919 pandemic. About 2% of the time influenza results in death, either directly or by causing pneumonia (inflammation of the lungs), but people who recover develop temporary immunity (weeks to years). Figure 6.7 shows the accumulated number of deaths from influenza and pneumonia in Chicago, September–October 1918. After 45 days more than 9,000 people were dead, and almost 50 times as many had contracted the disease, close to one-fifth of Chicago's population at the time. This figure shows that the doubling time, or the time it takes for the number of dead to double, comes to about 5 days in the early stages. Provided individuals who die are as likely to contract

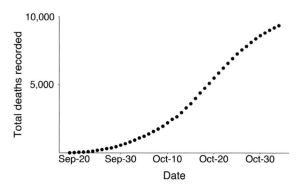

Figure 6.7 Cumulative number of deaths from influenza and pneumonia in Chicago, September–October 1918 based on daily records. The time taken to go from 500 to 1,000 deaths and from 1,000 to 2,000 deaths was approximately five days. (J. D. Robertson and G. Koehler. [1918]. Preliminary report on the influenza epidemic in Chicago. *American Journal of Public Health* 8:849–856)

the disease as those who do not, this should approximate the doubling time for the total number of people infected during the early stages. Across thirty-five U.S. cities, the doubling time is estimated to have been about 3 days. Following the HIV model, using the 3-day estimate and assuming exponential growth, we compute the per capita growth rate, $r = \ln(2)/t_{double}$, giving a value of $r = 0.25$ infections/infection/day.

Because an infected person clears the infection after about 6 days, a value of $r = 0.25$ infections/infection/day naively implies that one person infects about $6 \times 0.25 = 1.5$ others. However, an essential parameter needs to be added to the model. Someone who has contracted the influenza virus is only infectious for the last 4 days of the infection. The noninfectious early stage is the extra parameter (termed the latent period). The latent period for influenza is about 2 days. When this is added to the model, computer simulations show that a growth rate of $r = 0.25$ requires each individual to infect two others during the time they are infectious ($R = 2$). The implication is that, if about half the population was efficiently vaccinated, R would be less than 1 and an epidemic would be prevented.

The model we have introduced is termed the SEIR model:

1. *Susceptible*: The number of individuals susceptible to the disease. The population size of Chicago in 1918 was 2.5 million.
2. *Exposed*: Contracted the virus but not infectious. In the case of influenza, each infected individual is in the exposed class for an average of 2 days.
3. *Infected*: Infected and infectious. Infection typically lasts 4 days.

4. *Recovered:* Individuals who have recovered have acquired immunity and will not be infected again. About 98% of all individuals recovered in the 1918 pandemic.

6.7 Conclusions

Infectious diseases spread quickly and are most virulent when host populations are large and dense. This places common species, including humans, at risk. Diseases have also caused large reductions in the population size of previously common animals and plants. The take-home message from this chapter is that when numbers are low and individuals are spaced out (i.e., the population is at low density), individuals may either die or clear the disease before transmitting it, and the disease eradicated. This implies that the host species should not go extinct. That is all very well in principle, but sometimes the pathogen can persist in other hosts, or outside of the host altogether, or transmission probability is weakly affected by density. Such issues are taken up in Chapter 19, where we consider emergent diseases in humans, wild animals, and plants in more detail.

7

Evolution and Disease

Evolutionary principles explain why some diseases that have been long associated with humans, such as malaria, are so much more debilitating than others, such as the common cold. In this chapter, we show why. Specifically, this chapter:

(1) Examines host–parasite interactions to illustrate the general principles of evolution by natural selection, a framework that we will use throughout the rest of the book.
(2) Briefly describes how host immune systems have evolved as one means of combating parasites and pathogens.
(3) Considers how principles of evolution by natural selection can be used to understand why parasites and diseases vary so much in how sick they make the host.
(4) Illustrates these ideas by considering three diseases that continue to kill many people every year (AIDS, cholera, and malaria).

7.1 Evolution of Drug Resistance

Perhaps the most famous quote in evolutionary biology, and possibly in all of biology, is Theodosius Dobzhansky's contention that "nothing in biology makes sense except in the light of evolution." Evolution is the change in the average value of some trait, such as body weight, or the ability to fend off a parasite, from one generation to the next. Natural selection is the difference in the survival or reproduction of individuals resulting from differences in some trait such as body size or disease resistance. For example, because some individuals are larger than others, they may produce more offspring, in which case we would state that natural selection favors large body size in association with fecundity. Evolution by natural selection requires that the differences in the trait, in this case body size, are passed on to offspring—that is, offspring and their parents should be similar to each other, at least to some extent. In that case, if adults that have more offspring differ in body size from adults that have, fewer some of that difference will be passed on to their offspring (i.e., evolution has happened). Here, we illustrate how the biology of infectious disease

Ecology of a Changed World. Trevor Price, Oxford University Press. © Oxford University Press 2022.
DOI: 10.1093/oso/9780197564172.003.0007

makes sense when it is put into the Darwinian framework of evolution by natural selection.

Drug Resistance Arising in the Human Immunodeficiency Virus

The drug azidothymidine (AZT) was used in early attempts to control human immunodeficiency virus (HIV). The virus invades and destroys white blood cells that are an essential part of the human immune system, which provides a primary defense against parasites and pathogens. In animals, RNA is transcribed from DNA and the RNA is then translated into protein:

$$DNA \rightarrow RNA \rightarrow protein$$

HIV must work differently, for the starting point is RNA. When the virus adheres to a suitable white blood cell, it injects its RNA into the cell. The RNA is then translated into protein by the host cell. The first protein to be manufactured is an enzyme that causes the host cell to copy virus RNA into DNA. That DNA then inserts into the host's chromosome. Multiple copies of the viral RNA are subsequently transcribed from the inserted DNA, and then the proteins required to make the viral coat manufactured by the cell. So, the sequence is

$$RNA \rightarrow DNA \rightarrow RNA \rightarrow protein$$

AZT inhibits the first step in this chain, in which the RNA virus is transcribed into DNA. Because humans do not make DNA from RNA, the drug selectively slows down reproduction of the virus without seriously affecting the patient.

Mutations in the virus are common. Some of these mutations overcome the inhibiting effect of AZT. The viruses carrying the resistant mutation outreproduce the others, and resistant particles come to dominate within the host's body (Fig. 7.1). The resistant virus has had more offspring than the nonresistant form, and importantly, its offspring are also resistant. This is Darwin's theory of evolution by natural selection. First, the mutant form has a higher rate of reproduction (natural selection), and second, the mutation is inherited (offspring viral genomes are the same as their parent). In this particular case, the rapid evolution of resistance illustrates the general problem associated with using drugs to treat infections (Fig. 7.1). Higher doses of AZT are needed to achieve the same kill rate, but at high doses AZT starts to interfere with host cell replication.

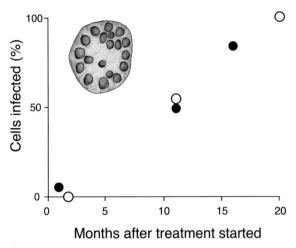

Figure 7.1 Two different patients infected with HIV (one is filled circles, the other, open circles) were placed on a course of the drug azidothymidine (AZT). HIV was isolated from each patient at different times from the beginning of the treatment. The virus particles were then inoculated on a plate of cultured human cells in the presence of 1 μl AZT and incubated for 3 days. On these plates, if a cell is infected, the infection spreads to nearby cells, which fuse to produce a single "giant cell" that can be easily observed (inset). The figure shows the number of giant cells present relative to the number before the drug was applied. After 20 months, the virus has evolved to become as infectious as it was prior to AZT application. (B. A. Larder, G. Darby, and D. D. Richman. [1989.] HIV with reduced sensitivity to zidovudine (AZT) isolated during prolonged therapy; example described in S. Freeman and J. C. Herron. [2004]. *Evolutionary analysis.* New York: Pearson)

7.2 The Immune System as a Defense against Parasites

Organisms are constantly being bombarded by parasites. If these parasites were always successful, the result would be rapid debilitation of the host. However, host species have immune systems that mount defenses and keep pathogen levels low. The importance of an immune system is apparent from our study of HIV. This virus is not in itself a cause of illness, but it destroys an essential component of the immune system, thereby weakening defenses against diseases such as those due to the tuberculosis bacterium.

The innate immune system, present in both plants and animals, is the first line of defense against foreign particles. For example, inflammation, making a wound turn red and eventually become covered with pus and scab, is a consequence of white blood cells engulfing and digesting foreign particles and then themselves dying. Besides this innate mechanism, animals also have an acquired immune

EVOLUTION AND DISEASE   67

system, which generates immunity lasting beyond the infection. As noted in Chapter 6, once an individual has had measles, they do not catch it again, whereas for other diseases such as influenza, immunity can be lost after some time.

The acquired immune system in humans is complex, and we describe only a part of it here, focusing mainly on showing just how complex it is. White blood cells, including B cells and T cells, are essential components. B cells detect and respond to foreign particles outside of cells, whereas T cells respond to an infection after it has been processed by certain host cells. Circulating B cells in the blood carry a unique protein receptor, called an antibody, on their surface. In humans, millions of different kinds of B cells turn over each day in the bone marrow. If a B cell carries an antibody that happens to be of the right configuration to bind a foreign particle (termed the antigen), the cell becomes activated, and a subclass of T cells (helper T cells) help B cells to produce an enormous amount of a soluble form of their antibody molecule, which is released in the blood. The antibodies bind the foreign particles, resulting in coagulation and subsequent engulfment by other immune cells. Help from T cells also allows for the production of long-lived, antigen-specific, memory B cells so that if the same foreign body were to invade in the future, the immune response would be quicker: hence the name "acquired immune system." The basis of a vaccine strategy is to trigger this type of response, often by injecting dysfunctional copies of the parasite/pathogen. Consequently, if an individual (or population) is exposed to the real pathogen in the future, they will already have existing immunity. In addition to helper T cells, other T cells directly detect infected cells and trigger their death. These cytotoxic T cells, as well as the helper T cells, form a memory population similar to the memory B cells.

We have given a lot of attention to the acquired immune response because it is a marvelous example of evolution by natural selection. Individuals with more efficient immune systems are able to survive and leave more offspring than those with less efficient immune systems. Because their offspring inherit the superior system, that system becomes more frequent in the population. Ongoing changes of this kind have resulted in the highly complex immune systems that we see today.

Successful parasites must evade their host's immune system. HIV escapes by hiding inside helper T cells. The liver fluke sequesters host molecules on its surface, thereby appearing to be part of the host. These are two ways parasites have evolved to overcome the host's immune system. They are also products of evolution by natural selection. Now we see a conflict. The host is evolving in order to defeat the parasite, and the parasite is always evolving to efficiently exploit the host. This is quite different from a drug such as AZT. The drug does not evolve to defeat HIV's resistance, which means the pathogen can potentially evolve to perhaps defeat it, in the absence of any retaliation.

7.3 Evolution of Virulence

Myxomatosis

The evolution of both the parasite and the host can be documented by studying virulence (kill rate) before and after an epidemic. Myxomatosis virus is the clearest example of how both host and pathogen have evolved (Fig. 7.2). In 1759, 24 rabbits were introduced from Europe into Australia. They rapidly increased in number and spread over half of Australia, destroying the habitats of native animals, and they also affected the economy of Australia by eating sheep food. By 1950, Australia may have had 600 million rabbits.

The myxomatosis virus is native to South America, where it is present in a different rabbit species. In 1950, some rabbits infected with myxomatosis virus were released into the wild in Australia. Initial infections resulted in a >99% kill rate, but over years the rate has declined, and more rabbits now survive an outbreak. One reason for the fewer deaths is that the rabbits have evolved to be more resistant. To show this resistance, after a series of epidemics at one

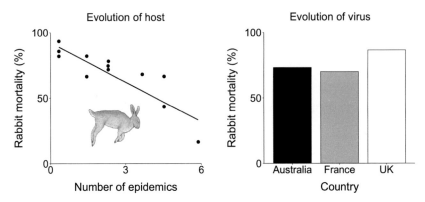

Figure 7.2 *Left:* Australian rabbit populations that have experienced multiple epidemics are more resistant to myxomatosis virus (rabbits were tested with the original virus that had been maintained in the lab). For this study, rabbits were taken from the wild and challenged with a virus that typically caused 70%–90% mortality. *Right:* Virulence in three countries in the 1960s. The virus had been introduced 10–15 years earlier, when it had a kill rate of close to 100% in all three locations. In the report, virulence was divided into five classes (<99%, 95%–99% mortality, etc.). Here, mean virulence in each class was computed by weighting by the number of individuals falling into that class. (F. Fenner and K. Myers. [1978]. Myxoma virus and myxomatosis in retrospect: The first quarter century of a new disease, pp. 539–570. In E. Kurstak and K. Maramorosch (eds.), *Viruses and Environment*. Cambridge, MA: Academic Press, right figure from p. 552)

location, surviving rabbits were brought into the lab and tested with the original myxomatosis strain. Less than 50% of the rabbits died in some tests (Fig. 7.2). The second reason for fewer rabbit deaths is that the virus also evolved. When domestic rabbits were injected with virus 10 to 15 years after myxomatosis had been present in the wild, the kill rate decreased from almost 100% to close to 70%, as also happened after introductions of the virus to France and to the United Kingdom (Fig. 7.2).

Thus, both the host and the parasite have evolved. While it is obvious that rabbits should evolve resistance, it is less apparent that the virus would evolve to be less harmful (i.e., kill the host less frequently). What is the explanation for this result? The answer likely comes from the difficulties of transmission. Mosquitoes transmit myxomatosis from one host to another, and mosquitoes become scarce in the winter. Those virus strains that kill their host quickly die off with it, whereas those that are less virulent enable the hosts to live long enough for a mosquito to visit them, and this increases the chance of transmission. In this way, natural selection favors those strains that are less likely to kill their host.

Parasite Control on Virulence

Much of an evolutionary biologist's work involves asking what sets the average value of any trait in a species through the study of trade-offs. For example, during a drought that caused high mortality of Darwin's ground finches on Isla Daphne Major, larger individuals had a greater chance of surviving, but the smaller ones subsequently reproduced more quickly. The balance between these selection events accounts for the average body size of the population. In the present case, the benefits to a parasite of being more benign are offset by costs, thereby determining level of virulence. Using this approach, one can ask how the balance can be shifted, causing parasites to become more or less virulent. At first sight, it seems that a decline in virulence should always be favored because the parasite keeps the host alive and active. Therefore, it makes most sense to ask: what is the countering advantage to an increase in virulence? At least three factors may give an advantage to a virulent parasite over one that is more benign:

(1) If large numbers of a parasite's offspring are present in the host, the chances are greater that one will be successfully transmitted. If the increased chances of being transmitted to another host outweigh the chances of killing the host before being transmitted, virulence will be favored. Here we are assuming that the number of parasites inside the host correlates with host debilitation.

(2) Parasites compete among themselves for resources within the host, and the more aggressive parasites inside the host may win over less aggressive ones, with side effects on host health.

(3) High virulence may be a result of the contest between the host immune system and the parasite. A parasite can die in two ways. First, the host dies. Second, the host's immune system clears the infection. A parasite may be favored to be virulent and even sometimes kill the host, if this means that it can more often defeat the host immune system and persist. For example, some host individuals may be weaker than others (e.g., older ones), and the ability of the parasite to persist in younger, healthier individuals may result in deaths when the parasite is in other individuals.

A useful approach is to start with a "parasite's view" of the world and assume the host can do little to prevent the evolution of a certain level of virulence. Parasites are likely to play the dominant role in determining levels of virulence for two reasons:

(1) Parasites often have many generations for every one generation of the host. Many have large population sizes and high mutation rates. These features imply that parasites show much more potential for new mutations to arise that enable exploitation of the host, and evolution can be more rapid in parasites than in hosts.

(2) Every parasite species depends on a host for its survival and reproduction, but many individuals of the host species may not be infected, so they survive quite happily. For example, in the case of the cuckoo (Fig. 6.1), every cuckoo is raised by another species, but most individuals of the host bird species are not parasitized and raise their own young. In this case, any adaptation the cuckoo makes to improve its chances of being successfully raised is likely to be strongly favored, but any adaptation the host makes to not raise the cuckoo will have much less advantage because it only benefits the individuals that actually encounter the cuckoo.

For these reasons, selection may be stronger and evolution happen more rapidly in parasites than in hosts. Thus, it is worthwhile for us to ask how much parasite evolution alone can explain different levels of virulence. We do this in the next sections by considering three of the main killers in humans. The factors leading to high virulence in these examples are human population density and mode of parasite transmission, which may be through direct contact, water, or an intermediate host.

EVOLUTION AND DISEASE   71

7.4 Virulence of Human Diseases

AIDS: Effects of Host Density

A possible example of spread of a virulent disease associated with an increase in ease of transmission is HIV, which still kills about 1 million people worldwide every year (Table 6.1). A particularly virulent form of HIV appears to have entered the human population about 100 years ago in Cameroon, West Africa. The date and location have been determined from the differences of the RNA sequence between chimpanzees at this location and humans. The strain has now spread across the globe.

The recent expansion of a virulent strain of HIV is associated with changes in human densities and customs, both of which resulted in increased chances of transmission. During the 1960s and 1970s, Africa underwent huge population growth and regional redistribution, resulting in mass migration to cities, particularly of men. Some prostitutes apparently had 1,000 sexual encounters in a year. As we noted in Chapter 6, this increased contact should in itself lead to rapid spread and an epidemic, but it may also have altered selection on HIV that favored increased virulence. A high number of virus particles circulating in the blood increases the chances of their being transmitted during contact. High numbers of particles should also increase virulence, making the individual more likely to die. However, as noted, provided the increased chances of transmission outweigh the chances of death, increased virulence gives the virus an advantage. Thus, increased human contact (1) enables the rapid spread of a virulent parasite that would otherwise remain at low levels or die out and (2) places selection on HIV to become more virulent. Note that although the logic is reasonable, it remains to be demonstrated that the virus has indeed evolved an increased level of virulence.

Malaria: Density and Mode of Transmission

A clearer example of virulence evolution comes from comparison of malaria species. Malaria is caused by a single-celled organism (a protist), rather than a virus. In humans, the disease is due to at least four different species: *Plasmodium falciparum, Plasmodium vivax, Plasmodium ovale*, and *Plasmodium malariae*. Mosquitoes transmit malaria from one human to another. *Plasmodium falciparum* is the deadly species, infecting up to 60% of blood cells, whereas the others infect just 2% and are milder. *Plasmodium falciparum* is found in areas of high human population density, where it outcompetes the other strains of malaria, but may kill the host. *Plasmodium vivax* (the main malaria in Asia) and *Plasmodium ovale* can remain latent in people and then reawaken years later to infect mosquitoes, whereas *Plasmodium malariae* remains at a low level of infectiousness for decades. The different strategies of the alternative species of malaria appear to reflect possibilities for transmission.

High host density likely favors the presence of the more virulent malarial species, *Plasmodium falciparum,* and mode of transmission helps explain why this malarial form is so deadly. Malaria killed more than 400,000 people in 2016, but the virus that causes the common cold has never been known to kill anyone. A compelling reason for the differences between malaria and the common cold lies in how they get from one host to another. The cold virus is passed on when an infected person sneezes on someone else. Hence, selection has favored virus particles that causes people to sneeze, but without debilitating them, so they regularly encounter as many other people as possible. The malarial parasite is transmitted by a mosquito, which means selection favors a strategy of debilitating the person, preventing them from swatting the mosquito. Indeed, when multiple diseases are compared, transmission through an intermediate host is associated with higher primary host death rates. If we define a virulent disease as one that regularly kills more than 10% of all infected individuals, one survey of human diseases determined that half of all diseases transmitted through intermediate hosts are virulent, whereas only a few (5%) transmitted in other ways are virulent.

An inadvertent experiment on malaria demonstrates how quickly virulence can evolve when transmission possibilities increase. *Plasmodium knowlesi* is another species of malaria. It occurs in monkeys (two species of macaque from South Asia) and causes only mild symptoms in humans. In the early decades of the twentieth century, doctors treated the sexually transmitted disease syphilis in humans (caused by a bacterium) by injecting *Plasmodium knowlesi* into the patient, thereby generating a mild fever and slowing reproduction of the syphilis bacterium. Because blood supplies from infected monkeys were in short supply, doctors injected blood from one human previously infected with *Plasmodium knowlesi* into another, then from this person into another, and so on. Consequently, the malarial parasite was selected to reproduce rapidly: any mutation that increases the growth rate of the parasite, and hence leads to higher numbers in the blood, would be more likely to be transferred to a new host. After 170 serial transfers, densities in the blood rose 500-fold. In the end, the malaria was causing as much debilitation as the syphilis.

Cholera: Transmission

A final example of how transmission affects virulence comes from diarrheal diseases. Cholera is a bacterial disease acquired from drinking contaminated water. It is truly a disease of the city, associated with high population density and unsanitary conditions, and it may actually have contributed to the collapse of some early civilizations. Over the past 200 years, seven major outbreaks (pandemics) have spread across the world. More than 100,000 people a year are still affected, mostly in Africa and Asia. Unsanitary conditions in Haiti following both the 2010 earthquake and Hurricane Matthew in 2016 resulted in cholera outbreaks; more than 3,000 people died in 2010.

When someone drinks contaminated water, only about one in 1 million cholera bacteria survive the stomach. The survivors lodge themselves in crevices of the gut and then produce a toxin that causes great thirst and a massive flux of water into the gut cavity. This movement of water rids the gut of other bacterial competitors and sends out bacterial progeny. Within hours of onset of the disease, a liter of stools can contain 100 billion bacteria. A liter of stools may be produced every hour.

Of the seven recorded pandemics of cholera, the latest has been of a different type, *el tor*, so named because it was first identified in El Tor, Saudi Arabia. This cholera variant is less virulent than the other strains. It spread out of Indonesia in the 1960s and 1970s and has replaced the original strain everywhere except in Bangladesh. *El tor* is less dependent on waterborne transmission, survives longer outside the body, and is more often transmitted by contact. It is likely to have become dominant in association with improved city sanitation. More generally, waterborne diarrheal diseases are more virulent than those that are transmitted by contact (Fig. 7.3).

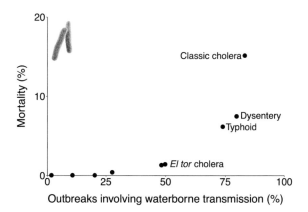

Figure 7.3 Outbreaks of diarrheal disease are more virulent when transmitted by water than by direct contact. Here, mortality is the fraction of people dying in an outbreak, in the absence of treatment. (P. Ewald. [1994]. *The Evolution of Infectious Disease.* Oxford: Oxford University Press.) Inset picture of the *Shigella* bacteria from the Centers for Disease Control, cdc.gov.

7.5 Conclusions

The examples presented in this chapter show that understanding how a parasite has evolved to pass its offspring from one host to another can explain many patterns of virulence. They also show that hosts are always under selection to reduce the negative effects of parasites. This generates an ongoing evolutionary arms race, whereby parasites are countering host adaptations to achieve an optimal level of virulence. One consequence of such a process is that when a parasite or pathogen encounters a host to which it has never been exposed, it can sometimes be exceptionally virulent. As we noted, the introduction of the myxomatosis virus into Australia had an initial kill rate greater than 99%. In this way, the movement of parasites and pathogens across the world has had major impacts on population sizes and extinctions, as will be documented in Chapter 19.

8

The Human Food Supply

Competition, Predation, and Parasitism

The large human population is straining the world's resources, especially the big three: energy, water, and food. The ecological and evolutionary principles covered so far in this volume have much to say about the last-named resource, given that our food consists entirely of other species. We all need to eat, and most of what we eat comes from crops, either directly (in 2017, 82% of all calories consumed) or indirectly from domestic animals, most of which themselves consume food grown as crops. The main wild food source is fish, which accounts for just 0.5% of all calories consumed, less than beer (1%). According to the United Nations Food and Agriculture Organization, more than 10,000 plant species are known to be edible, but only 150 to 200 species are regularly eaten. In fact, just two species, rice and wheat, account for about 40% of all calories consumed by humans. These two species, along with sugar cane and corn, are among the most common plant species on earth. Note that crops are also grown for purposes besides our consumption. Corn forms 5% of calories directly consumed, including extraction of fructose as a sweetener for foods (corn-based sweeteners are now more common than sugar in the United States). But corn is also the main ingredient of cattle feed (more than 30% of the total crop in the United States) and an important biofuel (about 40% of the U.S. crop). Likewise, soybeans are primarily grown for cattle feed and biofuel rather than for us to eat.

Threats to crops include all those pressures on a species we have discussed so far: competition, disease, and predation. This chapter considers each of these pressures in turn and asks how farmers defend against them. Weeds—defined as plants that are not valued in the place they are growing—compete with crop plants for resources such as space, light, and water. Animals eat plants. Plants also suffer from many diseases. Competitors, pests, and disease-causing pathogens are an ever-present threat to our food supply. As their populations increase and they keep evolving, we have to come up with new ways to combat them. The contest never ends. It is possible that we will become better at controlling crop enemies. It is also possible that we will experience a massive crop failure, as has happened at times in the past, such as the Irish potato famine in the nineteenth century, which is considered later in this chapter.

Ecology of a Changed World. Trevor Price, Oxford University Press. © Oxford University Press 2022.
DOI: 10.1093/oso/9780197564172.003.0008

76 ECOLOGY OF A CHANGED WORLD

This chapter therefore covers the impacts of:

(1) Plant competitors on crops.
(2) Pathogens (disease) on crops.
(3) Predators, notably insect pests, on crops.

Each section assesses both the direct impacts of these threats on crops and how competitors, pathogens, and pests are constantly evolving to counter our attempts to reduce these impacts.

8.1 Plant Competitors

Weeds are considered the biggest single cause of crop losses. For example, despite widespread removal by hand and application of herbicides, about 10% of rice worldwide is thought to be lost to competition from weeds. In one experiment, planting one-tenth the number of barnyard grass plants alongside rice plants resulted in a 50% reduction in rice yield. In experimental plots at the International Rice Research Institute in the Philippines, an absence of any weeding reduced yields by 60% or more (Fig. 8.1). Weeding early is particularly

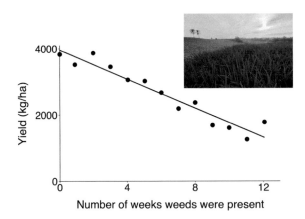

Figure 8.1 Yield of rice in experimental plots in the Philippines, when weeding was started a certain number of days after planting, in the 1975 wet season. In other plots where no weeding was carried out at all, yield was about 1,300 kg/ha. (K. Moody. [1990]. Yield losses due to weeds in rice in the Philippines. In *Crop loss assessment in rice*. Manila, Philippines: International Rice Research Institute, p. 196). Photo provided by the International Rice Research Institute, courtesy of Erin Razon.

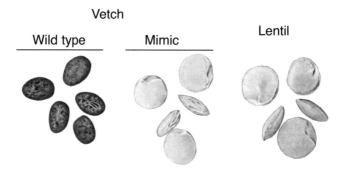

Figure 8.2 (Color plate 3) Vetch has evolved a mimic seed, resembling lentil seeds. (Redrawn from F. Gould. 1991. The evolutionary potential of crop pests. *American Scientist* 79:496–507).

effective, for rice plants that have grown can more easily outcompete young plants of other species.

Presumably since the beginning of agriculture, people have weeded by sight, and this has placed a great deal of selection on other plant species to resemble the crops that grow among them. For example, traditionally the seeds kept to be planted the following year were visually sorted from seeds of other species. This practice imposes selection on the weed seeds to look like the crop seeds, and there are some remarkable examples where they are very difficult to separate (see Fig. 8.2). Once planted, variants are also favored if they avoid being pulled out of the ground before they mature. Indeed, in the face of this selection pressure, barnyard grass has evolved in rice fields to become virtually indistinguishable from rice.

Over the past half century, mechanization and innovation have meant that removal by hand has been replaced by chemical herbicides. Some herbicides kill both unwanted plants and the crop, so they are applied before planting. Glyphosate (marketed as Roundup) is sprayed on growing plants. It is absorbed only through leaf and stem, and once in the soil it becomes bound to soil particles, so that the crop can subsequently be safely planted. Other herbicides selectively affect unwanted species and do not damage the crop. Atrazine is a very effective herbicide that inhibits photosynthesis. Corn carries an enzyme that rapidly detoxifies atrazine and is thus not susceptible. Atrazine is widely applied to cornfields in the United States, but it is not currently used in the European Union because of concerns about its effects on human health (Chapter 17).

The first cases of herbicide resistance arose in the late 1960s. In 2008, over 185 weed species had become resistant to one form of herbicide or another, including more than 60 species resistant to atrazine. The year 1996 saw the first

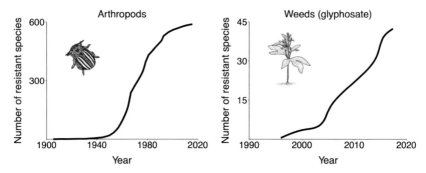

Figure 8.3 *Left:* Increase in the number of pest species (mostly insects) recorded as being resistant to a pesticide since 1910. *Right:* Increase in the number of weeds resistant to glyphosate (Roundup) since 1990. Illustrations are of the Colorado potato beetle, which has developed resistance to some insecticides, and the giant ragweed, which has developed resistance to glyphosate. (F. Gould, Z. S. Brown, and J. Kuszma. [2018]. Wicked evolution: Can we address the sociobiological dilemma of pesticide resistance? *Science* 360: 728–732)

Roundup-resistant weed. Now more than forty species show some resistance to the herbicide (Fig. 8.3), meaning that more herbicide has to be used to obtain the same effect.

8.2 Plant Diseases

All crop species are planted at high densities and usually in monocultures, which we have noted are exactly those conditions that promote the spread of disease. Viruses, bacteria, fungi, and other microorganisms attack plants. The famous Irish potato famine of 1845–1851 resulted from a blight (a kind of water mold related to diatoms) that infects both the leaves and tubers. The potato was brought to Ireland from South America in the 1500s and soon became Ireland's staple crop. Three hundred years later the blight came from Mexico. It can destroy the crop in 1–2 days. More than 1 million of Ireland's 8 million population died, and more than 1.5 million emigrated. Government policies also played a part in this famine, as potatoes continued to be exported from Ireland to England even when they were in critically short supply.

Current management of potato crops includes making sure only blight-free potatoes are used in planting; crop rotation; use of resistant strains of potato; and application of various fungicides. (Although the blight is not closely related to fungi, it was long thought to be related and the chemicals used to control it are still called fungicides.) In Ireland, for example, the crops grown as next year's seed potatoes are most susceptible to the blight because they are harvested late. They receive 8–15 sprays of fungicide to protect them. Clearly, it is an ongoing battle. In the northeastern United States, outbreaks have become more common because the mold can now persist through the winter on a native plant related to the potato, the hairy nightshade.

Presently, we are witnessing outbreaks of fungal diseases in all crops, including the two main staples, rice and wheat. Outbreaks of the fungal disease rice blast across the world may reduce annual crop yields sufficient to feed 60 million people. Two fungal diseases of wheat, stem rust and wheat blast, can cause some fields to have no production at all. Stem rust has affected crops throughout history, whereas wheat blast arose more recently. In the case of stem rust in Europe, scientists noted more than 100 years ago that wild barberry was the overwintering host. This discovery led to development of effective control measures based on removing barberry. Wheat strains resistant to stem rust were developed in the 1960s, but newly emergent fungal strains have caused major losses, especially in the Middle East. Factors driving the spread of emergent diseases are revisited in Chapter 19, with wheat blast as an example, but for now we reiterate that high-density monocultures of single varieties are especially at risk.

8.3 Pests

Herbivores include birds and mammals, but insect pests do far more damage to plants. Many of the spices we add to our foods are a manifestation of plant defenses, which have evolved to make the plant less palatable to insects. One good example is chili pepper, and another is ginger.

All crops have major insect pests. In the United States, corn has at least seven of these pests, plus many with less impact, the most important of which is the corn rootworm. Individual corn rootworms overwinter as eggs, the larvae feed on the roots, and the adults feed on the leaves and corn silk. One tactic used to deal with infestations is to starve the rootworm by rotating crops, for example, by alternating each year between soybean and corn. Amazingly enough, the rootworm evolved a way to skip a year too before it emerges (i.e., to spend almost 2 years as an egg) and thereby become synchronized with the crop. In Illinois, counties with more farms that practiced crop rotation had a higher fraction of the rootworm that follows the 2-year cycle (Fig. 8.4).

Insecticides include synthetic chemicals and those extracted from natural sources. The classic synthetic chemical is dichloro diphenyl trichloroethane (DDT) which was responsible for the decline of peregrine falcons and bald eagles, among many other species, starting in the 1950s (Chapter 17). The most widespread nonsynthetic pesticide is derived from *Bacillus thurigiensis* (known

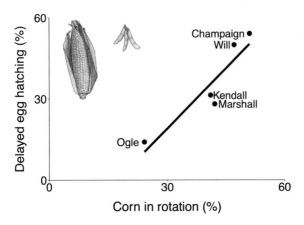

Figure 8.4 In 1986, Illinois counties with a higher proportion of farms rotating corn and soybean from one year to the next also had more corn rootworms that spent two years in the egg. County names are given. (E. Levine, H. Oloumi-Sadeghi, and J. R. Fisher. [1992]. Discovery of multiyear diapause in Illinois and South Dakota northern corn-rootworm (Coleoptera, Chrysomelidae) eggs and incidence of the prolonged diapause trait in Illinois. *Journal of Economic Entomology* 85: 262–267)

as *Bt*), a soil bacterium related to the species that causes anthrax in humans; because it is nonsynthetic it has been cleared for use in organic farming. *Bt* naturally infects many insects, perhaps quite selectively (i.e., one strain may affect only a single kind of insect species). The bacterium produces a toxin that kills the cells lining the insect gut. It is not known how virulent the bacterium is in nature, but *Bt* can be grown in the lab and applied as a spray at high concentrations. It has been used as an insecticide on cornfields since the early 1960s, and it has also been applied to cotton and other species. Many species of insects have evolved resistance and resistant corn rootworms have begun to appear. In the resistant form, a mutation in the protein to which the toxin normally binds prevents the toxin from binding. It is possible, at least in theory, to use more than one *Bt* toxin strain, which makes it harder for insects to evolve resistance. Also, one can maintain susceptible pests in the area by not using *Bt* on part of the crop. The hope is that pests that persist on this crop will cross with any pest carrying the resistant mutation coming from the bulk of the crop, and the offspring from this cross will be susceptible (Fig. 8.5).

Another approach to combating pests is to release an enemy of the pest in living form. Known as biological control, this approach may theoretically work more efficiently than pesticides because the enemy of the pest continues to reproduce and can even evolve to counter the pest's defenses. Biological control has often been used against invasive species. One reason why such species are so successful in

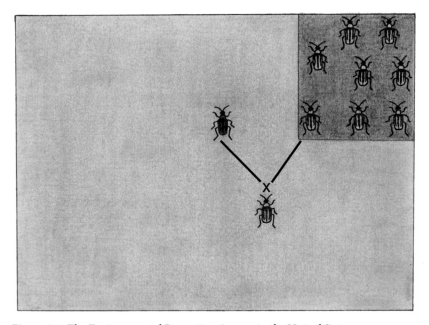

Figure 8.5 The Environmental Protection Agency in the United States encourages farmers to tackle resistant corn rootworm by leaving a part of their crop untreated with pesticides, or planted with nonresistant corn (dark shaded area), termed refuges, which allows beetles to develop. The bulk of the crop is treated, so the only survivors in the main field would be resistant forms. In the small area where the crop is not treated, susceptible beetles are common, so the resistant form will generally cross with a susceptible. Resistance genes are usually genetically recessive and are therefore masked in the offspring from the cross. Thus, a round of spraying or landing on a resistant plant will kill individuals that carry one copy of the resistance mutation. (https://www.epa.gov/regulation-biotechnology-under-tsca-and-fifra/insect-resistance-management-bt-plant-incorporated#refuges)

their new location is that they have left their natural enemies behind (Chapter 18). Therefore, bringing these enemies from the species' native area can help with control. The cottony cushion scale insect was introduced to California in the 1860s, apparently arriving from Australia on an acacia plant. Within 20 years it had virtually destroyed the citrus crop. The scale insect was not such a problem in Australia. In the late 1800s, a parasitic fly, which lays its eggs in the scale insect, and a lady beetle, which lays them on it, were discovered in Australia and thought likely to be controlling the scale insect population. A total of 12,000 flies and 500 lady beetles were successfully introduced to California, and within two years the scale infestation was greatly reduced, mostly by the beetle. Now the fly seems to be the biggest control agent on the coast, and the beetle inland.

8.4 Conclusions

Control mechanisms are essential but come with attendant side effects. Pesticides may not be selective and may even kill insects that eat the pests, thereby leading to secondary outbreaks. Biological control agents sometimes themselves become pests (Chapter 18). Pesticides and herbicides can affect humans too. In the next chapter, we consider how yields have been and are being increased despite these threats, and how some of the side effects can be reduced.

9

Food Security

Food production has increased substantially over the past 50 years, faster than human population growth. The average number of calories available per person has increased over this time by about 30% to 2,900 a day (Fig. 9.1; this includes food wasted in the home). The increase in the amount of food available per capita misses the point that many people remain undernourished. The United Nations defines undernourished individuals as those "receiving less than the amount of energy needed for light activity and attaining the minimum acceptable weight for attained height." The threshold per day vary across countries, depending on age structure and activity, but it is approximately 1,800 calories. According to the United Nations Food and Agriculture Organization (FAO), the number of undernourished people declined from more than 800 million in 2000 to 688 million in 2019, which represents about 9% of the world's population, but an increase of 60 million from its lowest level in 2014. It is expected to increase further following the COVID-19 pandemic, the war in Ukraine affecting food supplies, and the economic downturn. The major reason for the large number of malnourished is simply that people are too poor to buy food, with 900 million people thought to be living on an income of less than $1.90 a day.

The rest of this chapter addresses food security, which is defined as the state whereby "all people, at all times, have access to sufficient, safe, and nutritious food that meets their food preferences and needs for a healthy life." The FAO assesses food security through the use of a household questionnaire, and consequently estimates that as many as 2 billion people are food insecure. We also expect this figure to increase following the pandemic. The key to addressing malnourishment and food security is therefore raising the income of the poorest. That will inevitably lead to higher levels of consumption. If the human population grows over the next 30 years to add another 1 billion to 2 billion people, and individual consumption increases as people come out of poverty, the world's population might need to consume perhaps twice as many calories as it does today. We consider:

(1) Potential ways to increase the number of calories consumed.
(2) Traditional routes to crop improvement.
(3) The technology of genetic modification.

Ecology of a Changed World. Trevor Price, Oxford University Press. © Oxford University Press 2022.
DOI: 10.1093/oso/9780197564172.003.0009

84 ECOLOGY OF A CHANGED WORLD

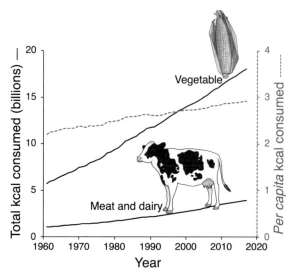

Figure 9.1 Number of calories consumed (including wasted at home). (Food and Agriculture Organization of the United Nations, http://www.fao.org)

9.1 Potential Routes to Increase the Number of Consumed Calories

Historical improvements in agricultural production over the past half-century have been accomplished by (1) farming more land (land devoted to crops has increased by about 9% over the past 50 years); (2) irrigating more land; (3) applying fertilizer; (4) using selectively bred crops; (5) applying herbicides and pesticides; and (6) introducing various intercropping methods (i.e., alternating plant species in time as well as space, which reduces competition as well as the impacts of specialist parasites and pathogens, which as we have seen are most likely to spread in dense monocultures).

Land under agriculture, land under irrigation, and intercropping are unlikely to contribute much more to increases in production. More forest could be cleared for agricultural land, but human settlements compete for such land and forests have important uses. Furthermore, even if land is added, other agricultural land is being lost (e.g., through desertification and urbanization); the amount of cropland increased by just 4% between 2000 and 2019. As we describe in Chapter 13, doubling food production solely by adding more land would use up the whole world. About 20% of all cropland is equipped for irrigation, which suggests potential for improvement, but water can be scarce. In 2019, 13 countries reported to the FAO their patterns of water use on cropland, and in total slightly less than half the land equipped for irrigation was actually irrigated. Furthermore, irrigation is not thought to be an important limit to increased crop yields in some locations, such as parts of the midwestern United States.

Altered Habits

How might the balance between future demand and future supply be met? We could be much more efficient with the food we presently grow. First, reduce waste. Between 30% and 40% of food is not consumed, either because it is wasted or is eaten by pests before getting to market (this being the main problem in developing countries), or because it is discarded in the home (accounting for more than half of all food loss in the United States). Second, consume less meat. As we noted in Chapter 7, many calories are lost as heat and waste during an animal's lifetime. Averaged across foods, about 11% of the calories provided by edible parts of a plant get to a person if they are first processed through an animal (estimates are 3% when it comes through beef, 12% for chicken, and 22% for eggs). One-third of all cereal production is fed to animals, which means 4 billion extra people could consume 2,700 calories a day if everyone in the world became vegan (Fig. 9.2). When we put these figures together, we see that complete elimination of waste and an absence of meat consumption would double the world's calorific intake, while at the same time reducing land use by as much

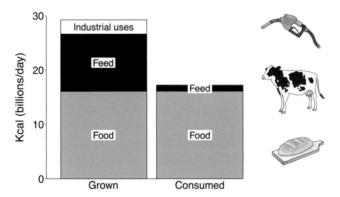

Figure 9.2 *Left:* Calories grown that could be used for human consumption (estimated for the year 2010). *Right:* Apportionment of those calories between meat and vegetables that humans actually consume. Numbers differ slightly from Figure 9.1 because the analysis was based on the 41 most common crops, forming about 90% of the calorific food supply. (E. S. Cassidy, P. C. West, J. S Gerber, et al. [2013]. Redefining agricultural yields: from tonnes to people nourished per hectare. *Environmental Research letters* 8: UNSP 034015)

86   ECOLOGY OF A CHANGED WORLD

as one-third. However, habits seem unlikely to change. Many people do not want to become vegans or eat up leftover food. Trends are in the opposite direction (e.g., Fig. 9.1).

A perhaps more realistic dietary change is to eat fish and other aquatic species instead of mammals and birds. This could be achieved through aquaculture. Aquaculture presently accounts for more protein consumed than either beef or wild-caught fish, but two-thirds of all production is in China. Across the oceans, plenty of suitable habitat is available for fish farms. It is estimated that aquaculture practiced over less than 1% of the productive marine inshore environment could generate the equivalent of all the seafood we currently eat. Fish and shrimp need to be fed, whereas farmed shellfish (e.g., scallops, oysters) do not because they obtain all their nutrients from filtering ocean waters. Estuaries, such as the River Thames beyond London, are now filling up with oyster farms. In the past, aquaculture has been a large polluter and has led to the destruction of coastal habitats, such as the case of shrimp farming in Thailand (Chapter 14). Furthermore, fish subjected to crowded conditions contract diseases, as is expected based on our knowledge of how parasites and pathogens spread. However, in the last 10 years new technology has reduced the environmental and health impacts of aquaculture.

The major advantage of raising fish over mammals is that a much larger fraction of what a fish eats is devoted to its growth—energy is not required to support a fish against the forces of gravity. Hence, a fish needs fewer calories than a mammal to produce the same quantity of calories available to us. Ongoing improvements in feeding regimes and diets means that, remarkably, Norwegian salmon are now fed only 20% more food mass than they themselves weigh at harvest. We eat about 70% of the salmon's mass, which implies that if we were to eat the food we give to a salmon, we would only consume twice the number of calories that we get from eating the salmon. Similar calculations imply that it is 6 to 15 times more efficient to eat salmon than beef and two times more efficient than eating chicken. This panacea, whereby a switch to seafood consumption comes at low cost to the environment, needs to be qualified because the fish do still need to be fed. Eighty percent of salmon feed now comes from crops, such as soybeans, which also form a cow's main diet, but the other 20% of the feed is from wild-caught fish oils and fish meal. Much research is being devoted to alternative fish food, such as farmed algae and fly larvae raised on biological waste. Freshwater carp, the dominant aquaculture species in China, are given supplements of recycled fish and other animal parts.

It appears that a switch from mammal and poultry to aquaculture-based diets might mean 2 billion more people could be fed with little increase in environmental impacts and no increase in land under cultivation. Consuming sustainably cultured shellfish such as scallops from the northeastern United States is now thought to have a lower environmental impact than consuming chicken eggs.

Increased Production

Even if we do not change habits, yields can be increased on land that is currently cultivated. Given the current technology, we could close the yield gap, which is defined as the extent to which yields could be increased given current technology (Fig. 9.3). For instance, a study of more than 1,200 farms in Zimbabwe showed how farmer education plus limited use of nitrogen fertilizer increased corn yields by 40%. In addition, we can improve production on current lands through ongoing advances in technology.

Assuming that new technology continues to increase crop yields at a similar rate to what it has been over the past 20 years, one can ask how technological advances, the closing of the yield gap by technology transfer, and the distribution of additional fertilizer across the globe could contribute to increased yield. In Figure 9.4 the height of the bars shows land that would need to be cleared if technological advances continued at the same rate, and the yield gap was closed by technology transfer. The height of the internal back bars considers impacts of technology, closing of the yield gap, and the addition of 2.5 × the quantity of fertilizer. In that case, almost no land would need to be cleared double crop yield.

Figure 9.3 (Color plate 4) The extent to which crop yields of the three major grains (wheat, corn, rice) were below the maximum possible in the year 2000, given technology and fertilizer available at that time. Mapped areas contain 98% of the crop. (N. D. Mueller, J. S. Gerber, M. Johnston, et al. [2012]. Closing yield gaps through nutrient and water management. *Nature* 490:254–257)

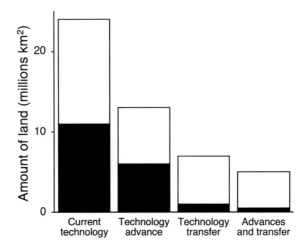

Figure 9.4 Amount of land needed to double the world's food supply under different scenarios. The calculations are based on 2006, when humans added about 100 million tonnes of nitrogen as fertilizer. The filled (black portion) of each bar is the land that would be needed if we upped the fertilizer added to 250 million tonnes as well. For reference, the United States and China together cover 19.4 million km². (D. Tilman, C. Balzer, J. Hill, et al. 2011. Global food demand and the sustainable intensification of agriculture. *Proceedings of the National Academy of Sciences* 108:20260–20264)

Of course, this analysis comes with many uncertainties and if implemented, would come with many unwanted side effects (e.g., fertilizer is a major pollutant—see Chapter 17). On the positive side, it is possible that technology will increase yields at a faster rate than predicted. Most crops have not been subject to the same intensive research as corn, rice, and wheat, and most effort has been placed on improving yields in temperate regions. The remainder of the chapter describes technological advances, with a focus on genetic improvement, which is being revolutionized due to the introduction of genetically modified foods.

9.2 Genetic Improvement

Improved farming techniques and the application of pesticides and fertilizers have gone hand in hand with selective breeding to improve crops. For example, over the past 50 years, corn yield in the United States dramatically increased (Fig. 9.5). Much of the genetic improvement is in the way crops respond to increased fertilizer and other innovations. A comparison of old and new varieties of corn suggests that at least 50% of the yield gain can be attributed to genetic improvement, with the rest due to improved farming methods and the

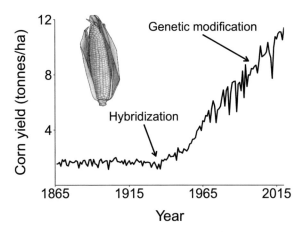

Figure 9.5 Corn yield (tonnes per hectare) in the United States since 1861. (US Department of Agriculture (http://quickstats.nass.usda.gov/)) The first hybrid corn was planted in 1933, and GM corn in 1996 (A. F. Troyer. 2006. Adaptedness and heterosis in corn and mule hybrids. *Crop Science* 46: 528–543). The term *hybrid* here means crosses between different varieties of corn.

use of fertilizers, pesticides, and herbicides. In the future, genetic improvement seems likely to play an even larger role in increasing yields, as well as combating threats from weeds, pests, and pathogens.

Since the advent of agriculture, humans have likely been selecting for crop improvement, at first subconsciously and more recently by directed breeding programs. The obvious thing to do is to take individuals that carry an exaggerated form of a desired trait and breed them together. Known as artificial selection, this approach forms the human complement to natural selection. If one breeds two individuals with the same extreme trait together, each one is likely to be carrying different genes affecting the trait of interest. Bringing these genes together then creates an offspring carrying an even more extreme version. For example, cabbage, kale, cauliflower, broccoli, kohlrabi, and brussels sprouts were all derived from wild cabbage. Brussels sprouts originated in the 1700s from breeding together those plants that had large buds on their stem.

Artificial selection results in rapid change if individuals have large genetic differences (if all individuals were genetically identical, no response would be possible). To speed up responses to selection, genetic differences have been created by generating mutations (e.g., by UV radiation) and by selectively transferring genes from elsewhere, including genes from other varieties or even other species. For example, the wheat used for bread is a hybrid of three species. Two of these species hybridized before domestication, and then this hybrid itself

Figure 9.6 Part of "The Harvesters," a painting by Pieter Bruegel the Elder, which depicts a scene in Holland in 1565. Compare this with a recent picture of wheat harvesting. Although people are taller now than they were then, they are not that much taller! Instead stem length has been shortened, mainly by crossing with a short stem variety, followed by artificial selection and choosing to propagate short-stem individuals with high yield. (*Left*: public domain, *Right*: purchased from https://depositphotos.com)

hybridized with a third species in the farmers' fields. It has been improved by selection and by crossing with other varieties. For example, a variety of wheat was first brought from Scotland to Canada in the mid-1800s, where it was susceptible to late frosts. A variety from the Himalaya was brought in 50 years later, and frost resistance genes successfully crossed into it. In another example, in the late 1950s, a research team introduced two dwarfing genes into the standard Mexican wheat. The effect was to make the wheat stem substantially shorter and more uniform in height (Fig. 9.6), thereby reducing the plant's nutritional investment into the stem rather than seeds. This new variety has better nutrient uptake and improved response to fertilizer, and it does not easily get blown over. The result was a 60% increase in wheat yields in India. This one innovation was a major factor in the Green Revolution, which substantially reduced starvation and malnourishment in the 1970s. A more recent example (from 2006) of inter-variety transfer is the introduction of the *sub1* gene by crossing from one rice cultivar (resistant to submergence) to another more productive form. This cross has higher rice yields whenever floods occur. The Gates Foundation has sponsored the planting of new varieties of rice such as this, and by 2014 more than 50,000 km^2 of cropland were being planted with *sub1* varieties in South and Southeast Asia.

9.3 Genetic Modification

While hybridization and selection have improved crops substantially, new technologies enable the introduction of genes from distantly related organisms. In

FOOD SECURITY 91

this way, many of the traditional methods have been augmented and superseded by genetic engineering, whereby single genes are artificially introduced, permanently modifying the organism.

To genetically modify an organism, one needs to identify a gene of interest. The gene coding for the *Bt* (*Bacillus thurigiensis*) toxin (Chapter 8) is one such example, and it has been inserted into soybeans, corn, and cotton. To illustrate the general principles, consider one of the earliest methods of performing this procedure. The bacterium *Agrobacterium tumefacians* naturally infects plants, where it transfers some of its DNA into the host cell's DNA. This makes the plant cell grow as a tumor, and manufacture special sugars that form the bacterium's main food. To insert the *Bt* gene into the plant's DNA, the researchers synthesized a stretch of DNA that included (1) the *Bt* gene and a genetic sequence from the plant to turn the gene on, (2) a gene that confers resistance to a toxin, and (3) short sequences from the bacterium on either side of the construct, which enable integration into the plant's DNA. Small pieces of leaf inoculated with bacteria carrying the sequence are grown on a medium containing the toxin. Those leaves that have integrated the bacterium into their DNA develop as a tumor, but the others die because of the toxin. The growing tumor can be transferred to another dish, which contains a plant hormone, stimulating development into a plant. In this way, genetic modifications are permanently produced. New, more advanced methods all basically use a similar approach—that is, integrating a gene of interest permanently into the plant or animal's genome.

The main pest-resistant GM crops on the market are corn and cotton. When susceptible insects eat *Bt* crops, they die. In the United States, more than 95% of all corn planted is now *Bt* corn. One advantage is that spraying can be greatly reduced or even eliminated, so insects that are not consumers of the crop can persist. A second benefit is that the *Bt* toxin is both inside and outside the plant, so an insect pest that eats the plant finds no relief. By 2015 nearly all of China's cotton crop was *Bt*. Benefits include both increased yields, which can be substantial, and the decreased need to use pesticides. The resulting reduction in spraying with noxious chemicals has been shown to increase farmer health in China.

Another type of genetic modification is herbicide resistance. For example, Roundup is absorbed through the leaves and prevents protein synthesis. At first, researchers tried to generate crop plants that produce excess quantities of the enzyme that Roundup interferes with. This method did not work. Instead, the breakthrough came with the discovery of a soil bacterium that carried an enzyme that was very similar to the one in plants but that was not affected by Roundup. The gene was transferred from bacteria to plant, and it worked. Most soybeans grown in the United States are Roundup-resistant. Weeds do not have to be removed prior to planting, which reduces runoff and erosion, and Roundup only needs to be applied when weeds become a problem. *Bt* crops and

Roundup-resistant crops are remarkable examples of transfer from one kingdom (bacteria) to another (plants), and certainly these improvements would be impossible without GM technology. Agricultural companies now produce two-in-one, *Bt* plus Roundup-resistant strains, and over 80% of the planted corn and cotton in the United States is both Bt- and Roundup-resistant.

GM applications continue to expand. Enzymes that lead to the synthesis of anthocyanin have been introduced into tomatoes, turning them purple (anthocyanins make blueberries blue). Anthocyanins are antioxidant compounds, with many health benefits. The eventual idea is that tomato ketchup will become healthier to eat. Another innovation has been the introduction of enzymes that synthesize omega-3 fatty acids into a plant. Omega-3 fatty acids are good for health and previously mostly derived from fish. The new plant meal has successfully been used to augment fishmeal in aquaculture.

The Politics of GM

The practice of genetically modifying foods has met with strong resistance in some parts of the world. For instance, essentially no genetically modified foods are grown in the United Kingdom. Objections to the use of GM include the strong selection it places on pests and pathogens to become resistant; moral arguments about the consequences of transferring genes from one species to another; health concerns; and the possibility that GM genes may spread into native forms of the same species. Other criticisms of GM have more to do with agribusiness and the profit motive in general. For example, in India an improved (non-GM) cotton required the purchase of an associated fertilizer. When crops failed in drought years, farmers were unable to repay their debts. As a consequence, more than 3,000 farmers in the state of Andhra Pradesh have committed suicide over the past 30 years. In the United States at least one agricultural company has been patenting its seed and prosecuting farmers who retain seed from one year to the next.

Technological improvements are varied, and some are more environmentally and people friendly than others. More generally, as technology has progressed, farms in the developed world have become larger and more efficient, and food has become cheaper. The result has been strong competition between developed and developing countries for products, causing hardship for small farmers who develop cash crops for export. Canola (rapeseed) oil has been genetically engineered with the addition of a single gene to produce lauric acid, although apparently it is not yet commercially viable. Lauric acid was formerly only found in high quantity and quality in tropical coconut and palm oils.

9.4 Conclusions

While changed habits—less meat and less food waste—could ease the pressure on the environment, we appear to be moving in the opposite direction. Therefore, technological improvements that increase crop yields are essential if we are to avoid the complete destruction of nature. Part 2 of the book (Chapters 10–22) emphasizes this message by considering the amount by which humans have already impacted the world.

PART 2
THE THREATS TO BIODIVERSITY

10

Prediction

The next section of the book uses the ecological and evolutionary principles covered so far to evaluate the future of biodiversity. We wish to predict how various threats will impact biodiversity over time, and start with a consideration of the issues associated with prediction. One often sees statements about the future written as facts. One regularly quoted example is that the world's population will increase to 9 billion by the middle of the century. Such an outcome clearly depends on multiple factors, including birth rates, wars, and disease, and hence it is uncertain. In this chapter we delve more deeply into uncertainty. We ask why predictions fail, how uncertainty can be quantified, and how to use predictions to decide on action to prevent undesirable outcomes. The chapter:

(1) Describes why predictions are so difficult to make.
(2) Explains how models are used to make predictions.
(3) Considers why predictions often do not come true and what this means for planning into the future.

Much of the chapter draws on predictions about climate change because they illustrate many of the more general points, have been well explained in the literature, and are a major concern to many. However, the general principles apply more widely and will be frequently referred to in the following chapters on other threats.

10.1 Uncertainty in Prediction

Uncertainty can be quantified as the probability of a future event. For example, the weather forecast comes with statements such as "the probability of rain tomorrow is 30%," which implies that meteorologists believe that current conditions will lead to rain 3/10 times, whereas 7/10 times they will not. The goal of much of forecasting is to make such predictions as precise as possible. In fact, predicting the weather is a recent success story. Uncertainty has been reduced, and a five-day forecast now is as accurate as a one-day forecast was in 1980.

To make predictions such as these we need a model in hand, so the assumptions are clear. For example, if we toss a coin, the probability of heads

Ecology of a Changed World. Trevor Price, Oxford University Press. © Oxford University Press 2022.
DOI: 10.1093/oso/9780197564172.003.0010

98 ECOLOGY OF A CHANGED WORLD

or tails is the same. In both cases the probability equals 0.5, and the assumption is that the coin is fair. In reporting its evaluations of global climate trends, the Intergovernmental Panel on Climate Change (IPCC) rates it "very likely" if the probability of an outcome is greater than 0.9 and "virtually certain" if the probability of an outcome is greater than 0.99. However, even predictions with high probabilities come with a set of assumptions. In the case of climate change, one assumption the IPCC makes is about how carbon emissions will change into the future.

Predicting a particular event in the future is often difficult. Why? The essential reasons are as follows:

(1) Many different factors affect a particular outcome, and many are unknown or difficult to measure (termed *value uncertainty*).
(2) The factors interact with each other in complicated ways; that is, the effect of one factor on an outcome such as a disease causing a certain number of deaths depends on the state of another factor, such as how warm it is (termed *structural uncertainty*).
(3) Unanticipated events. Predictions are based on prior information, and the future is not the same as the past. Value and structural uncertainty come with clearly stated assumptions, which include the absence of various events that could affect the outcome (e.g., volcanic eruptions affecting climate).

In the case of the weather, the availability of abundant satellite data has reduced value uncertainty. Structural uncertainty has been reduced in two ways: (1) by asking what happened in similar conditions in the past (if it rained today, what are the chances of it raining tomorrow?); and (2) by following basic mathematical principles, achieved by advances in our understanding of atmospheric physics. In the following section, we consider value and structural uncertainty, how they lead to the placement of probabilities on future events and how uncertainty may be reduced. We then discuss some examples.

Value Uncertainty

Consider a hypothetical example of 100 towns and suppose that each year half of these towns experience a fire, causing on average $1 million in damage (Fig. 10.1). If one town is more or less the same as another, we might ask why it is that a fire started in a particular place. Presumably, this outcome reflects an unfortunate concatenation of events, each of which is difficult to predict individually. For example, a town may have experienced a fire because on a day

PREDICTION 99

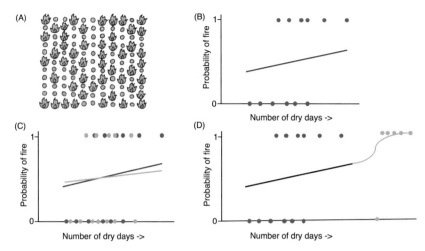

Figure 10.1 (A) 50 towns out of 100 experience a fire in a year. The question is, can we predict which ones? (B) This plot reports occurrences in the past for a hypothetical sample of 12 towns. The probability values are either 1 (fire happened) or 0 (fire did not happen). A line can be fit to estimate the probability that a fire happens as a function of the number of dry days. The probability increases with the number of dry days. Note that the probability itself does not increase a great deal (in this example, from about 0.45 on the left to about 0.55 on the right). (C) Part of value uncertainty. A different sample of towns results in a different estimate. Combining the 24 towns would improve the estimate, but not a great deal: much value uncertainty results because of other factors which are not measured. (D) Example of structural uncertainty. Even if we have a good estimate of how the number of dry days affects the probability of a fire, extrapolating beyond the range changes the relationship between extra dry days and fires. Note that an unusual combination of unmeasured variables can also create structural uncertainty, even if each variable is within the present range.

when it had not rained for some time, someone dropped a cigarette onto a piece of dry paper at a location some distance from the fire station and no passersby noticed it.

At present, we are stuck with the prediction that there is a 50% chance that the town you live in will experience a fire, which means that we might as well toss a coin to forecast the fire risk. How might we improve this prediction? Although it may be impossible to trace the events that led up to someone stepping out for a smoke and dropping a cigarette, we might be able to predict that, based on geography and climate, some towns are going to suffer more dry days than others. And if we look back in the past, we should see an association between the number of dry days and the chances a town experienced a fire (Fig. 10.1B). In this way, we

100 ECOLOGY OF A CHANGED WORLD

can assert that a certain number of extra dry days increase the probability of a fire by a certain amount.

What we have done is make a prediction about the future, based on knowledge of the impact of a certain factor in the past. First, we estimated how much an increase in the number of dry days affects the probability of a fire from a limited set of observations. This estimate in itself is uncertain. If we add more data or if we collect a similar amount of data again, we would get a different estimate of the relationship, leading to a different expectation (Fig. 10.1C). That means that in another similar data set, the line would have a different slope. An estimate of what would happen based on the first line would give a different expectation than an estimate based on the second line. While inferring the effect of dry days on fires in towns requires the use of past data, in other cases one can appeal to the basic laws of physics. For example, once could model the impact of a neighboring mountain on precipitation in a certain town.

Much uncertainty remains. Even if we accurately estimate the relationship between dry days and the probability of a fire by collecting large amounts of data, because of multiple other factors, our ability to predict the probability of a fire increases just a little. Observe that in Figure 10.1B the probability of a fire goes from about 0.45 in wet years to about 0.55 in dry years, which is hardly a big increase. Some factors are difficult to measure accurately, and others are impossible to measure at all, such as whether a passerby would happen upon the lit cigarette. Poorly measured factors, unmeasured factors, and unknown factors that affect an outcome are termed statistical noise. Such noise makes it hard to quantify the effects of the factors we do measure. In this example, perhaps a little more can be done to improve the prediction for an individual town. For example, the number of smokers in the town might be added to increase predictive ability. The fact that not all factors can be measured precisely, or sometimes at all, is what the IPCC calls value uncertainty. If value uncertainty is large, it is difficult to make precise statements about future events:

> Ecology and economics have much in common. They both involve highly complex systems whose behavior is affected by countless different forces. This is why economists are no better at predicting the future than fisheries managers. . . . There are just too many unknowns. Putting terms into the models for all of these unknowns doesn't help much if you can't predict or accurately measure the values plugged into the model. (Roberts, 2007)

This perceptive statement does belie the fact that even small increases in our ability to predict an outcome can be used to make stronger general statements: if the probability of a fire in every town goes from 0.5 to 0.55 as a result of an increase in dry days, the probability is greater than 0.99 that more fires will happen in a given year (see the section for this chapter in the online Appendices). For this reason, we can make strong statements (e.g., about climate, such as the average day will be warmer in the future) even if an outcome of a single event, such as temperature on a particular date in 2025 compared with the temperature on the same date in 2015, is less certain. In a similar way, scientists were able to make a definitive, near-certain statement that a pandemic would occur in this century but were unable to predict that it would occur at the end of 2019.

Structural Uncertainty

Structural uncertainty is uncertainty in the way different states of one factor alter the effects of another. We are trying to extrapolate into the future, based on what has occurred in the past, and often beyond the range of the observations in the past. Therefore, even if we accurately know how much an increase from 10 to 20 dry days a year affects the probability of a fire, it is generally not the case that an increase from 20 to 30 dry days would have the same incremental effect. This is the meaning of the common statement that responses are nonlinear (Fig. 10.1D).

Responses are often nonlinear because of positive and negative feedback loops. The exponential growth equation (Chapter 2, Fig. 2.1) is one demonstration of a positive feedback loop, generating a nonlinear increase with time. Even with constant birth and death rates a growing population grows at an ever-faster rate because at each time interval more individuals are reproducing (positive feedback). In this case, eventually some negative feedbacks kick in, such as competition for food causing an increase in death rates, and in a full model we would need to know how they would come into play. Feedback loops are generally complex because multiple factors interact, creating all kinds of positive and negative feedbacks. A remarkable feature of complex feedbacks is that a small perturbation can be greatly amplified by subsequent feedbacks, making predictions impossible. In the extreme, we get what is known technically as chaos and colloquially as the butterfly effect. As originally stated, a flap of a butterfly's wings is sufficient to create a small change in air circulation that may eventually result in a change of where a tornado hits land, or even whether it occurs at all. This phenomenon is often promoted as a classic example of why weather is so difficult to predict more than a few days in advance.

Interactions between different factors can result in a great deal of structural uncertainty. In the fire example, suppose that a drought causes an increase in smoking cigarettes through the stress that hot, dry days cause. However, this increase in smoking is only apparent if the drought is prolonged (e.g., over 60 days). Then the probability of a fire versus the number of drought days would change slope as time goes on, increasing at a more rapid rate after 60 days than before.

102   ECOLOGY OF A CHANGED WORLD

Note, too, that even if several factors all vary within their normal range, and for some reason all attain extreme values at the same time, their summed effect is identical to moving beyond the normal range and can result in qualitatively different outcomes than what is predicted from past associations.

Unanticipated Events

Value and structural uncertainty can be reduced, increasing the reliability of predictions. Still, making a prediction depends on the assumptions that go into the model. Unanticipated factors may come into play, affecting outcomes in multiple ways. Some of our unmeasured factors may be very rare. A town may get hit by lightning, thereby starting a fire in a once in a millennium event, but we would not consider this possibility in making a prediction. In consideration of biodiversity, myriad possibilities exist (e.g., a predator may invade a community). The purpose of a model is to predict what would happen, or to understand what has happened, given the stated assumptions. It is up to the researcher to evaluate whether the assumptions are reasonable.

10.2 Examples

Disease Spread

In late 1995, the number of syphilis cases in Baltimore almost tripled. The Centers for Disease Control, noting that use of crack cocaine increased at the same time, suggested a causal link based on increased risky sexual behavior that increased transmission rates. Others noted that the city had cut back on its health clinics at the time, and patient visits declined by nearly one-half, so people carrying the disease were infectious for a longer period of time. Yet a third explanation is that the city demolished several inner-city housing projects, translocating people carrying the disease to other areas. Although presented as alternatives, none of them by themselves may have been sufficient to trigger the epidemic. Instead, all of them acted together, in an unfortunate confluence of events, which led to a sustained period during which the infection rate exceeded the rate of recovery.

This is an example of structural uncertainty, where three factors together create a qualitatively different outcome than any of them would if the others had not changed. It emphasizes the important point that we saw with the hypothetical smoking example—notably, it is rarely just one thing that causes a major change, but rather a concatenation of multiple events, each of which may be independent of the other. The principle has become known as a "tipping point" or "perfect storm," which are the titles of two popular books published at the end the last century.

Climate Change

Climate change is considered in detail in Chapter 15. Here we focus on predictions of future climate, illustrating how uncertainty can be modeled. The IPCC has produced six general reports dating back to 1990, with the latest published in 2022. The reports have moved from models with much value uncertainty (a great deal of statistical noise) to models that try to account for as many factors as possible. Such model building requires an understanding of how the different influences on climate interact. For example, water vapor is an important greenhouse gas contributing to warming, but the amount of water in the atmosphere depends directly on the temperature and therefore on other greenhouse gases, notably CO_2. The relationship between temperature and the water-holding capacity of air is accurately known and easy to place in the models. Other factors are much more difficult. The effect of ocean circulation patterns was only explicitly modeled in the third report—before that it was just considered a source of statistical noise.

Both value and structural uncertainties remain in models that predict the progression of climate. Value uncertainty is present because it is impossible to include all variables or to measure them accurately at the present day. Structural uncertainties create great difficulties because of the many feedback mechanisms in the climate system that amplify (positive feedback) or diminish (negative feedback) warming, which in theory can result in quite different outcomes under small differences in the assumptions. For example, as the climate warms, snow and ice melt. Melting exposes land and water, which are darker than snow and ice. These darker surfaces absorb more of the sun's heat, causing more warming, which causes more melting, and so on, in a self-reinforcing cycle. The effect of clouds is a particular issue. Clouds typically cover two-thirds of the earth's surface. Despite consisting of water vapor, which is a greenhouse gas, clouds have an overall cooling effect on the planet because they reflect some of the sun's radiation. A reduction in cloud cover should thus lead to an increase in temperature, and the increase in cloud cover should lead to a decrease in temperature. However, many nuances influence the relationship between cloud cover and temperature change, such as the height of the clouds above land, the latitude that the clouds are present, and the cloud type. In the IPCC's fifth report, different models of the effects of cloud radiation were described that result in predictions that differ by up to 5°C over the next century. However, all agree

104 ECOLOGY OF A CHANGED WORLD

that the effects of clouds will be to generate positive feedback. For example, one prediction is that clouds will on average be higher in the sky, which results in greater warming.

The IPCC summarizes models of how factors interact as they change over time, derived from both previous observations and principles of physics. Each model uses equations to project climate forward from a set of initial conditions, based on parameters whose values are inferred from past measurements. Value uncertainty is added to the model, both as uncertainty in the initial conditions and as statistical noise (adding random values to the estimates, such as cloud cover, at each time step). Structural uncertainty arises because different models come with different assumptions (e.g., about the effect of clouds). Because of the large number of factors interacting in complex ways, computer simulations are needed to project outcomes. The result is a range of possible outcomes in temperature increase, with the more extreme values considered unlikely.

One way to check a model's value is to see how well it reconstructs the past. Past temperatures are derived from land and sea recordings, averaged across regions of equal area across the world and, in the most widely reported compilation, omitting data from regions without any monitors. Figure 10.2 compares the results of different models that include different assumptions, starting points, and noise. One can see that some runs predict more warming than has happened and some less, but the average of all models agrees rather well with the trajectory of past temperatures. One needs to be somewhat careful in this assessment because models that only included the effects of greenhouse gases predicted more warming up to 2010 than actually happened. As noted in Chapter 2, one purpose of a model is to find out why predictions do not provide a good match to data. For example, the major volcanic eruption of Mount Pinatubo in the Philippines in 1991 released many particles into the atmosphere. These particles reflected the sun's radiation, which causes cooling. The model runs in Figure 10.2 include this eruption. The implications are that greenhouse gas warming would have been larger and have fitted model predictions better if Mount Pinatubo had not erupted, but if we wanted to improve our prediction of climate, we would have to predict volcanic eruptions.

From the years 2000 to 2010, the average of all models together overestimated the observed warming, which fueled the arguments of climate-change deniers. One possible contribution is that the additional climate stations were added to the data set, and it has been hard to evaluate how these have affected the overall measurements. Remaining discrepancies appear to result from (1) cooling from minor volcanic eruptions, (2) a greater than expected decrease in heat coming from the sun (the sun fluctuates in the amount of radiation emitted), and (3) climate fluctuations that have no directional trend and operate on timescales of

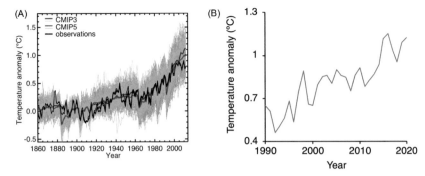

Figure 10.2 (Color plate 5) (a) Thick black line: Annual deviations from the average of 1880–1919 for global mean surface temperature up to 2005 (in places one sees more than one line because three different assessments have been made). Thin lines represent the average of different models of temperature change, accounting for both human and natural causes, with the shading around them different run of the models. (taken from Figure TS.9, Panel [a], T. F. Stocker, D. Qin, G.-K. Plattner, L. V. Alexander, S. K. Allen, N. L. Bindoff, F.-M. Bréon, et al. [2013]. Technical Summary. In: *Climate Change 2013: The Physical Science Basis. Contribution of Working Group I to the Fifth Assessment Report of the Intergovernmental Panel on Climate Change*, edited by T. F. Stocker, D. Qin, G.-K. Plattner, M. Tignor, S. K. Allen, J. Boschung, et al. New York: Cambridge University Press, 33–115. doi:10.1017/CBO9781107415324.005; www.ipcc.ch/report/ar5/wg1/technical-summary/)
(b) Global temperature 1990–2020. C. P. Morice, J. J. Kennedy, N. A. Rayner, et al. [2012]. Quantifying uncertainties in global and regional temperature change, using an ensemble of observational estimates: The HadCRUT4 data set. *Journal of Geophysical Research* 117:D08101. doi:10.1029/2011JD017187; www.metoffice.gov.uk/hadobs/hadcrut4/. This data set is based on averaging climate records across 5° x 5° grids across the world (111*111 km at the equator, less at higher latitudes), for which there are any temperature records at all. It includes the Atlantic and Indian oceans, but not much of the Pacific, the extreme north, and parts of Africa, South America, and central Asia. A large upgrade in the number of stations in 1990 means we have a more accurate estimate since then.

approximately 15 years. The main driver of this internal variability is that of el Niño and la Niña events, driven by oscillations in atmospheric pressure altering current flow in the Pacific Ocean (often termed ENSO, for El Niño—Southern Oscillation). Large el-Niño events happen when cold upwelling shifts further west in the Pacific, resulting in ocean warming along the South American Pacific coast and drought over large parts of the tropics worldwide. The large el Niño of 1998 pushed temperatures higher, which then relaxed. Thus, the scientific consensus is that the models predicted more warming than was observed, not because the models were inaccurate, but because several sources of statistical noise (creating value uncertainty) all operated in the direction of a cooling influence. If this is the case, these sources should average out over time, and the warming trend over the next 10 years would be higher than expected. That expectation has been borne out by the data, which show that 2011–2020 was a period of exceptionally rapid warming (Fig. 10.2; 2016 was another el-Niño year).

Figure 10.3 depicts the projections made in the IPCC's 2014 report. The IPCC narrows down predictions by making greenhouse gas emissions one of its model's assumptions, rather than adding this to the model as an extra uncertainty. Shadings about the different lines indicate uncertainty (Fig. 10.3). The shaded intervals are obtained by running the models many times and recording

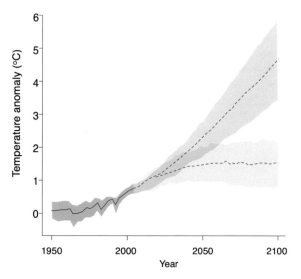

Figure 10.3 Predictions of climate change. The solid line is past global surface temperatures, and dashed lines are predictions under two extreme scenarios of emissions. The shading gives the 90% confidence limits, reflecting the uncertainty of past estimates and future predictions. Multiple models are run, with different starting conditions; 5% of the time outcomes lie above the shaded area, and 5% of the time outcomes lie below. Illustrated is the worst-case scenario where coal output increases at the rate it was in the 1990s (CO_2 concentrations at end-century >900 ppm), which is now thought very unlikely, and a best-case scenario, whereby emissions peaked in 2020 and started to decline (CO_2 concentration at end-century would be 420 ppm, see Fig. 16.1). (National Center for Atmospheric Research NCAR GIS Program. 2012. Climate Change Scenarios. https://gisclimatechange.ucar.edu/gis-data, accessed August 23, 2020)

all outcomes. A 95% confidence limit, for example, is obtained by finding the temperature above which the top 2.5% of all predicted temperatures fall, and the temperature below which the bottom 2.5% of all predicted temperatures fall. The implication is that 95% of the time we construct such a model, these upper and lower bounds will encompass the true outcome, given the assumptions of the model (see the online Appendix for further elaboration on the meaning of confidence limits).

10.3 Planning for the Future

In the prologue to his famous work, *The Population Bomb* (1968), Paul Ehrlich wrote, "the battle to feed all of humanity is over. In the 1970's and 1980's hundreds of millions of people will starve to death." This prediction was not fulfilled, partly because the Green Revolution introduced technological improvements in agriculture, such as increased wheat productivity due to artificial breeding for shorter stems (Chapter 9). Ehrlich's statement might have been better stated as "we predict with high probability that mass starvation will happen, under the assumptions that no novel technology is invented and no other major factors change." In any event, failed predictions such as this have been used as ammunition against environmentalists' "doom and gloom prognoses." Is this criticism reasonable? More generally, why don't predictions always come true?

One reason predictions of environmental deterioration do not come true is that they lead to remedial action. For example, the decline of the tiger population in India led to establishment of reserves in strategic areas. Tiger populations in India now appear to be holding steady or even slightly increasing. In another example, Rachel Carson, in perhaps the most consequential environmental book of the twentieth century, *Silent Spring* (1962), drew attention to the damaging effect of pesticides and predicted the loss of many bird species. This work led to ban of the pesticide DDT in the United States (Chapter 17). Subsequently, several threatened species, most notably the United States' national bird, the bald eagle, have recovered from their low numbers. Finally, chlorofluorocarbons (CFCs) were formerly used as refrigerants and in air conditioners. In extreme cold, they catalyze the conversion of ozone to oxygen in the stratosphere. One consequence is the presence of an ozone hole above the Antarctic, which is especially pronounced at the end of the Antarctic winter. Ozone at high altitudes absorbs harmful ultraviolet radiation. CFCs were banned under the Montreal Protocol in 1989, an action that is likely in the long run to prevent many premature deaths. In 2019, the ozone hole was the smallest it has been since its discovery in 1985, although this may reflect the warmth of the year rather than a strong effect of the ban.

Although remedial action can prevent unfavorable outcomes, predictions do not come true primarily because they are difficult to make. They come with both an associated probability and underlying assumptions. Unless the probability is 1.0 and the assumptions hold, some predictions will certainly not come true. If a prediction has a greater than 0.99 probability of being correct, we should always act to prevent the outcome, providing the benefits of action outweigh the costs of inaction. What about the situation in which 50% of the time the prediction turns out to be true? We can illustrate this type of cost-benefit assessment by returning to the analogy of fire risk in a town. What is the cost compared with the benefit of building a fire station in a town, given a certain probability that a fire will start?

As graphed in Figure 10.4, the cost of maintaining a fire station each year is the same whatever the probability of a fire, but the average benefit that accrues over a large number of years increases from 0 when a fire never happens (probability = 0), to a large number, clearly exceeding the cost, when a fire is certain to happen (probability = 1; in Figure 10.4, the point on the right-hand axis is the cost of the fire). The point at which the cost and benefit curves cross is the probability at which the cost of maintenance of the fire station over a very long time exactly matches the benefits from having the station, because of the fires that have occasionally occurred. In the left-hand graph of Figure 10.4, this point occurs at

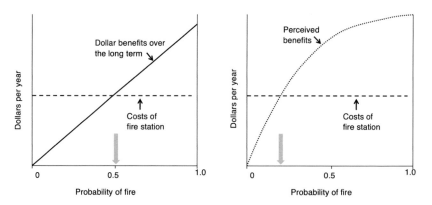

Figure 10.4 *Left*: If the costs of maintaining a fire station are half the costs of a fire, and if the probability of a fire in a year = 0.5 (arrow), the costs exactly match the benefits over the long run. Under this criterion, the station should always be built if the probability is greater than 0.5 and never built if the probability is less than 0.5. *Right*: The utility curve translates the perceived benefits of a fire station into dollars. The utility curve indicates that even if a fire's probability = 0.2 (arrow), the fire station should be built. The second curve reflects a "risk-averse" scenario and implies that one is prepared to pay a regular amount to avoid a major disaster, even if the dollar cost of the disaster averages out to be greater than the expense. This type of risk aversion is the basis for insurance.

a fire probability of 0.5 because the benefit (i.e., the savings from rapidly putting out the fire) is twice the cost of maintaining the fire station.

Humans are typically risk averse, willing to pay more than the expected (or average) benefit benefits to avoid low-probability events. Our total insurance premiums for example, must on average be higher than our claims, or else insurance companies would not turn a profit. The utility curve replaces the benefit curve in describing how much extra we are willing to pay. The point at which the utility curve crosses the cost curve gives the value above which we are prepared to pay to cover the probability of a fire. In Figure 10.4, we still build a fire station when the probability of a fire is about 0.2. Perhaps this scenario can be viewed as if there is a fire we lose everything, whereas if there is not a fire, we lose just a little.

Without factoring in risk aversion, a cost-benefit analysis can be used to determine how much we should invest to stop a rare harmful event from happening. If a fire costs $1 million and the probability of a fire is 0.01 per year, without risk aversion then an investment of $1 million $\times$ 0.01 = $10,000/year indicates an exact balance between costs and benefits. So even with low probabilities of bad events occurring, we should take some action, and with risk aversion, we should invest even more to prevent such events. However, political realities can create risk-prone behavior, where long-term benefits are heavily discounted against short-term costs (Chapter 14).

10.4 Conclusions

Despite the popularity and high impact of many research papers asking what the world will look like in the future, predictions regarding biological systems will always be difficult to make. This is the case not only because of the many factors involved, but because many low-probability events cannot be modeled, any one of which could alter a trajectory. Nevertheless, we can assign probabilities to events if accompanied by clearly stated assumptions of the model (which include an absence of these low probability events). Perhaps the main take-away from this chapter should be that, as models improve and data are added, very small improvements in prediction can have far reaching implications: when costs to the adverse outcome are large, the expected cost (probability $\times$ cost) is also large. This is clearly true for climate change, where models have predicted with a 50% probability a 3°C rise based on one reasonable emissions scenario (Chapter 16). A 3°C rise would be devastating. Such a prediction, even though it is to the level of a coin flip ($P = 0.5$), is the reason why efforts are being made to reduce the use of fossil fuels (Chapter 16).

11

Human Population Growth

On November 1, 2021, the U.S. Census Bureau estimated a world population of more than 7.8 billion people, a very large number. If one were to start counting people at the rate of one person every second, it would take 247 years to count everyone. Compare human numbers with those of our closest relative, the chimpanzee, with a population size of perhaps 400,000 (0.005% that of humans). Genetic evidence suggests that only 100,000 years ago, ancestral humans had a population size substantially smaller than that of chimpanzees.

The human population is large and it is continuing to grow rapidly, by about 12 people every 5 seconds, or 200,000 a day. The birth rate, b, in 2021 was about 0.02 births/individual/year, and the death rate about 0.01 deaths/individual/year, making the population growth rate about 1%. That is almost half the peak human growth rate estimated for 1968 (Fig. 11.1). Because the number of people added to the population is the product of the growth rate and the population size at the beginning of the year, the number of individuals added each year continued to increase for another 20 years after 1968, even as the growth rate declined. However, since 1988 the number of individuals added each year has also decreased, and the annual increase is now about 85% of the 1988 peak (Fig. 11.1). In this chapter we explore patterns of human population growth in more detail. Specifically, the chapter:

(1) Describes the history of human population growth.
(2) Asks why the rate of growth is slowing.
(3) Considers some projections.

11.1 History of Human Population Growth

It is thought that humans and chimpanzees last shared a common ancestor (which lived in Africa) about 6 million years ago but perhaps as much as 8 million years ago. A major step forward in human evolution was our progression to movement on two feet, an advance that was refined following the shift to exploitation of open savannah rather than forests about 4 million years ago. Modern humans are recognized from at least 300,000 years ago, with well-developed language probably dating at least to that time. They may have moved out of Africa in

Ecology of a Changed World. Trevor Price, Oxford University Press. © Oxford University Press 2022.
DOI: 10.1093/oso/9780197564172.003.0011

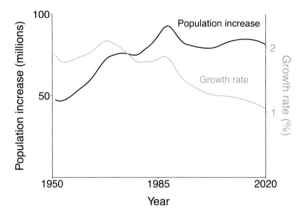

Figure 11.1 Human population growth and growth rate (expressed as a percentage). The period is from July 1 of the year indicated (the first data are 1950–1951). Per capita growth rate peaked in 1968, and population increase peaked 20 years later, in 1988. (United Nations, http://esa.un.org/unpd/wpp/DVD)

various waves beginning as long as 180,000 years ago, replacing, and now known to have hybridized with, the earlier colonists, the Neanderthals. Humans reached Australia 50,000 years ago, and the Americas (across the Bering Strait) probably about 16,000 years ago, but perhaps as much as 30,000 years ago. Culture started to change quickly beginning about 70,000 years ago, associated with clothing, sophisticated hunting, and, eventually, cave paintings, the oldest of which are known from Sulawesi in Indonesia, dating to almost 46,000 years ago. The origin of agriculture in the Middle East dates to about 10,000 years, as we emerged from the last Ice Age, and is associated with a shift from hunter-gatherer societies to permanent settlements. Subsequently, agriculture originated at several other locations elsewhere in the world, apparently independently. One societal innovation, the brewing of beer, began about 7,000 years ago, in the Middle East.

Human Population Growth

If we look at estimates of population size throughout the past 2,000 years, growth appears to have been exponential (Fig. 11.2). But how well does this model actually fit human population data? In Chapter 2, we introduced the model of exponential growth:

$$N_t = N_o e^{rt}.$$

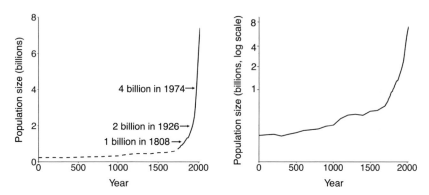

Figure 11.2 *Left:* Estimates of human population size over the past 2,000 years. *Right:* The same curve, where population sizes have been log-transformed. (K. K. Goldewijk, A. Beusen, J. Doelman, et al. 2017. Anthropogenic land use estimates for the Holocene –HYDE 3.2. *Earth System Science Data* 9, 927–953; https://easy.dans.knaw.nl/ui/home)

N_o and N_t are the population sizes at some time and t years later, respectively, and r is the intrinsic growth rate, which can also be regarded as the fractional increase from one year to the next. We can rewrite this equation by taking the natural logarithm of each side:

$$\ln(N_t) = \ln(N_0) + rt$$

If birth rate and death rates stay constant in an exponentially growing population, a plot of the logarithm of population size, $\ln(N_t)$, against time, t, should be a straight line with slope r. The plot of log of human population size against time, however, curves upward (Fig. 11.2), implying that the growth rate (r) has itself been increasing over the past 2,000 years. Models fit to data that assumed the increase in r would continue at this rate led to the projection of a human population size that by 2023 would be expanding away from the earth at the speed of light! Such a rate of increase obviously could not happen, but it is of interest in showing how difficult it can be to make predictions, and in showing that birth and/or death rates had to change. This model, appropriately termed the Doomsday Curve, is described further in the section on this chapter in the online Appendices.

A good way to see that the rate of increase was itself increasing is to examine doubling times. With exponential growth, the time it takes for a population to double in size is the same, whatever the population size. In fact, the human population may have reached 1 billion in about 1808, 2 billion in 1926 (118 years later), and 4 billion in 1974 (just 48 years after reaching the 2 billion mark; see Fig. 11.2). Joel Cohen in *How Many People Can the Earth Support?* (1995) proposes that the rapid shortening of doubling times happened at three demographic transitions in the last 10,000 years (see Table 11.1).

According to Cohen, the first demographic transition was associated with the origin of agriculture starting about 10,000 years ago, which allowed human societies to develop and sustain higher densities than were possible in previous hunter-gatherer societies. A second demographic transition came in the eighteenth and nineteenth centuries and was associated, at least in part, by the spread of agricultural products between continents (e.g., potatoes from South America became a staple in Europe and the temperate regions of Asia), leading to generally increased nutrition. The third demographic transition occurred in the mid-1900s as a result of improvements in human health, which lowered mortality rates. For example, the death rate, d, in Sri Lanka fell by almost 50% in the 5 years after the Second World War as a result of various health initiatives. Among these initiatives was the spraying of pesticides against mosquitoes, which contributed to a decrease in the incidence of malaria (though the actual contribution of spraying is disputed and may have been less than 15% of the total). Mortality from disease continued to decline, at least until recently. The number of people who died from communicable diseases in 2016 was half the number recorded in 2000 (Table 6.1). The current pandemic has altered that picture somewhat, and in 2020 COVID-19 accounted for more deaths than from all other diseases combined.

Earlier, we showed that the simplest model of population growth with constant birth and death rates leads, often quite rapidly, to impossibly large population sizes. In the late eighteenth century, Malthus suggested that overpopulation results in war, pestilence, and famine (Chapter 2). Indeed, war, pestilence, and

Table 11.1. Four Proposed Demographic Transitions through Recent Human History

	Date	Population (millions)	Doubling time (years)	
			Before	After
Agriculture starts	10,000 BP*	5	~150,000	~2,200
Agricultural interchange	~1750	750	~1,000	~120
Public health	1950	2,500	87	36
Fertility	1970	4,000	34	>60

*BP: years before present. (J. Cohen. [1995]. *How many people can the world support?* New York: W. W. Norton.)

famine have been common throughout history. The two world wars of the first half of the 20th century killed more than 80 million people; one recent estimate of the effects of the influenza pandemic in 1918 is at least 17 million dead but may be as much as 50 million; and famine across much of the tropics between 1875 and 1895 killed tens of millions.

More recently, high human population densities have clearly contributed to episodes of mortality, but these have had relatively small impacts. In 1994, genocide in Rwanda resulted in the deaths of 800,000 people. Jared Diamond in his book *Collapse* describes farmers trying to survive on plots of land as small as a small house (less than 300 m^2), and property disputes that were settled by murder. Although nominally a tribal conflict, genocide was probably triggered by these desperate conditions of existence. With respect to diseases, by the end of 2021 AIDS had killed more than 36 million people and COVID-19 more than 5 million. As described in Chapter 7, HIV likely spread as a result of rapid urbanization in Africa and associated high human densities. Similarly, the spread of COVID-19 reflects high human densities driving ease of transmission. Finally, famines in northern Africa regularly occur and are partly a consequence of many people living in drought-prone areas.

Increased death rates are an expected consequence of high population densities. But despite the events we have documented above, these factors have had only small impacts on recent population growth, less than those large-scale mortality events of 80–150 years ago, including the world wars, disease pandemics, and mass starvation. The death rate has been decreasing rather than increasing (Fig. 11.3): on average, people are living longer. Remarkably, what is happening is that the birth rate is declining all over the world over, with few exceptions. The birth rate is declining faster than the death rate (Fig. 11.3), which explains why the population growth rate is slowing (Fig. 11.1). This slowdown represents Cohen's fourth demographic transition, and the first that has resulted in a lengthening of doubling times (Table 11.1).

11.2 The Demographic Transition to a Low Birth Rate

A small contribution to the declining growth rate is that women delay having children until they are older (see the online Appendix to this chapter), but the main reason is that women are having fewer children. The fertility rate is the number of children a woman entering reproductive age expects to have if current practices hold and she survives to old age. It is calculated based on the birth rate for each age class. For example, if in a particular year, 0.4 of all 20-year-old women had a child and 0.8 of all 21-year-old women had a child, but no one else did, we would infer that a woman entering reproductive age will on average have

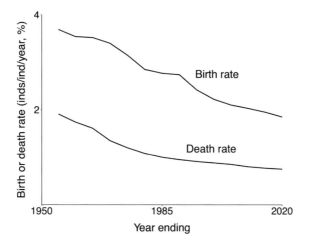

Figure 11.3 Birth and death rates averaged over 5-year periods. In 2019 the birth rate, $b = 0.0185$, the death rate, $d = 0.0075$, so the per capita rate of increase, $r = 0.0109$, or about a 1% increase in population size. (United Nations, http://esa.un.org/unpd/wpp/DVD)

$0.4 + 0.8 = 1.2$ children in her lifetime, if conditions do not change. Therefore, 1.2 would be the fertility rate for that year.

The global fertility rate halved between 1960 and 2020 (from 5 children per woman to 2.5 per woman; Fig. 11.4) but is not yet quite at the level that would produce long-term population stability, which would be 2.1 (one male, one female, with 0.1 to account for death before completing the reproductive stage of life). Only a few countries in Africa and west Asia have yet to show a slowing in the fertility rate. Currently, Niger in Africa has the world's highest fertility rate, with more than 7 children per woman. In 2015, countries with fertility rates below 1.5 included several from eastern and southern Europe, Korea, Japan, and China. Some parts of east Asia have fertility rates around 1. The Chinese city of Shanghai (population 23 million) had a fertility rate of only 0.88 in 2008, and it may have declined further.

Why Is the Fertility Rate Declining?

Having fewer children would seem to be contrary to Darwin's principle of evolution by natural selection (Chapter 7). Darwin's principle implies that individuals that have the most surviving offspring increase their genetic representation over generations. This seems to imply that high reproductive rates would be favored because individuals with large families would leave more descendants than individuals with small families. However, the operative word in the formulation is

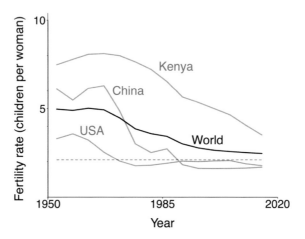

Figure 11.4 Fertility rates for 5-year periods estimated for the world and three countries; X axis labels indicate the beginning of the period. The dashed line is the rate which should eventually lead to a stable population. (United Nations, http://esa.un.org/unpd/wpp/DVD)

surviving. In nature at least, many species produce few young (e.g., elephants), which they can better provision, rather than large numbers of less well-nourished young, all of which might perish. This is not to say that one could have predicted the current decline in the birth rate, but there should be no expectation from Darwin's theory that, once improved health care and sanitary conditions allowed many more children to survive, women would continue to have as many babies.

The proximate causes of the demographic transition to low birth rates are now becoming understood. The decline in the fertility rate generally results from an increase in wealth and longevity (Fig. 11.5). The main driver is thought to be education, particularly of girls, coupled with the availability of contraceptives. The importance of education can be seen not only by comparing countries that have not gone through the demographic transition with those that have (Fig. 11.4), but also within the same country. In the years between 1992 and 2011, Ethiopian women without any education had about six children, whereas those with 12 years of schooling initially had about two, declining to 1.5 by the end of the period. Similar findings have been reported for other African countries.

Presumably, individuals choose to have fewer children for many different reasons, but perhaps one major contribution is a vision of the benefits of investing in offspring quality (e.g., their education) rather than quantity. In Kenya, the more immediate reasons for having fewer children are diverse. They include limited private holdings (crowding and costs), availability of contraceptives, direct education, and a TV soap opera that has promoted small family size. In China, the

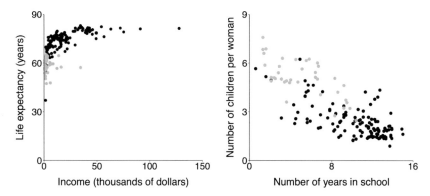

Figure 11.5 Across 167 countries, life expectancy at birth is associated with per capita income *(Left)*, and fertility is associated with education *(Right)*. Fertility and life expectancy are based on birth and death rates from 2010; number of years in school is for women aged 25 to 34; and income is in dollars, based on gross domestic product in 2010 (corrected for inflation and purchasing power to 2005 USA values; see Chapter 12). Gray points are sub-Saharan African countries. (www.gapminder.org).

transition may have been aided by the government's one child per family policy initiated in 1979 and not officially abandoned until October 2015. However, the largest decline in fertility rates likely occurred before that. Comparisons of provinces with different income levels show the clear effects of education and wealth.

Effects of the Demographic Transition on Population Growth

The progression of the current demographic transition to low fertility can be visualized using a population pyramid, which diagrams the number of individuals of each sex, grouped into age classes. Figure 11.6 shows pyramids for females in three countries. A triangular shape characterizes a rapidly growing population that has not undergone the demographic transition, as illustrated for the east African country of Burundi. The youngest age class represents the number of children born, and the main reason the next oldest age class has fewer individuals is that fewer individuals were reproducing in the earlier time step. The top of the pyramid represents both this effect and increasing mortality with age. As the demographic transition to lower fertility rates takes place, the number of individuals in the youngest classes may be equal to or less than those in older classes (as seen here for China). If less, a bulge gradually moves up the pyramid, which in turn becomes more cylindrical. A stable population would have roughly equal numbers in age class until mortality steeply increases in old age.

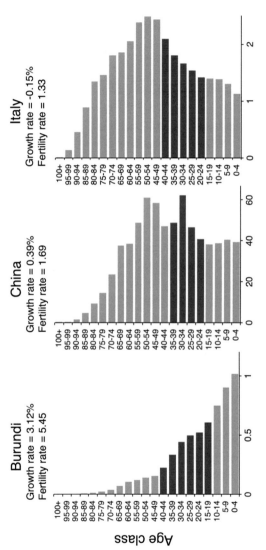

Figure 11.6 Population pyramids for three countries. Graphed are the number of females grouped into 5-year age classes for the 12 months preceding June 30, 2020. For each country, the % increase in population size over the previous year and the fertility rate (averaged over the preceding 5 years) are indicated. The black bars span the age group contributing 90% of the babies (98% in the case of Italy). China is undergoing the demographic transition, with a fertility rate below 2, but a positive growth rate. The continuing increase in population size comes from the legacy of 25–29 years ago, so that many women are now of reproductive age and give birth, which outweighs the number dying, given there were fewer people born >60 years ago. (United Nations, http://esa.un.org/unpd/wpp/DVD)

Even if every woman from now on were to have only two children, the human population would continue to increase because, as a legacy of past high birth rates, a very large number of women are entering reproductive age. Thus it is important to realize that it takes time for population growth rates to slow. Indeed, averaged between 2016 and 2020 the fertility rate in China was 1.69, but the population continued to grow at a rate of 0.4% per year. More individuals are having a child in a given year than are dying. The demographic transition has proceeded further in Italy, where the population declined by 0.15% between 2019 and 2020. In the United States, the fertility rate in 2020 was about 1.8 children/woman. The population increased by about 0.4% because the birth rate still exceeds the death rate and also because of immigration. According to the U.S. Census Bureau, births now account for about 90% of people added and immigrants about 10%.

11.3 Population Size Projections

Predictions are always difficult, but population projections do have some built-in features that help reduce uncertainty. In particular, if we know the number of people alive today who are 10 years old, we can make a reasonable estimate of the number of people 40 years from now who will be 50 years old. For sure, there will be no more. This of course applies to any age cohort.

The major uncertainties in the projections are what will happen to fertility and, less critically, how longevity and patterns of migration will change. The United Nations Population Division makes predictions for both the birth and death rates of each country according to that country's progress in making the demographic transition and adding in uncertainty by estimating past variation within the country, and differences between countries. The UN expects that fertility will decline substantially during the demographic transition, followed by a slight uptick, as has been observed in some countries. On this basis, under the most likely model, in 2100 the fertility rate for those countries that have completed the demographic transition is modelled to be 1.78. In this scenario, the UN gives a probability of 0.5 that the population will be greater than 10.8 billion in 2100, when it will still be slowly increasing. However, the model also embraces the possibility (a probability of 0.1) that the population will be less than 9.5 billion and declining (Fig. 11.7). Other modelers have used predictions about changes in education and wealth to predict fertility rates. Two examples are shown in Figure 11.7 (no confidence limits were placed on these predictions, which were made in 2010). Given a scenario of rapid development, the population in 2100 would be smaller than it is today, with a fertility rate of 1.33. We are presently exceeding the population sizes predicted under that favorable scenario (Fig. 11.7).

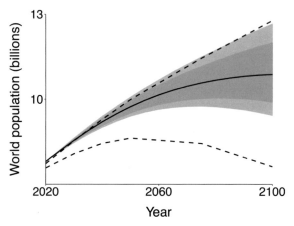

Figure 11.7 Population predictions under different assumptions. The black line gives the United Nations 2019 predictions, based on the progress of the demographic transition (http://esa.un.org/unpd/wpp/DVD). Gray areas give the 80% (inner) and 95% confidence limits; that is, the true outcome should lie somewhere between limits constructed in this way 80% and 95% of the time, respectively, given the model's assumptions. The dashed lines are predictions made in 2010 from models based on changes in education and wealth, with no estimates of uncertainty. (W. Lutz, W. P. Butz, and S. KC. [2014]. et al. 2014 (eds.). *World population and human capital: Executive summary.* The Wittgenstein Center for Demography and Global Human Capital; http://www.iiasa.ac.at/publication/more_XO-14-031.php). The lower dashed line assumes that progress toward development continues as in the previous decades, but with reductions in resource use and dependency on fossil fuels. The upper dashed line considers a world separated into regions characterized by extreme poverty and some pockets of moderate wealth. In this case, the emphasis is on security at the expense of international development.

11.4 Conclusions

The population in 1900 was 1.65 billion, which is the same number of people added to the world in the 20 years between 2000 and 2019. The rapid increase in human population size over the past 120 years has been driven by reductions in mortality associated with technological advances in medicine and food production. The huge number of people now present in the world is contributing to nature's decline. But this is not the only factor in the increasing human impacts on the environment. Concomitant with the population increase, and also associated with technological advances, people have greater spending power. The increase of personal wealth is the subject of the next chapter, after which we move on to how people and their money have affected and are affecting the state of nature.

12

Growth of Wealth and Urbanization

People worldwide are on average getting richer. As shown in Chapter 11, this trend has been accompanied by an increase in life expectancy, education, and smaller families. Eventually, a stable, smaller population size should mean good news for the environment and subsequent human generations. However, increases in spending power are presently putting enormous stresses on the environment. This chapter investigates the growth of income and its impacts. It also briefly touches on urbanization: the fact that an increasing fraction of the population lives in cities, which some have argued is as important as wealth and population growth with regard to the long-term health of the planet. The goals are to:

(1) Define income.
(2) Ask how income and expenditure are increasing.
(3) Consider how technology could lessen environmental impacts.
(4) Consider the influences of urbanization.

12.1 Income

The usual measure of income is gross domestic product (GDP), which is the total amount of money spent by people, the government, and businesses on end products (e.g., a car, a cup of tea) and services (e.g., the person who repairs the car or who serves the tea in a restaurant) over a given time interval, typically one year. End products are those that are not sold on. Because income and expenditures should approximately balance, GDP is also estimated as total earnings through salaries and business profits (investments are ignored in the accounting). Net domestic product is lower because of capital depreciation (e.g., the machines used to refine tea may wear out and may not be replaced). Note that GDP is not a measure of how much is owned (i.e., the capital of the country and its people, termed wealth), but instead a measure of cash flow.

Measures of GDP must be adjusted to account for vagaries in exchange rates, which means that the cost of the same item varies across countries. Purchasing power parity (PPP) presents GDP for a country in units of a currency, adjusted to the cost of a set of goods for a specified country, typically the United States. For example, if the set of goods costs $20,000 in the United States and $30,000

Ecology of a Changed World. Trevor Price, Oxford University Press. © Oxford University Press 2022.
DOI: 10.1093/oso/9780197564172.003.0012

in another country, the latter country's GDP will be downvalued by one-third. A second adjustment is needed to deal with inflation, which is the increase in money spent for the same set of goods over time. To account for inflation, GDP is corrected to a specified year, based on what it would cost to purchase the set of goods in that year. For example, if goods cost 2% more than in the previous year, we would multiply earnings in the second year by 100/102 to make it comparable to the first. The approach requires that the goods being compared in each year are identical. This is not easy because many items are constantly being improved (e.g., cellphones from 2005 are hardly comparable to those of today). For this reason, comparisons are usually done from one year to the next in order to give an annual inflation rate. The set of goods used is carefully chosen by a group of economists, and others might be used. For example, reflecting recent globalization, *The Economist* magazine has introduced the Big Mac Index, where the basket of goods is a big Mac hamburger.

As another indicator of inflation we might use a chicken. Assume that the chicken we wish to eat was of the same quality in 1919 and 2009. In 1919, a typical chicken cost $1.23, whereas in 2009 it cost $2.86 (inflation of 2.86/1.23 = 230%). If earnings increase at the same rate as inflation, per capita GDP will stay steady. But earnings have increased by over 7 times. A good way to see how real earnings have increased is to note that in 1919 in the United States, it took over 2 hours and 30 minutes for the average worker to earn enough money to buy a chicken, whereas in 2009 it took just 30 minutes. Hence, on the basis of chickens, assuming the amount of time working has not changed, per capita GDP in the United States increased fivefold between 1919 and 2009, which is quite close to the increase based on a more continuously adjusted set of goods (Fig. 12.1). In Figure 12.1, dollars are adjusted to match earnings in the year 2011, so if per capita GDP were to stay constant, the actual dollars received in the pay packet would be less in 1919 (by 1/2.3 = 43% of the 2011 figure) and greater in 2019 (by 16%; annual inflation in the United States has varied from 0.7% to 3% over the 10 years 2009–2019). In general, in presenting GDP, one needs to indicate the baseline year.

12.2 Growth of Income

Per capita GDP for the United States over the past 200 years has increased approximately exponentially (Fig. 12.1). Growth in GDP results from increases in productivity (i.e., the goods and services produced by a typical person). Productivity itself increases partly because of improved capital investments, such as keeping machines better maintained and replaced more quickly, but

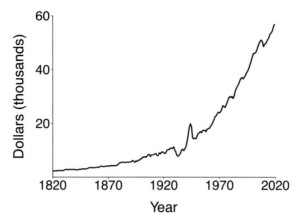

Figure 12.1 Estimated per capita GDP in the USA over 220 years, adjusted for inflation to 2011 U.S. dollars. Note the economic recessions of 1929–1939, and 2008. (www.gapminder.org up to 2015, and 2016–2019 from http://databank.worldbank.org)

mostly because of advances in technology (e.g., faster microchips, chainsaws instead of hand saws), and better training and education. As productivity increases, the same goods can be sold for less, and in a competitive market they surely will be sold for less. For example, people now spend less than 10% of income on food, compared with more than 40% in 1900. This, of course, frees up money to be spent on other things.

Figure 12.2 shows the growth in per capita GDP for the world and separately for Europe, India, China, and the United States from 1990 to 2020. On average across the world, per capita GDP increased by about 2.3% per year, and assuming exponential increase, we can use the formula introduced in Chapter 2:

$$t_{double} = \ln(2)/r$$

to deduce a doubling time of around 30 years. Can this progress be continued? It is regularly stated that growth must slow because resources are finite. For example, William Jevons argued in 1865 that the deeper seams being mined for coal in the United Kingdom would soon make coal an unprofitable source of energy, thereby leading to a recession. Instead, in the face of demand from consumers, technological improvements increased profitability. A different example comes from whaling, where natural resources were indeed used up. For almost 400 years up to the end of the 19th century, oil for heat and light mostly came from whales and seals. Millions were slaughtered, and many species have not

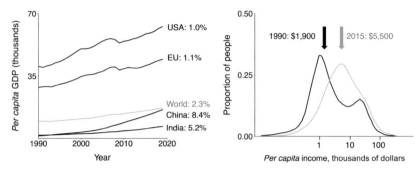

Figure 12.2 *Left:* Growth in per capita annual GDP, 1990–2013 (adjusted to 2011 US dollars); the percentages give the average over the 20 years 2000–2019. *Right:* Distribution of individual incomes in two years (based on household surveys, also adjusted to 2011 dollars). Income within countries became slightly more inequitable, but income between countries became more equitable, for example due to economic growth in China and India, and this dominates the pattern. The arrows separate the top 50% from the lower 50%. The top 5% of all people had an income of more than $38,000 in 1990, $52,000 in 2015. (*Left:* http://databank.worldbank.org, *Right:* V. Jordá, and M. Niño-Zarazúa. [2019]. Global inequality: How large is the effect of top incomes? *World Development* 123:104593)

recovered (Chapter 21). A whaler in the late 1800s reflected on the economic consequences:

> At one time it was almost thought the world would stand still if the supply of fur of seals and bone and oil of whales should cease. The supplies did cease, but the world still goes on, and what was half a century ago so highly valued is now scarcely missed. Science and nature have ministered to man's necessities, and a far better oil for illuminating purposes ... has been supplied in such abundance that the homes of the poor are supplied with a better, cheaper, and more healthy illuminant than whale oil. Thus, we have an assurance that the needs of man will always in some form be supplied by a bountiful providence. (C. Roberts, 2007)

By the middle of the nineteenth century, increased demand for fuel for night lights had far outstripped the ability of whale oil to provide, stimulating the development of fuel alternatives. At first, these were derived from alcohol derivatives, and later they came from petroleum. A government tax on alcohol drove the switch to petroleum. We see a similar combination of market forces and government interventions (taxes, subsidies, and legislation) driving innovations in the energy market today. Over the past 20 years, technological breakthroughs

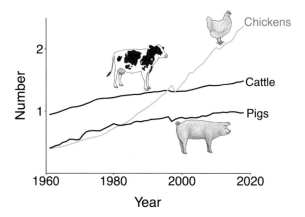

Figure 12.3 Global increase in cattle, pigs, and chickens. Units are billions for cattle and pigs, and tens of billions for chickens. Between 2000 and 2018, the number of cattle increased by 13%, pigs by 9% and chickens by 64%. (From UN Food and Agriculture Organization; http://faostat3.fao.org)

have led to new methods for extracting fossil fuels, such as gas by fracking and oil from tar sands (Chapter 16). In summary, unlike human population growth, which has slowed, we see no evidence for a slowing in the growth of real income. That is the reason the number of chickens in the world is increasing exponentially (Fig. 12.3).

In a capitalist system, financial gains are driven by the investment of wealth (i.e., capital) to create more wealth. That requires consumers to consume more products, which is achieved partly by people's wishes, amplified by advertising and improved quality of product. As our wealth increases, we find new ways to spend it, and consumption continues to expand. We eat more meat (Fig 12.3), trade more goods internationally, eat out more, and travel more (Fig. 12.4); the figures in this chapter all show the trend of an increase in some indicator of consumption over the past 20+ years, often at an increasing rate. Economists note that within countries, wealthier people do eventually spend a smaller proportion of their income on goods, investing the rest, and they consequently suggest that individual consumption levels off with wealth, slowing environmental impacts. However, these investments drive economic growth within countries. Looking between countries, environmental impacts have so far have increased proportionately with per capita GDP, and show little sign of leveling off (Fig. 12.5). The exception seems to be the city of Singapore. However, in this case it is possible that adding migrant workers to the equation would lower average per capita income, more than altering per capita environmental impacts.

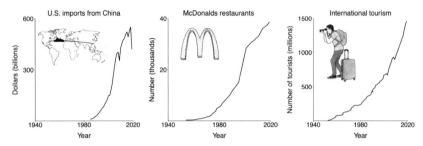

Figure 12.4. Examples of rapid increases in consumerism and globalization since 1945. Imports from the US Census Bureau (adjusted here to 2019 dollars), the number of McDonald's restaurants (which now has franchises in 118 countries) and international tourists. (*Left*: census.gov/foreign-trade/balance/c5700.html; *Center and Right*: W. A. Steffen, A. Sanderson, P. D. Tyson, et al. 2004. *Global change and the earth system: A planet under pressure*. Berlin: Springer-Verlag, with recent years added for McDonald's from statista.com/topics/1444/mcdonalds/ and for tourism from unwto.org/global-and-regional-tourism-performance)

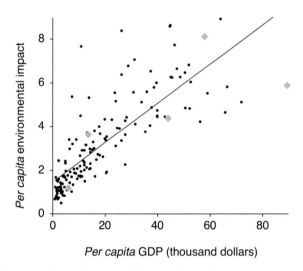

Figure 12.5. Per capita environmental impact plotted against per capita GDP (2020 dollars), both measured in 2016. Environmental impact is roughly the amount of productive land (in hectares) utilized to support one person (discussed in Chapter 13), but here it includes land that would need to be planted as forest to capture released carbon. Two small countries with very high per capita environmental impacts (Luxembourg and Qatar) are not shown. The grey diamonds are from left: India, China, UK, USA, Singapore. (www.footprintnetwork.org/en/index.php/GFN/page/footprint_data_and_results/, databank.worldbank.org)

12.3 Impacts Reduced by Technology

If consumption does continue to increase with wealth, human impacts on the environment will not necessarily increase as well. Such impacts are the result of three contributions: population size, per capita consumption, and technology. Consumption inevitably stresses the environment (as detailed in later chapters, the threats of climate change, overharvesting, pollution, habitat loss, invasive species, and disease all become magnified as consumption increases). However, technological improvements can have both negative and positive effects. Positive effects include (1) increasing the efficiency of production that lowers the environmental impacts of a particular set of goods, achieved coincidentally with a reduction in price, which makes them immediately attractive to the consumer; and (2) lowering the environmental impact of the product even in the face of price increases:

Increased Cost Efficiency

The same goods may be produced at less cost. Obvious ways to reduce costs are through using less energy or land for the same return, e.g. miles travelled or crops harvested. In such cases, the profit motive and conservation goals appear to be aligned. Unfortunately, however, lower costs have so far had little effect on reducing consumption. As we have already seen, people find ways to spend savings, thereby negating the energy or land saved. For example, in Chapter 9 we showed that technological improvements mean that we can produce more food per hectare than we could 50 years ago. Rather than reducing land cultivated, these improvements stimulate increased consumption through price reductions in a competitive market, which means that people eat more, and more often substitute meat for vegetable diets.

Increased cost efficiencies can sometimes create an overshoot, whereby total use of the product increases so much that its environmental impact is greater than before. This overshoot is termed Jevons' paradox, after that same Englishman who was concerned about the loss of coal. He pointed out that increasing efficiency in the steam engine led to an increase in the use of coal, not a decrease, as one would expect if one continued with the same lifestyle. In the United States between 1975 and 1996, energy efficiency (energy produced/dollar) increased by 33%, but each person nonetheless consumed 6% more energy. For example, as cars have become more efficient in fuel consumption, we have bought larger ones and driven them further. In richer nations, this may be changing. Rapid increases in energy efficiency over the last 10 years have been associated with a flatlining of energy consumption in the United States and a slow decline in Europe. In this case the increase in GDP is leading to expenditure

128 ECOLOGY OF A CHANGED WORLD

on other items, many with negative environmental effects, such as electronics (promoting mining for metals), plastics, and clothing.

Technology Can Reduce Impacts on the Environment

The previous sections suggest that the conservation of resources associated with reduced production costs may lead to little or no small environmental improvement, especially as incomes increase from low levels. One alternative is to make equivalent products that are less harmful to the environment. Such an approach may often increase costs and consequently be undesirable to both consumer and producer. That is well illustrated by arguments from various entrenched forces, such as oil companies, which still oppose reductions in fossil fuel production. The cost of transition toward renewable forms of energy may be high, in light of the immense investments already established in technology and equipment for oil, gas, and coal extraction. In 2006, a paper entitled the *Economics of Climate Change* (also known as the *Stern Report),* noted that declines in CO_2 emissions of greater than 1% per year may slow economic growth and potentially lead to economic recession, because of the substantial infrastructure invested in fossil fuel use, the slow development of alternative energy sources, and difficulties in changing people's habits. However, the paper emphasized the importance of making such a transition, with the summed costs to the future lower the sooner the transitoin is made, and the real possibility that rapid development of alternative energy sources could stimulate growth. Between 2010 and 2020, development of renewables (primarily wind and solar) led to cost decreases of 50% (offshore wind) to 85% (solar installations). In 2022, we were at the point where in many places in the world the building of new coal-fired power plants is less profitable than the installation of renewables, and even the retiring of some inefficient coal-fired plants for renewables would be economically profitable. Still wind and solar provided only about 5% of the world's energy (oil 31%, coal 27%). An ongoing switch to renewable energy sources should stimulate economic growth first, because government investment itself stimulates growth, and second, because of the inevitable further reduction in costs of renewables driven by a competitive market (Chapter 16). Incentives to switch to less environmentally damaging products may be coupled with other market forces. People may demand more environmentally friendly products and are prepared to pay more for them. Beyond energy, Chapter 14 considers the example of organic food, and Chapter 22 highlights sustainable fisheries.

The increase in the use of renewable energy has been brought about largely through government incentives (e.g., subsidies for renewable energy), rather than government regulations, or taxes to make fossil fuel less competitive.

GROWTH OF WEALTH AND URBANIZATION   129

A government intervention associated with legislation and fines that led to technological innovation for the good of the environment came from efforts to curtail air pollution in the United States. In 1970, the Environmental Protection Agency (EPA) was established to oversee the implementation of new regulations devoted to reducing pollution, such as the Clean Air act, which was also introduced in that year. In 1993, the EPA set limits on the amount of sulfur dioxide (a major component of acid rain) that coal-fired power stations could emit. They put aside 2.8% of the total limit on emissions for power stations to purchase as credits, if they planned to exceed their individual limit. A starting price at $250 per ton of sulfur dioxide was established, with the expectation that companies would bid against each other and the price would rise. In fact, credits in the first year went for $136, and companies worked to reduce emissions. By 1996, credits went for less than $70 and in 2020 for less than 10 cents. Sulfur dioxide emissions are now 20% of what they were at their maximum in the 1980s. Auctions still operate, but continue mainly to allow new companies to start up. This is a nice example of market manipulation, where government legislation (1) revealed the true cost of controlling pollution, which was originally stated by the power companies to be more than $250 (perhaps as a political ploy to prevent legislation) and (2) lowered costs by stimulating invention and creativity. All this seems to be a great success. However, companies took the cheapest method available to lower emissions, which was to absorb the SO_2 using limestone and discarding the sludge produced into pits. This is now leaking into groundwater in some locations, including near my home in Indiana, generating yet new problems of pollution.

12.4 Urbanization

The European Commission defines urban centers as communities with at least 50,000 people living in adjacent 1 km^2 grids, each of which has more than 50% of the ground built up and/or is occupied by more than 1,500 people. In 2015, 13,135 cities met this criterion. Urban centers are occupied by 48% of all people but they cover just 0.4% of the world's land. When towns and suburbs are included, more than 80% of all people live in urban settings. This percentage is predicted to increase steadily, thanks to internal population growth, migration from the countryside, and expansion to include neighboring communities. Per capita income is higher in urban centers than elsewhere, which implies that pressures on the environment from cities are currently relatively high as well, but the demographic transition toward smaller families is proceeding more rapidly in cities. In the developed world, larger urban centers have higher per capita income than the smaller centers but require less per capita infrastructure (e.g., roads), suggesting reduced impacts on the environment as cities grow. However, as a generalization, suburbs currently have the highest environmental impact in terms of per capita infrastructure and energy use than both urban centers and rural areas.

Concentrations of people at high densities are reportedly beneficial to the environment by relieving pressures on wild places, but we have no clear way to assess this effect, given that a world without urbanization does not exist. Instead, discussion of biological conservation and cities has included the following topics: (1) The direct impact on the environment. While the area occupied by cities is small relative to cropland, cities are often placed in biologically rich areas and have effects that stretch far beyond the city (e.g., the influence of the Amazonian city of Manaus on fish stocks can be detected up to 1,000 km away); (2) both water and air pollution; (3) the failure of urban dwellers to experience, appreciate, and value biodiversity; and (4) biodiversity loss in the city itself (cities contain few species). Some benefits of cities do accrue to surrounding wildlife. Cities managed for greenspace can preserve species that are absent from surrounding areas, as is argued for bees, critical plant pollinators that have been severely affected by pesticides in croplands.

12.5 Conclusions

While per capita GDP is one measure of wealth, an environmentalist is more concerned with total GDP (the product of population and per capita GDP) as this, along with how income is spent, is what impacts natural systems. As shown in Figure 12.6, world GDP has recently increased by about 3.4% annually, a doubling time of about 20 years, with the causes more or less equally split between more people and increases in income. That the growth of income is a

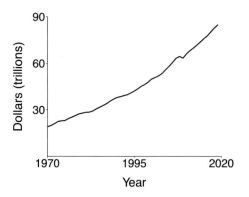

Figure 12.6 World GDP 1970–2019 (2010 dollars). Between 2000 and 2019 GDP increased by 70%, for an annual growth rate of 3%. (databank.worldbank.org)

GROWTH OF WEALTH AND URBANIZATION   131

major contributor to environmental degradation will become apparent in the next chapter. There, we will show that if everyone had the current consumption patterns of the United States, with current production patterns, 1.3 worlds would be required to support humanity.

The underlying principle of economics is that of matching supply to demand. When something is common or not in demand, it has low value; when something is scarce and in demand, its market value increases. Consequently, environmental degradation may sometimes raise GDP because more can be charged for the scarce resources it creates. For example, clean water may become a commodity, resulting in people now charging for packaged water, increasing incomes. This outcome is of course counter to the idea that greater wealth improves quality of life. A decrease in quality of life with an increase in GDP is sometimes termed Lauderdale's paradox. It is why some have argued that a cessation of economic growth is a viable alternative to growth in a more environmentally friendly way, but we see little evidence that this is happening.

Despite these issues, economic considerations also give us a reason for optimism. Wildlands are an increasingly scarce resource, and so, as people become wealthier, the demand to visit parks and reserves is increasing. In 2018, Yellowstone National Park in the United States received more than 4 million visits. Visits to Corbett National Park in India are now capped at 300,000 a year, despite high demand. Given the decline in wild areas, and the increase in demand for access to them, we anticipate that the transition to a wealthier society will lead to increased economic benefits from natural areas. Such economic benefits can exceed those benefits of the land if it were to be used for other purposes, ultimately leading to wildland expansion and making an important contribution to the preservation of nature. For example, a natural area near Corbett is being rapidly developed for ecotourism, with the number of visitors increasing from 2,000 in 2012 to more than 200,000 in 2019. (Possibilities such as these are developed further in Chapter 14.)

13

Habitat Conversion

We now move on to consider COPHID in detail (Climate change, Overharvesting, Pollution, Habitat loss, Invasive species, and Disease), though not in that order. For each threat, we consider a pre-human background as baseline, changes that have happened, their physical and biological underpinnings, and changes that may happen under different scenarios.

It makes most sense to start with habitat conversion because it is a result of appropriation of land by humans and therefore connects us to the earlier chapters on agriculture. Habitat loss is also seen as the largest single driver of species declines. Further, evaluations of the effects of humans on the environment are usually made in a two-step process. The first is land-use change, converting habitats from their natural state, and the second is loss within habitats due to one or more of the other threats. Hence, the goals of this chapter are to ask:

(1) How much of the world's plant production is being usurped by humans?
(2) How much land is being used by humans?
(3) What is the present rate of habitat conversion, emphasizing tropical forests?

13.1 Appropriation of Plant Productivity

We start by asking how much annual natural plant growth has been diminished. This statistic is analogous to GDP in that it assesses annual impacts rather than standing natural or monetary wealth.

The quantity of carbon vegetation assimilated at a particular location over a given timespan, usually a year, is termed net primary productivity (NPP); the units are often grams of carbon/m^2/year. Productivity is "net" rather than "gross" because plants themselves burn off some of the carbon they assimilate, which is not included in the accounting. The statistic that captures the effects of humans on NPP is termed Human Appropriated Net Primary Productivity (HANPP), calculated as the difference between the NPP that would be observed in the absence of human influences and that which is left on the land after we have harvested crops, allowed domestic animals to graze, have built over the land, and so on.

Ecology of a Changed World. Trevor Price, Oxford University Press. © Oxford University Press 2022.
DOI: 10.1093/oso/9780197564172.003.0013

Human appropriation of net primary productivity occurs for three reasons. The first is a result of land-use change ($NPP_{pristine} - NPP_{actual}$). For example, a parking lot has zero plant growth, so in this case the HANPP would be identical to $NPP_{pristine}$. Most changes in land use reduce NPP (e.g., by conversion of forest to grazing land or cities), but in some cases they can raise NPP (e.g., by irrigation of deserts). The second contributor to HANPP is the amount of plant growth in a location that is removed by harvesting or grazing. The third is from human-set fires, both deliberately in order to clear croplands after harvest or forests for

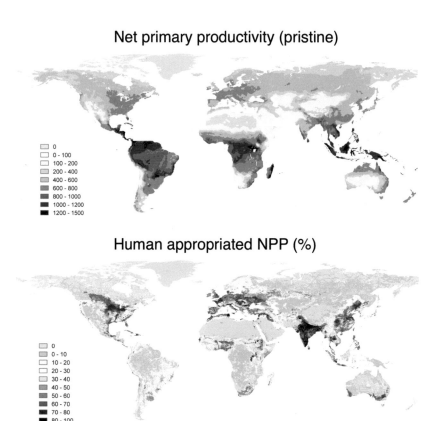

Figure 13.1 (Color plate 6) *Above:* Estimated net primary productivity ($NPP_{pristine}$) as it would be in the absence of humans. The units are grams carbon/m²/year. *Below:* Human Appropriated Net Primary Productivity (HANPP) as a percentage of $NPP_{pristine}$ for the year 2000. (H. Haberl, K. H. Erb, F. Krausmann, et al. 2007. Quantifying and mapping the human appropriation of net primary production in earth's terrestrial ecosystems. *Proceedings of the National Academy of Sciences* 104:12942–12945; data at https://boku.ac.at/wiso/sec/data-download)

134 ECOLOGY OF A CHANGED WORLD

agriculture and accidentally through burns of natural areas. The equation used for calculating HANPP in a given location is therefore:

$$\text{HANPP} = \text{land use losses} + \text{amount harvested or grazed} + \text{amount lost to fire}$$

or

$$\text{HANPP} = (\text{NPP}_{\text{pristine}} - \text{NPP}_{\text{actual}}) + \text{amount harvested or grazed} + \text{amount lost to fire}$$

To measure HANPP, we need to estimate each of the quantities on the right side of the equation:

(1) $\text{NPP}_{\text{pristine}}$: An estimate of what terrestrial Net Primary Productivity would be across the world in the absence of humans is shown in the upper panel of Figure 13.1. This estimate is calculated from a model that relates temperature, water, and geochemistry measurements to NPP. The values that go into the model are estimated from direct measures of productivity in some locations. The warm and wet tropics are the most productive places in the world, whereas cold and dry areas have much less plant growth. Indonesia seems to have slightly higher primary productivity than the other tropical regions, which may be because recent volcanic activity has raised available nutrients.

(2) $\text{NPP}_{\text{actual}}$: This is the measure of plant carbon uptake that actually happens in a location, based on information from the United Nations Food and Agriculture Organization. The FAO divides the world's land surface into a few broad classifications, notably cropland, grazing land, forest, and built-up land. For grazing lands, $\text{NPP}_{\text{actual}}$ is calculated using the same model as that for $\text{NPP}_{\text{pristine}}$, after accounting for the effects of fertilizer, vegetation change, and the like. For croplands, $\text{NPP}_{\text{actual}}$ is calculated from direct measurements of plant growth.

(3) For grazing lands, the amount removed is obtained from the known consumption rates of animals. For croplands, the amount harvested is directly measured.

(4) Fires: The amount lost to human-set fires can be directly calculated, based on satellite imagery.

Using these methods in the year 2000, humans appear to have appropriated about one quarter of global terrestrial NPP, a large quantity given that we are only one among millions of animal species dependent on plants. Overall, harvest contributed 53%, land-use changes 40%, and fires 7% to the appropriation. The lower panel in Figure 13.1 shows the global distribution of HANPP as a fraction of NPP$_{pristine}$, excluding the effects of fires. In some places such as India and Illinois, well over 50% of the NPP was appropriated. In a few desert locations, NPP has increased over the pristine value because of irrigation. While no one has updated the estimate across the globe in this way, according to the Living Planet report, HANPP increased from the year 2000 to the year 2020 by 18%.

13.2 Appropriation of Land

It is one thing to estimate annual appropriation of primary productivity, but more ambitiously, we can estimate appropriation of land, marine (continental shelf), and freshwater bodies by humans. This is analogous to the standing wealth of a nation, whereas the HANPP measure would be analogous to GDP. An organization named the Global Footprint Network has attempted to do this, also using the United Nations' FAO database.

According to the Global Footprint Network, the productive area available to humans totals 119 million km^2 (Table 13.1), a sum that is obtained by adding continental shelf and freshwater bodies to cropland, grazing land, forest, and built-up land (low-productivity areas such as deserts, high altitudes, and polar regions are ignored). Given a global human population of 7.9 billion in 2021, this is just 1.5 ha of productive land for each person.

Table 13.1. Land Categories and Their Assigned Calorific Values Relative to Fishing Grounds

	Area (millions km$_2$)	Proportion	Relative Value	Proportion after Weighting*
Cropland	16.0	0.13	6.8	0.34
Built-up land	1.67	0.01	6.8	0.04
Forest	39.0	0.33	3.4	0.41
Grazing land	34.0	0.29	1.2	0.13
‡Marine and inland water	28.3	0.24	1.0	0.09

*For example, the relative contribution of cropland after weighting by calorific value is $(16 \times 6.8)/(16 \times 6.8 + 1.67 \times 6.8\ 39 \times 3.4 + 34 \times 1.2 + 28.3 \times 1) = 0.34$

‡Marine refers to continental shelf only. About half was considered used. (B. Ewing, D. Moore, S. Goldfinger, et al. *The Ecological Footprint Atlas 2010.* Oakland, CA: Global Footprint Network. https://www.footprintnetwork.org/content/images/uploads/Ecological_Footprint_Atlas_2010.pdf).

136 ECOLOGY OF A CHANGED WORLD

Next, land-use categories are weighted to account for their potential value to humans based on the number of calories they could provide (Table 13.1). The method follows FAO measures of "crop suitability," which rank different areas of the world according to the number of calories per hectare they produce. In this accounting, fishing grounds generate the lowest value per hectare, and the value for other categories is expressed relative to that for these areas. The Global Footprint Network considers that 20% more calories come from the same area of grazing land, thereby giving grazing lands a value of 1.2. Croplands get the highest ranking because people tend to cultivate land where plants grow well. Built-up land gets the same value as cropland because cities emerge in those areas too. Forest land gets a ranking between grazing and cropland because it covers land that could be converted to one or other of these categories (e.g., converted Siberian forests are suitable for grazing, whereas Indonesian forests are suitable for oil palm plantations). These methods are regularly updated, but the current weightings remain similar to those presented in Table 13.1.

The next step is to work out how much of each category is being used. Cropland, cities, and grazing land are by definition fully exploited by humans, but this is not the case for forests or fishing grounds. In the 2016 accounting, about one-half of fishing grounds and one-third of forests were considered exploited. Exploited forests include plantations. They are used primarily to provide timber and paper products rather than calories.

Estimates of the proportion of the world devoted to each land-use category, after weighting by suitability, are given in Figure 13.2. Under present production patterns, we could maximally double our calorific production by converting all forests into food production, but doing so would eliminate the use of forests for paper and timber. These measures should correlate well with impacts on nature because the calorific value of land is closely correlated to plant productivity. Thus, we would conclude that about one-half of all pristine productivity ($NPP_{pristine}$) was on land that is currently devoted to feeding us (Fig. 13.2).

The one-half figure is an average estimate for the world and does not account for considerable variation in resource use across different geographic regions. One can calculate appropriation on a per country basis by accounting for each country's imports and exports, and considering the resources each country uses rather than the resources it produces. The United States appropriates about twice as many resources per person as the world average, so we would need to convert all forests to food production if everyone were to have the United States' present-day consumption patterns, given the present state of technology (Figure 13.3). While these conclusions are sobering, in Chapter 9 we noted that one could double calorific production without increasing land use by either (1) changing habits, such as altering diets and reducing food waste, or (2) altering farming practices, including the optimal distribution of fertilizer and technology worldwide, and improved technology.

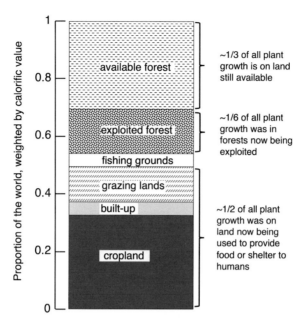

Figure 13.2 (Color plate 7) Exploitation of the planet by humans, after weighting land by its calorific value (Table 13.1), estimated for the year 2011. See Table 13.1, first column, for the actual amount of land devoted to each of these uses. Here we assume calorific value to humans translates into growth of native plants. (global footprint network; https://www.footprintnetwork.org/resources/data)

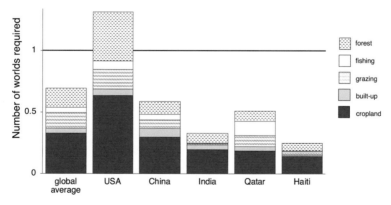

Figure 13.3 (Color plate 7) Requirements from the planet if everyone in the world had per capita consumption patterns of the five countries listed, based on 2011 data. The gray line is one world. In 2011, Qatar had the highest average income in the world (rightmost point in Fig. 11.4, *Left*), and Haiti the lowest life expectancy (lowest point in Fig. 11.4, *Left*). Qatar now somewhat exceeds China. It is a desert country whose excessive use of energy results in a low land use footprint (global footprint network. https://www.footprintnetwork.org/resources/data)

13.3 Impacts on Habitats

The appropriation of land for food has had a major effect on natural ecosystems. An ecosystem is the community of species, plus all associated nonbiological factors such as nutrients and pollutants. All ecosystems have deteriorated from human exploitation. For example, the United States (lower 48 states) has lost an estimated 50% of its wetlands. Tallgrass prairie once covered large parts of the Midwest in the United States and Canada, most of which has been converted to corn and soybean production. For example, the US state of Iowa, has only 0.1% prairie left.

Forest cover about 8,000 years ago may have been close to what it would be without human effects, but by the early 1990s it had been reduced by about half (Fig. 13.4). Forest loss continues and is increasingly well documented through national surveys, as reported by the FAO and through remote sensing by satellite. Remote sensing regularly records the state of the land in areas as small as 30 m × 30 m grids over the entire world. In Figure 13.4 (top right), each grid is colored based on the extent to which it was covered with trees greater than 5 m in height in the year 2001. More than 3 million km^2 of forest loss between 2001 and 2015 (9% of that standing in 2000) is attributed in equal proportions to one of four causes: (1) forestry, (2) shifting agriculture, (3) wildfire, and (4) permanent land-use change for commodity production (notably, soybeans in Brazil and palm oil in Indonesia). Forest converted into commodities represents permanent loss, whereas forest lost

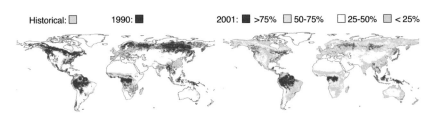

Figure 13.4 (Color plate 8) *Left*: Forest inferred for c. 8,000 years before present and in 1990. *Right*: Average cover in 3km*3km grid squares estimated by remote sensing across the world in the year 2001. *Color plate 8 contains an additional figure documenting forest change from 2001 to 2019.* (United Nations monitoring program; http://www.unep-wcmc.org/resources-and-data, M. C. Hansen, P. V. Potapov, R. Moore, et al. [2013]. High-resolution global maps of 21st-century forest cover change *Science* 342: 850–853; https://glad.earthengine.app/view/global-forest-change#dl=1;old=off;bl=off;lon=20;lat=10;zoom=3)

due to the other three causes should regrow. Today, the most pristine forests are located mainly in the boreal regions of Canada and Russia, the central Amazon, and the Congo (Fig. 13.4, color plate 8). However, people have lived in the tropical regions for many years, and in Amazonia a relatively high abundance of useful fruit trees is thought to represent their selective cultivation and preservation.

The rest of this chapter focuses on tropical rainforests, because they contain a large fraction of all the world's species of animals and plants. The humid tropics contain more than 35% of all forest, covering an area of 11.5 million km^2.

The main locations of tropical rainforests are in South America, southeast Asia, and west Africa (Fig. 13.5).

The Amazonian Rainforest

The forest of Amazonia contains about 5.5 million km^2, about half of all remaining tropical rainforest. It was largely untouched until the 1970s, when road construction into the area led to settlements and clearing. Large swaths of land have been cleared mostly for cattle and increasingly for crops, much of which are soybeans to be used as cattle feed and much of which is exported. Brazil holds about two-thirds of current forest areas. The progression of forest removal is associated with logging companies taking out the best trees, thereby opening up the area with additional roads, often followed by the setting of fires. Beginning in 2004, the annual rate of deforestation in Brazil decreased

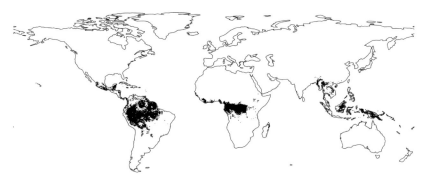

Figure 13.5 Moist tropical rainforests in 1990. (United Nations monitoring program; http://www.unep-wcmc.org/resources-and-data)

(Fig. 13.6). The reasons likely include reduced profitability of exports (as the Brazilian currency has become stronger), enforcement of environmental laws, improved forest monitoring by satellite, incentives for utilizing already deforested lands, expanded protected areas, and some private-sector companies realizing the public relations value of a lighter footprint. The rate of deforestation in Brazil recently increased, in keeping with a new political climate. About one ha/minute was being cleared in May 2019, although this has decreased more recently (Fig. 13.6).

As clouds move over South America from the Atlantic, rain that falls is recycled by evaporation and falls again. In forested regions, evaporation from trees accounts for as much as one-third of the Amazon's rainfall. Models imply that deforestation of the Amazonian region would reduce total precipitation by at least 15%, and increase temperatures by 2°C. Hence, deforestation not only creates climate change by the release of CO_2 but also exacerbates it through reduced precipitation recycling.

Southeast Asia and Oceania

The island of Borneo consists of three countries, which have about 50% of their forest cover remaining, down from 75% in 1973 (Fig. 13.7). The rate of forest loss

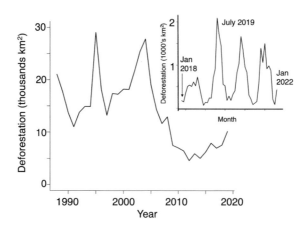

Figure 13.6 Estimated forest lost annually (July to June accounting) in the Amazonian region of Brazil, based on Brazil's National Institute for Space Research monitoring program, called DETER. The insert shows monthly amounts from January 2018. (Collated by mongabay.com; https://rainforests.mongabay.com/amazon/deforestation-rate.html)

has been 30% lower in this century than it was in the last 25 years of the twentieth century, a slowing of loss that resembles that in Brazil (Fig. 13.6). For example, despite being a protected area, Gunung Palung National Park was considered seriously threatened in 2002, when conversion to oil palm proceeded apace around its perimeter. At that time, illegal logging and poaching led to the suspension of tourism in the. The formation of the Gunung Palung Orangutan project in 2000 led to the emergence of a network of conservation advocates throughout the region (Chapter 28 describes some of their actions.) Logging continues in the park but has slowed substantially; it currently contains about 2,500 orangutans (the total number of orangutans in Borneo is unknown but may be in the region of 50,000). A slowing of forest loss across Borneo may be attributable to several of the same reasons that apply to Brazil, including lobbying for enforcement and protection, and the increased use of non-forest land for industrial plantations of oil palm (see Fig. 24.1).

Papua New Guinea (the western part of New Guinea in Oceania) is exceptional because it still has 83% of its forest cover and has had a low, but constant, rate of forest loss, at about 400 km^2/year since 1973 (about 0.1% of the standing forest and 10% of the average in Borneo). This picture is changing, with exploitation increasing (see Fig. 28.2).

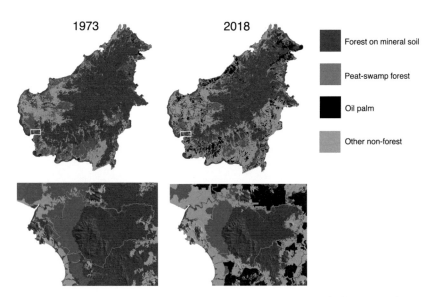

Figure 13.7. (Color plate 9) Borneo land use in 1973 and 2018. The lower graphs show an expanded version of the white rectangle, which covers Gunung Palang National Park and its surroundings. (Compiled by the Center for International Forestry Research; https://nusantara-atlas.org)

Congo Basin

The Congo Basin contains about 2 million km^2 of forest (5 times that of Borneo), more than half of which is present in the largest country, the Democratic Republic of the Congo. Best estimates are that this country lost about 2.5% of its forest areas between 2002 and 2014, driven by small holdings and commercial logging. Industrial agriculture seems to be currently less of a threat than in other regions, but it is increasing and commercial logging is intensifying (see Fig. 28.2).

13.4 Conclusions

This chapter has highlighted the large amount of land that humans have appropriated for agriculture, grazing, forest products, and cities; we are in a very different world than that of 100 years ago. Recognizing the benefits of conserving land both now and into the future, multiple initiatives, both private and public, are ongoing. However, laws are not always easy to enforce. A route to conservation beyond legislation is to increase the economic value of forests, while at the same time minimizing the impact of their use. That is the focus of the next chapter.

14

Economics of Habitat Conversion

People who own forested land may not clear it because they value their time more than their money or because they simply like trees. However, when forest or other natural habitat *is* cleared, it is nearly always for material benefit. This applies whether the forest is cut down by a multinational company to plant palms for oil or by poor farmers who grow crops for subsistence, as well as cash crops, such as coffee, cocoa, and the timber itself. This chapter addresses the costs and benefits of habitat conversion. We seek to determine whether the benefits of conservation can exceed, or be made to exceed, the benefits that come from other uses. The key question is whether the short-term benefits can be made to exceed the costs for the interested party, rather than for the future of humans and for nature, because decisions are rarely made in terms of public or long-term benefits.

If the goal is to maximize profit, it may be better to engage in high-intensity agriculture across an entire landscape rather than conserving any natural land. Indeed, this strategy has dominated many places, such as the corn and soybean fields of the Midwest in the United States and the wheat belt in India, both of which have led to the almost complete loss of original grasslands. Economists emphasize the benefits of such a strategy, where each geographical location is devoted to maximizing the productivity of that region. For example, in some midwestern United States, retaining woodlots and harvesting wood from them as a source of energy is less efficient than importing energy from regions specialized in producing it. Analogously, division of labor among people with different skills is championed as a major contributor to rapid economic growth, and is often the justification for free trade agreements.

Even if the goal is to maximize profits, the clearing of forest or natural habitat is not necessarily the optimal strategy because it ignores the value of wildlands. For example, forests in upstate New York provide clean water and recreation to neighboring communities, resulting in economic benefits that may exceed the advantages of clear-cutting. A patch of wild land surrounded by agricultural or urban land may be especially economically valuable. As we noted in Chapter 12, the rarity of an in-demand commodity means that we can charge more for it. Growth in wealth has led to ecotourism as a burgeoning industry, and tourists are increasingly willing to pay to see exceptional wild places or to hunt a wild

Ecology of a Changed World. Trevor Price, Oxford University Press. © Oxford University Press 2022.
DOI: 10.1093/oso/9780197564172.003.0014

144 ECOLOGY OF A CHANGED WORLD

animal. In some cases, then, both economic and conservation interests may be aligned.

In principle, it would seem straightforward to assess the costs and benefits associated with clearing or preserving a piece of land, and in the first example that we consider below, this is the case. In practice, however, it is often not so easy to do this. The goal of this chapter is to ask why, considering:

(1) Potential benefits from conservation, with an example from New York City's water supply.
(2) The difficulties of placing a dollar value on recreational benefits.
(3) The difficulties in evaluating future value (the question of discounting).
(4) The difficulties in evaluating costs and benefits across society, not just to the decision maker (the question of externalities).
(5) How costs and benefits may be altered to improve prospects for conservation over alternative use of the land.

14.1 New York City's Water Supply

About 90% of New York City's water supply comes from six large reservoirs in upstate New York. In the early 1990s, the quality of the city's drinking water fell below the standards required by the U.S. Environmental Protection Agency, due to sewage inflows and to fertilizer runoff. In 1996, New York City began buying land in and around the reservoirs to restrict land use, moving it away from agriculture and other forms of development and toward forest cover. The city also subsidized the construction of better sewage treatment plants in the area. Water quality improved. In April 2007, the EPA concluded that water supplies remained sufficiently clean that they did not need to be filtered, with further review in 2017. The city has saved the cost of building a filtration plant, estimated at more than $6 billion, and running costs of more than $100 million a year. The city continues with the program, planning to set aside $300 million per year to further its conservation and treatment programs around the reservoirs.

This example describes a straightforward cost-benefit analysis that aided conservation goals. However, even in this case, it is apparent that not all costs and benefits have been included in the analysis. For example, communities surrounding a park may get clean water but may also have coyotes coming out of the park to eat their chickens. Hence we should also consider benefits that are less easy to assign dollar amounts, an issue addressed in the next section.

14.2 Value for Tourists

If a national park is set aside for public use, with low entrance fees, it is difficult to ascribe a dollar value to the benefits people receive from recreational use. One could ask people who do not go to a park how much they are prepared to pay to keep it in existence (termed the no-use value because the person asked does not use it). This is clearly an uncertain accounting approach, albeit resulting in the valuation of U.S. national parks at $33 billion solely for this reason. Alternative approaches assess expenditures by people who visit parks which include travel expenses per se, hotel stays, entrance fees, etc, and are termed travel costs.

The travel cost approach has been widely applied, but assessment is not straightforward. For example, controversy has arisen over whether cost estimates should account for the salary that people forgo when they take time off from work to travel go to a park. Nevertheless, assuming one can work out the expenditures of each visitor, one can then place a lower bound on the park's value by summing across all visitors. This is considered a lower bound because people may be willing to pay more in entrance fees to visit the park. The consumer surplus is the difference between what people pay and what they are willing to pay. Estimates of such a surplus are useful because that helps quantify the park's value in a way that is commensurate with other values that may be placed on the land (e.g., grazing). Knowing the consumer surplus could also be used to estimate how much more one could charge for entry if revenue is an important consideration.

Travel costs can be used to assess the consumer surplus. One approach is the zonal travel cost method. People who live farther away from a park must spend more to visit it, which means that typically the farther away from a park that someone lives, the less likely they are to visit it. Suppose, for example, that we divide distance from the park into three concentric zones such as 0–20 km, 20–40 km, and 40–60 km. Assume that each zone contains the same number of people and the same average income. A plot of a number of visitors from each zone against the average cost of a visit is termed the trip-generating function. In the hypothetical example of Figure 14.1, more people visit from the nearest zone, presumably because it is cheaper for them to do so (Fig. 14.1A). In the example, 300 people visit from zone 1, 200 from zone 2, and 100 from zone 3, for a total of 600 visitors. Total expenditures by visitors are $34,000, giving a minimum estimate of the park's value. From this example, one can see that some visitors from zone 1 must value the park more highly than they are paying because visitors from zones 2 and 3 are identical in every way to those from zone 1, except that they live farther away. In other words, if some people in zone 1 moved to zone 2, they would continue to visit the park. Hence, we can infer that they are prepared to pay more, even if they remain in zone 1.

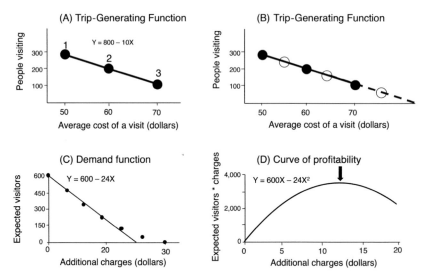

Figure 14.1 Estimating the value of a reserve to visitors using the travel cost method. (A) 300 visitors come from nearby locations (zone 1) spending $50 to do so; 200 visitors come from more distant locations, spending $60; 100 visitors come from even further away, spending $70. The value of the park based on this expenditure is 300 × 50 + 200 × 60 + 100 × 70 = $34,000. (B) The trip-generating function is used to calculate the number of visitors predicted for a given cost increase. For example, 450 visitors are expected if costs are increased by $5, which is the sum of the number predicted at $55, $65, and $75 (open circles). (C) The demand curve plots the expected number of visitors for a given cost increase (e.g., $5: 450 visitors, $20: 120 visitors). In this example, the demand curve is not a straight line because beyond $80 no one is expected to come to the park. We approximate a straight line fit to get to (D) The product of the demand curve and additional costs can be used to estimate maximum revenue the park could generate.

The trip-generating function can be used to ask how many people would visit if costs were increased (e.g., by increasing entry fees). The line in Figure 14.1B predicts how the number of visitors would decline if entry fees were increased by $5. Instead of 600 visitors, only 450 visitors would come (250 from zone 1, 150 from zone 2, and 50 from zone 3). The plot of the number of visitors for different costs is termed the demand function (Fig. 14.1C, where the circles indicate $5 increments). Summing the number of visitors prepared to pay each incremental value gives the total consumer surplus (450 people will pay $5 more, and of these 450, 300 will pay another $5, etc.). In the example, the consumer surplus comes to about $5,000 (see the section in the Appendices for this chapter), which is an additional 15% over the actual expenditures of $34,000. As entry fees increase, more revenue accrues to the park per person, but fewer people visit, so there is an optimum entry fee that maximizes the product of the entry fee and the number of visits. Figure 14.1D plots how increased charges translate into additional income. With a $12.50 increase in charges, only 300 visitors would come to the park, but this increase would maximize income, gaining an additional $3,750.

The practical application of methods such as these involves many additional considerations. For example, one needs to account for differences in income and the numbers of people in each zone, and the relationship between travel cost and number of visitors is generally not a straight line. Actual expenditures by visitors are often relatively high, even without consideration of consumer surplus; after applying the travel cost method, annual visits to Australia's Great Barrier Reef are valued at more than $1 billion. Across India's flagship national parks, estimates of consumer surplus add about 5% to values on the order of $1 million a year and generally show that entrance fees could be raised if maximizing profits were the goal.

This section shows, using examples, that it is always going to be difficult to measure all costs and benefits of conservation. These difficulties are exacerbated in two important ways: considering benefits that extend across many years and assessing impacts on others besides those who make the decisions. We consider each of these in turn, illustrated by example.

14.3 Discounting: Sustainable Harvesting in the Amazon Region

One of the earliest studies of the benefits that accrue from conservation considered the value of the Amazon rainforest. Researchers compared clear felling for cattle ranches against the benefits of sustainably harvesting goods from the forest. These goods include timber, as well as fruit, honey, fungi, and rubber.

Benefits and costs need to be evaluated into the future (i.e., what ranchers expect to gain over their lifetime). Such projections require placing a dollar amount on future benefits. The future is nearly always valued less than the present. That is because if one could receive $100 and invest it at an interest rate of 5%, the following year one would have $105. Hence, by waiting a year to receive the $100, one has lost $5. We define the discounting rate, d, as the interest rate on the future valuation that is required to reach today's value. For example, if one receives $100 today and the discounting rate is $d = 0.05$, then one would value receipt of $100 next year as worth $95.2 because 95.2 + (0.05 × 95.2) = 100. More generally we would write:

$$V_1 \times (1+d) = V_0$$

where V_1 is the valuation at time $t = 1$ and Vo the valuation at time $t = 0$. We turn the equation around to calculate future value in the following year as:

$$V_1 = V_0 / (1+d)$$

One possibility is to equate the discounting rate directly with the bank's interest return: an interest rate of 5% means that investing $95.2 today would give you $100 next year. However, the discount rate d is based on the return you would want to get next year, which may often be greater than the interest rate. You would probably want to figure in inflation, as well as issues such as how much you need the cash right now rather than next year. How one discounts when faced with a certain situation can vary a great deal (see Figure 14.2 and the section in the Appendices for this chapter).

A typical discount rate is set at 5% ($d = 0.05$). As shown above, this leads to a valuation of $100 in the next year as $95.2. Valuation for the year after that would be 95.2/1.05 – $90.7, that for the third year, 90.7/1.05 = $86.4, and so on (Fig. 14.2). You may recognize this as a geometric series, which we described for population growth in Chapter 2, but here we have declining numbers toward 0, at a rate of

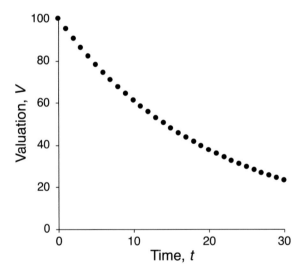

Figure 14.2 Discounting of an annual income of $100, at 5%. The values are obtained as $V_t = V_0/(1+d)^t$, where t is time and d is the discounting rate. The interpretation is that V_t (i.e., the discounted value for that time) is the investment one would need to make to get to $100, given an annual interest rate = d. Note that at time $t = 30$ intervals, even with a low discounting rate of 5%, $100 is only valued at $23, which by 50 years is $8 and by 100 years 80 cents.

ECONOMICS OF HABITAT CONVERSION   149

Table 14.1. Cost-Benefit Analyses of Two Options for the Exploitation of 1 Hectare of Peruvian Forest in 1987

	Clear-cut		Sustainable harvest	
	Recurring benefits	Value	Recurring benefits	Value
Timber sales	0	†1,000	†310	‡513
Fruit	0	†400	†300	*6,300
Latex (rubber)	0	†22	†22	*462
Cattle ranching	148	*3,108		
Total value		4,530		7,275

†Benefits that accrue in the first year of exploitation of the forest (total $1,422 clear-cut, $632 sustainable harvest). *Fruit, latex, and cattle ranching are discounted at the rate of 5%/year ($d = 0.05$). ‡Some wood is harvested every 20 years, which at a 5%/year discount is equivalent to $d = 1.53$. (C. M. Peters, A. H. Gentry, and R. O. Mendelsohn. Valuation of an Amazonian rainforest. *Nature* (1989) 339: 655–656).

$1/(1 + d)$, rather than an increase toward infinity. If one sums such a geometric series across all years into the future, the complete return is given by the formula $V_o \times (1 + d)/d$. If $V_o = 100$, with $d = 0.05$, the total value is $2,100.

Assuming a discounting rate of 5%, Table 14.1 gives valuations of clear-cutting and sustainable harvesting for rainforest in Peru in 1987:

Clear-Cutting

After transport and labor costs, clear-cutting 1 ha of forest for timber, fruit, and rubber results in an immediate profit (see the † numbers in Table 14.1). If the land is subsequently converted to cattle pasture, one gets additional revenues from cattle grazing every year. This revenue is estimated at $148 per year, which, when summed into the future at a discounting rate of 5%, comes to $3,180 although this ignores the costs of maintaining the pasture (e.g., fencing). When added to income from the timber, fruit, and rubber sales at the time of the clear-cut, the benefit totals $4,530 in 1987 dollars.

Sustainable Harvesting

The evaluation assumes that, for sustainability, all the rubber and three-quarters of the fruits could be harvested each year, leaving the rest of the fruits to

150 ECOLOGY OF A CHANGED WORLD

regenerate and provide resources for wild animals. In addition, every 20 years some timber could be sustainably extracted. Putting this together and summing into the future, the value of 1 ha is $7,275 (Table 14.1), which is 60% more than clear-cutting followed by conversion to pasture. Thus, according to this analysis, sustainable harvesting is a better bet economically. It is also better for the environment, though to a lesser extent than leaving the forest untouched (see Chapter 24, Fig. 24.2).

This example illustrates the difficulties associated with future valuations. For a variety of reasons, a developer may discount the future completely. That could be because a forester plans to move to another location the following year, because a farmer absolutely needs the income gained by clear-cutting to survive to the following year, or because a local villager expects their land to be usurped by the army sometime soon, as recently happened at a location in northeast India. With complete discounting of the future (i.e., d set to be a very large number), clear-cutting is the economically favored solution, $1,422 versus $622 (the summed † entries in Table 14.1).

The term opportunity cost is widely used in the economics and conservation biology literature but perhaps would be more easily understood if the term benefits forgone was used instead. In the above example, it is the cost of not exploiting the forest. If one were to completely leave the forest alone (the best strategy for conservation), the opportunity cost for a company contemplating clear-cutting is $4,530, whereas for a farmer who wishes to take up sustainable harvesting, the opportunity cost is $7,275. The section in the Appendices for this chapter discusses the opportunity cost concept further.

14.4 Externalities: Shrimp Farming in Thailand

The 1980s in Thailand witnessed the rapid development of shrimp farming, which involved clear-cutting mangrove forests to create marine ponds. Between 1980 and the early 2000s, 20–50% of all mangrove cover in Thailand was lost. About 40% of that loss has been attributed to the development of shrimp farms. In an economic analysis, the benefits to a developer of converting mangrove to a shrimp farm are more than 15 times greater than sustainably harvesting the forest (Table 14.2, top two lines). Thus, unlike the Peruvian rainforest example, a shrimp farmer has no obvious economic incentive to preserve the mangrove.

However, costs and benefits to clearing the mangrove do not fall solely on the farmer. Other parties are affected. A more complete accounting assesses these externalities, defined as the costs and benefits of an action that affect individuals other than the perpetrator. In this case, externalities include the value of mangroves as a fish and shellfish nursery for the fishing industry, which

ECONOMICS OF HABITAT CONVERSION   151

Table 14.2. Cost-Benefit Analysis of Converting Mangrove to a Shrimp Farm in Thailand*

	Development of farm	Conservation
Farmer	9,632	584
Local fishermen (mangrove as a fish nursery)	0	987
Local community (storm protection)	0	10,821
Thai people (taxes provide business subsidies)	0	8,412
Restoration costs	0	9,318
Total value	9,632	30,122

*U.S. dollars per hectare (1996 values), with a discounting rate of 10% per year over 9 years, after which time the farm is assumed to be not viable. The opportunity cost of conservation to the farmer = 9,632–584 = $9,048. The opportunity cost of development to all = 30,122–9,632 = $20,490. (TEEB–The Economics of Ecosystems and Biodiversity for National and International Policy Maker—Summary: Responding to the Value of Nature 2009, p. 12. www.teebweb.org/media/2009/11/National-Executive-Summary_-English.pdf. The original source is N. Hanley and E. B. Barbier (2009). *Pricing nature: Cost-benefit analysis and environmental policy.* (Elgar. Northampton, MA, USA).

supports a large population of inshore fishermen, and the value of mangroves as providers of storm protection. The number of coastal "natural disasters" more than tripled between 1975 and 2005 in Thailand, which may partly result from the loss of mangrove. The value of mangroves as storm protectors came under increasing scrutiny after the devastating 2004 tsunami. An additional externality is that of subsidies provided by the government for developing a business (e.g., to reduce the cost of such items as equipment), but come out of the Thai population's taxes. These externalities fall on different segments of the Thai population. For example, the loss of fishing nurseries mainly affects local fishermen, but government subsidies affect everyone who pays tax. All externalities share the feature that the costs do not fall entirely on the shrimp farmer. By one account, when externalities are quantified, the costs are more than two times the benefits (Table 14.2).

This example not only illustrates the value of taking externalities into account but also shows how difficult it can be to place dollar values on them. Two different approaches have been used to estimate the costs associated with storm protection. The first method evaluates the cost of replacing mangrove with the construction of an artificial barrier. The second method examines the costs associated with past storm damage and multiplies these costs by the predicted frequency and severity of storms in the future. The two methods yield wildly

152 ECOLOGY OF A CHANGED WORLD

different estimates of the benefits of conserving mangroves for storm protection. The benefits based on the second method (the figures shown in Table 14.2) is just over 10% that of the first method.

Until about 15 years ago, shrimp farms and other aquaculture endeavors had large environmental impacts. In Thailand, intensive methods have resulted in outbreaks of shrimp diseases. Partly for this reason and partly because of economic analyses such as the one outlined above, a new generation of more environmentally friendly, sustainable aquaculture projects is coming online. Chapter 9 described some recent advances.

14.5 Conservation

One way to improve the prospects for long-term conservation is to manipulate the costs and benefits in the short term, so that individuals who act in their own immediate interest also act in environmentally friendly ways. This can be done by using market forces or by resorting to government intervention. With respect to the market, some consumers are interested in buying from ecologically sound sources. Shade coffee in Central America (coffee grown under a natural forest canopy) and organic coffee (coffee grown without synthetic pesticides) are both environmentally less damaging than coffee grown in the standard commercial way, and people will pay more for these products. Note that in the first case, the coffee drinker feels better by contributing to the public good, but in the second case they may be directly concerned with their health.

Government intervention schemes operate on a grander scale and can be used to manipulate the market to benefit the entire population. CO_2 release into the atmosphere is an excellent example of an externality, with the benefits from burning a gallon of oil accruing to one or a few individuals, but with the costs shared across the world's population. Because CO_2 emissions have global impact, control requires international cooperation. In 2005, the United Nations introduced the Collaborative Programme on Reducing Emissions from Deforestation and Forest Degradation in Developing Countries (REDD) as a way to slow carbon emissions. The REDD program was originally devoted to funding reforestation because trees capture carbon. In 2010 the scheme was modified to REDD+, which adds in conservation (i.e., not cutting down trees, as a means of preventing additional carbon release). By the end of 2019, a total of 65 tropical and subtropical countries had become engaged in REDD+, about half of which had received funds to conserve or restore forestland. REDD+ is controversial because locals have ceded control of their traditional lands. The international meeting in Glasgow in November 2021 on climate change contained proposals to protect large swathes of land and ocean, with the genuine concern that this would lead to the usurpation of many local people.

Including carbon storage in cost-benefit analyses of the clearing of tropical forests can heavily influence the conclusions drawn from such analyses. One study of tropical woodlands in Paraguay considered benefits in terms of (1) sustainable harvesting of plants and animals; (2) people prepared to pay to search for medicinal plants; and (3) aesthetic values (assessed by the no-use method involving how much people are prepared to pay for the conservation of areas even if they themselves will never visit the area). The study then added the benefit of carbon locked up in trees, which was given a value of ~$9/tonne of carbon. That valuation is based on (1) estimates of damage expected from climate change and (2) proposed costs under various trading schemes, whereby a polluter can pay someone else to obtain carbon credits. Assuming some organization, such as the United Nations, is indeed prepared to pay this price, carbon storage overwhelms all other benefits at $378/hectare/year (2004 dollars, Fig. 14.3). Timber harvest ($28/ha), existence value ($25/ha), wild animals for meat ($16/ha), and prospecting for medicinal plants ($2/ha) together contribute less than one-fifth in total. When carbon is not included, it appears economically more profitable to convert forest to agriculture. However, when carbon is included, it becomes economically wiser to purchase forest and preserve it, except in a few places where purchase costs are high (Fig. 14.3).

14.6 Conclusions

By definition, conservation takes a long-term view. The aim is to preserve nature for future generations. A strong discounting of the future therefore works against the goals of conservation. Even a low discounting rate of 5% per year means that one would value income 100 years from now as worth less than 1% of the income today. Failure to include all externalities also typically works against conservation because the costs are generally distributed across more people than the benefits. This means that a developer is more likely to pursue their own interests, as in the shrimp farm case. The difficulties of discounting and externalities highlight a major point made by conservationists, which is general to the problem of predicting future value. In a world with a higher standard of living, people will likely value nature far more than they do at present. Even now, many pay large amounts of money to go to a park or to see a whale. As demand increases and parks and whales become scarcer, the future value of natural amenities may far exceed a 5% interest rate. Perhaps we should be counting up (formally known as negative discounting).

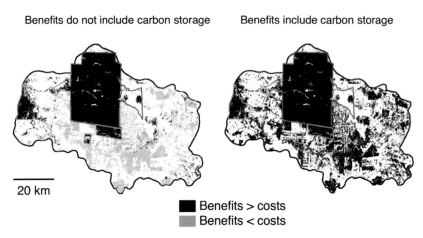

Figure 14.3 (Color plate 10) Benefits and costs of conservation across the upper watershed of the Jejuí River in eastern Paraguay. The opportunity cost of conservation is estimated from the cost of purchasing land, which depends on soil quality, topography, and who owns it. *Left:* The benefit of retaining forest from harvesting and aesthetic value alone. *Right:* Harvesting, aesthetic value, and carbon storage. Black: benefits of conservation outweigh costs. Gray: costs of conservation outweigh benefits. White: not forested nor considered. In the continuous black areas of the left graph, a park and indigenous areas are already protected, and hence the cost of purchasing the land for conservation is absent or low. (R. Naidoo and T. H. Ricketts. 2006. Mapping the economic costs and benefits of conservation. *PLoS Biology* 4: e360)

Cost-benefit analyses come with many uncertainties and sometimes result in an economic solution that favors cutting down trees (e.g., Fig. 14.3). As a result, some argue that we should not be using such analyses at all when deciding whether to conserve or develop. Ultimately, however, a wild land that tangibly pays for itself is likely to be best protected, and cost-benefit analyses indicate how the costs might be reduced or the benefits increased so that this actually happens.

15

Climate Crisis

History

Climate is best described as average weather over a certain timespan, and on average it is getting warmer. Global warming is a clear threat to humanity as well as to nature. For this reason, it is under greater political scrutiny than habitat loss, which makes it more promising that something will be done to bring it under control. On the other hand, if we do not do something about climate soon, all the other threats we are considering will become less relevant because of the great changes we expect from this one factor alone. This chapter:

(1) Considers the history of climate up to the present, so we can place what is happening now in context of what it was like in the past.
(2) Summarizes our understanding of the link between CO_2 and climate change.
(3) Describes the detectable effects of recent climate change on weather and species.

The next chapter describes predictions and potential remedial action to slow and reverse CO_2 build-up.

15.1 History of Climate

We are getting an increased understanding of what climate was like far into the past, based on analyses of fossils. One route is to estimate deep ocean water temperatures from oxygen isotope ratios in the shells of a group of single-celled organisms known as Foramanifera (or to researchers colloquially as "forams"). Oxygen comes in two main isotopes ^{18}O and ^{16}O, with the heavier ^{18}O carrying two extra neutrons. At lower temperatures, Foramanifera deposit relatively more ^{18}O than ^{16}O into their shells. The quantity of ^{18}O in bottom-living Foramanifera over the past 65 million years has progressively increased, indicating that the world has been cooling down (Fig. 15.1). The actual temperature has been calibrated based on laboratory experiments with living Foraminfera. A correction also needs to be made for the past 30 million years, since permanent ice sheets

Ecology of a Changed World. Trevor Price, Oxford University Press. © Oxford University Press 2022.
DOI: 10.1093/oso/9780197564172.003.0015

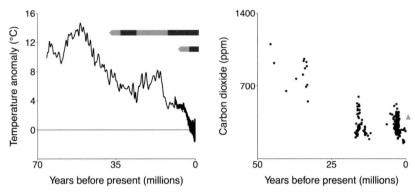

Figure 15.1 *Left:* Estimate of deep ocean temperatures over the past 65 million years. Until the formation of the ice sheets (indicated with lines at top, Antarctic above, Arctic below), this estimate is based directly on oxygen isotope ratios and magnesium/calcium ratios in Foramanifera, both of which are affected by temperature and are quantified with laboratory experiments on present-day Foramanifera. Ice sheets preferentially tie up ^{16}O; because this is the lighter isotope, water with this isotope evaporates more readily and then comes down as precipitation. That makes the ocean more concentrated in ^{18}O. A correction has been made to account for this, with ice volume based on estimates of sea level. The Antarctic ice sheet was much reduced 25–15 mya, associated with relative warmth (lighter grey). *Right:* CO_2 estimates based on boron isotope ratios in Foraminifera over the past 50 my. CO_2 concentration in 2019 is indicated with the grey triangle. (*Left*: B. S. Cramer, K. G. Miller, P. J. Barrett, et al. [2011]. Late Cretaceous–Neogene trends in deep ocean temperature and continental ice volume: reconciling records of benthic foraminiferal geochemistry ($\delta^{18}O$ and Mg/Ca) with sea-level history. *Journal of Geophysical Research* 116:1–23. *Right*: G. L. Foster, D. L. Royer and D. J. Lunt [2017]. Future climate forcing potentially without precedent in the last 420 million years. *Nature Communications* 8:14845)

have been present. Ice preferentially incorporates ^{16}O, with the consequence that the ocean has a higher concentration of ^{18}O than it would in the absence of ice (Fig. 15.1).

Superimposed on the general cooling trend over millions of years, climate cycles are triggered by variation in the tilt of the earth's axis (with respect to the plane defined by its orbit around the sun) and variation in the axis about which the earth rotates, which cycle over 41,000 years and 23,000 years respectively. These interact with variation in the shape of the earth's orbit around the sun, which cycles on the order of hundreds of thousands of years, to drive changes in the strength of the seasons and in mean temperatures; a steeper tilt of the earth's axis makes for cooler summers and hence less recovery from the winter ice. Temperature records have been derived from $^{18}O/^{16}O$ ratios in air bubbles trapped in polar ice cores. Relatively more ^{18}O is found in cold periods because ^{16}O is preferentially tied up in ice, meaning the relative concentration of ^{18}O in the air increases, just as it does in the ocean. Figure 15.2 shows temperature estimates from an ice core from Antarctica extending back 400,000 years. Several thousand-year long interglacial periods punctuate glacial periods at intervals of approximately 100,000 years. The most recent were 130,000, 240,000, 340,000, and 410,000 years before present (Fig. 15.2). Globally, temperatures are estimated to have been similar to the present day during the last interglacial, and about 7°C cooler during the last glacial maximum. Just 18,000 years ago, Chicago was under more than 1.5 km of ice.

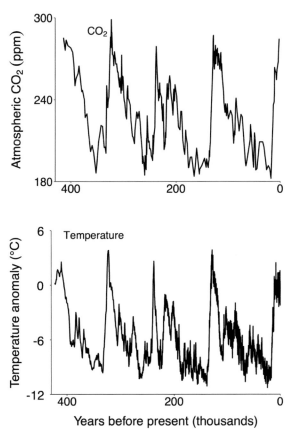

Figure 15.2 Surface temperature and CO_2 over the past 400,000 years, inferred from an Antarctic ice core (J. R. Petit, J. Jouzel, D. Raynaud, et al. [1999]. Climate and atmospheric history of the past 420,000 years from the Vostok ice core, Antarctica. *Nature* 399:429–436).

158 ECOLOGY OF A CHANGED WORLD

Climate fluctuations also occur on shorter timescales. In glacial times, especially in the northern hemisphere, extremely rapid increases in temperature, sometimes by as much as 16°C over a few decades, were followed by thousands of years of recovery (these changes still left Greenland quite cold, from –52°C to –36°C). The causes of these rapid changes remain unclear but likely reflect altered ocean currents. Global climate changes over the last 1,000 years include the "medieval warm period" (950–1100) associated with a major drought in the U.S. Southwest, followed by "the little ice age" from the 1500s to the 1800s: a cooling of perhaps 1°C below early twentieth-century levels in the northern hemisphere. Regional examples include a warming of 4°C in the 1920s along the north Atlantic coast of North America and the dust bowls of the North American prairies associated with a drought in the 1930s.

Climate is currently changing fast and perhaps shifting to a new, permanently hotter, regime at an unprecedented rate in earth's recent history. Up to 2022, the years 2016, 2019, and 2020 were the three hottest ones since records began in 1880 (Fig. 10.2). However, the earth is not getting warmer uniformly. Land is warming up faster than the oceans, with a rise of 1.5°C (averaged over the past 10 years) against the 1880–1920 baseline, compared with the global average of about 1°C (Fig. 10.2). The polar regions are heating up the fastest.

15.2 Greenhouse Gases

The recent increase in temperature is a result of human activities, mostly the release of greenhouse gases into the atmosphere via the burning of fossil fuels. Coal, oil, and gas that have stored carbon over hundreds of millions of years are now releasing their carbon back into the atmosphere over tens of years. The quantity is large: a single gallon of gas is thought to be derived from 90 tonnes of plant material (marine phytoplankton).

Heat accounting of the planet works as follows:

(1) About 30% of the sun's heat energy is reflected, for example, by clouds, small particles (aerosols), and ice.
(2) Earth absorbs the rest and to be in balance must emit the same amount of heat, which it does as long-wavelength radiation.
(3) Water vapor and CO_2 are the main gases in the atmosphere. They absorb long-wavelength radiation and reemit it, warming the lower atmosphere and ground. In this way, these greenhouse gases act as a 'blanket' for long-wavelength radiation, just as glass in a greenhouse keeps plants warmer. CO_2 is the critical greenhouse gas because the quantity of water vapor in the air is directly correlated with temperature. With no CO_2, the earth

would completely freeze over. Methane is also an important greenhouse gas and may have contributed as much as one-quarter to the warming we have recently experienced (coming from domestic animals, rice paddies, and leakages during gas extraction and combustion), but it breaks down to CO_2 within 10 years or so and then contributes only a small proportion of all released CO_2.

Presently, the earth is in an energy imbalance. More energy is being absorbed than emitted. Because of this imbalance as well as the present concentration of CO_2, the earth would continue to warm even if we entirely stopped emitting greenhouse gases in 2021.

Estimates of past CO_2 levels come from multiple sources, including isotope ratios and the cellular structure of fossil plants, but they are still quite uncertain. One method uses boron isotope ratios in Foramanifera, which correlate with ocean acidity, which in turn correlates with atmospheric CO_2 levels. Best estimates suggest that the last time CO_2 was as high as it is now was at least 3 million years ago, but perhaps as much as 20 million years ago.

Drawdown in CO_2 from the atmosphere results from its capture by marine organisms making calcium carbonate (e.g., corals, Foramanifera, mollusk shells). The capture is in turn driven by extensive weathering of basalt rocks, which release the calcium ions needed to form the shells, as described by the Urey equation:

$$CaSiO_3 + CO_2 \leftrightarrow CaCO_3 + SiO_2$$

or in words:

Calcium silicate + carbon dioxide $\leftrightarrow$ calcium carbonate + silicon dioxide

The silicon dioxide is inert. Calcium carbonate gets buried on the ocean floor, locking away CO_2, in a two-step process:

(1) Weathering on land releases calcium and bicarbonate ions (Ca^{2+}, HCO_3^-) that are then washed into the ocean. Weathering is accelerated in warm, wet climates, and because of continental movements, large areas of basalt have been present in the tropics over the past 66 million years.
(2) Marine organisms synthesize calcium carbonate shells from these products, which then get buried on the ocean floor.

The removal of CO_2 is widely seen as the cause of cooling over the past 66 million years. The role of weathering in the drawdown of CO_2 generates an important long-term thermostat. The idea is that when the earth warms, increased

160 ECOLOGY OF A CHANGED WORLD

evaporation and associated rainfall accelerate the weathering of rock, increasing the rate at which CO_2 is removed. On the other hand, as the earth cools, the amount of weathering is reduced, so CO_2 rises as it is continuously released into the atmosphere. This is a feedback loop that promotes climate stability over hundreds of thousands of years. It is not relevant to climate change over the next thousands of years, but provides an ultimate control on temperature. Without it, the earth would not be habitable.

Release of CO_2 occurs by a reverse of the Urey reaction at high temperatures in the earth's crust. A slowing of tectonic plate movements between 15 and 5 million years before the present has been suggested to have led to a slowing in the subduction of carbonates and hence the rate of release of CO_2, contributing to accelerated cooling over that period.

The Arctic and Antarctic ice cores give accurate estimates of CO_2 concentrations in air bubbles over the past few hundred thousand years. They reveal a major surprise, which is that CO_2 concentrations increase as temperature increases when the earth comes out of a glacial period (Fig. 15.2). CO_2 levels arc high when temperatures are relatively high and low when temperatures are relatively low. Unlike the long-term thermostat described above, this is a positive feedback, whereby a warming temperature results in release of CO_2, generating more warming. CO_2 concentrations went from about 200 ppm (parts per million) in the last glacial maximum to 280 ppm before human contributions. About half of the total warming when coming out of a glacial period has been attributed to increased solar radiation from the sun, which triggered the warming, plus other positive feedbacks (notably, as ice disappears, less heat is reflected). The other half of the warming results from released CO_2. The carbon isotope composition in calcium carbonate deposited at the time shows that the CO_2 must have been released from the ocean, and not land. A warmer ocean can hold less CO_2 anyway, and altered ocean circulation patterns likely contributed in an important way. Whatever the cause, this kind of positive feedback loop, which accelerates warming, has warnings for today.

Daily measurements initiated in Hawaii in 1958 show increase in CO_2 levels in the atmosphere over the past 60 years (Fig. 15.3). The world is breathing. Every spring, CO_2 levels are relatively high, as plants in the northern hemisphere have not been growing during the northern winter. The northern temperate regions are much larger than the southern temperate regions and have much more forest (see Fig. 13.4). Every fall, CO_2 levels are relatively low because plant growth through the summer takes up carbon. The breathing world illustrates just how important plants are as a storage system for CO_2. The greater the number of trees, the less CO_2 is in the air.

In 2019, humans released about 11.75 billion tonnes of carbon. More than 1.8 billion tonnes came from land clearance, and the rest from fossil fuels, made up

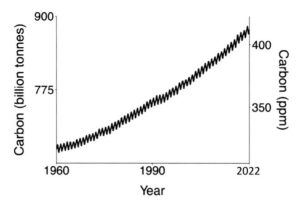

Figure 15.3 Atmospheric CO_2 measured at the Mauna Loa observatory in Hawaii, 1960–2022 (January). Note the 25% increase over the past 50 years, as well as the annual fluctuations. Normally, one sees this graph as ppm (parts per million, *right axis*), which is translated here also into billion tonnes carbon (1 ppm = 2.13 billion tonnes) to compare with other figures. (National Oceanic and Atmospheric Administration: https://gml.noaa.gov/ccgg/trends/data.html)

of coal (40%), oil (35%), gas (20%), and cement production (4%). Just under half of this remained in the atmosphere (Fig. 15.4). Measurements and models of carbon uptake by the ocean imply that about 40% of the rest was taken up there, leaving the remaining 60% to be taken up on land. Increased CO_2 stimulates plant growth, as should warmer temperatures and a longer growing season at higher latitudes, and increased nitrogen fertilization. Based on measurements of tree growth, tropical forests in the 1990s removed 17% of CO_2 emissions. However, in Amazonia higher temperatures and droughts have resulted in water becoming more limiting than CO_2, and additional tree death means that the amount of carbon stored is decreasing (Fig. 15.4). The eastern part of Amazonia has warmed rapidly and suffered more from droughts and deforestation than the west; direct measurements of atmospheric CO_2 using aircraft indicate it is now a net source of carbon. In total forest loss and the subsequent reduced uptake by trees had lowered the total intake from tropical forests by half between the end of the last century and 2015. Carbon storage by forests is still increasing through temperate forest uptake, but one of the greatest concerns is that this too could start to decrease, and land could then become a net emitter rather than an absorber of carbon.

Ultimately, much of the extra carbon in the atmosphere will be taken up by the ocean, but this takes time. Halting emissions entirely at the end of 2021 would cause atmospheric CO_2 to decline from 415 ppm to 350 ppm by the end of the century. Some of the absorbed CO_2 is dissolved in the ocean as gas, but a

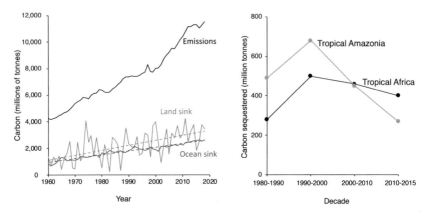

Figure 15.4 *Left:* Total annual emissions from human causes (including burning fossil fuels and deforestation) and estimates of the amount taken up by land and by the ocean. Dashed lines show general trends. *Right:* Average annual carbon sequestration by tropical forests on two continents. (*Left:* P. Friedlingstein, M. W. Jones, M. O'Sullivan, et al. [2019]. Global Carbon Budget 2019. *Earth System Science Data* 11: 1783–1838; *Right:* W. Hubau, S. L. Lewis, O. L. Phillips, et al. 2020. Asynchronous carbon sink saturation in African and Amazonian tropical forests. *Nature* 579: 80–87)

small amount reacts with water to produce carbonic acid, which in turn reacts with calcium carbonate to produce bicarbonate, summarized as:

$$CaCO_3 + CO_2 + H_2O \leftrightarrow 2HCO_3^- + Ca^{2+}$$

Or in words:

Calcium carbonate + carbon dioxide + water ↔ bicarbonate + calcium ions

As CO_2 increases, the equilibrium moves to the right side of the equation, causing calcium carbonate to dissolve off the ocean floor. After a few thousand years, only about 10% of the released CO_2 should be in the atmosphere, even if the terrestrial carbon sink ceases. Absorption of the rest awaits the long-term weathering cycle.

15.3 Present-Day Effects

Across the world, temperatures have increased by about 1°C with respect to the 1880–1920 average (Fig. 10.2). Some effects of this increase in temperature are (1) Heat waves, such as, record-breaking April and May 2022 temperatures in India, and record-breaking May 2019 temperatures in several southeastern U.S. cities; three runs of exceptionally high temperatures in June, July, and August 2019 in Europe; and temperatures exceeding 50°C in Pakistan. (2) Sea level has risen by a little more than 20 cm over the past century, in part due to the thermal expansion of a hotter ocean (recently about 2 mm/ year) and in part to runoff from glaciers and ice sheets (1.5 mm per year). Between 1992 and 2018, loss of ice from Greenland alone caused a 1 cm rise. (3) More ice from the Arctic Ocean is lost each summer. This melting does not in itself affect sea levels because floating ice already displaces water. (4) The years 2005 and 2020 had the most ever hurricanes recorded in the Atlantic.

In 2005, 2010, and 2016, major droughts in the Amazon Basin, attributed at least in part to high sea-surface temperatures (in the Atlantic for the first two and in the Pacific for the third event, associated with a strong el Niño event, Chapter 10) resulted in the deaths of billions of trees. The drought in 2005 was labeled a once-in-a-century event when it happened, but the following droughts have been even more extensive. During a drought, carbon that would be sequestered in normal years is not absorbed because trees fail to grow. Further, as the dead trees rot and decompose, they release additional carbon. The estimated atmospheric effects of the 2010 drought were equivalent to a release of about 2.2 billion tonnes of carbon in total, which was more than China emitted that year (1.72). This is another dramatic example of a positive feedback loop whereby released CO_2 will lead to more warming and more droughts. In other effects, trees that are water stressed reduce fruit production, with ramifying consequences for species that consume fruit (see Figure 24.1).

Based on a 2003 analysis of birds, insects, and plants primarily studied in Europe and North America, plants are flowering earlier and birds are breeding earlier (on average, one day earlier every four years). Sea level rise has been invoked as the cause of extinction of one mammal species (the Bramble Cay mosaic-tailed rat) that used to live on a small Australian island and whose habitat has been lost. This species could not escape from its island, but the most obvious changes in species in response to a warming climate are shifts in their geographical and altitudinal ranges. In the 2003 compilation, northerly limits of species ranges had moved to higher latitudes at a rate of 0.6 km per year. Along mountainsides, recent evaluations demonstrate average upslope movement of more than 2 m per year, twice as fast as fifteen years ago. Upward movements puts species on mountaintops at risk. Indeed, a resurvey of a 1,415 m ridge rising out of the Amazon rainforest in Peru found that four common high-elevation bird species in 1985 had disappeared in 2017. Further, species that were present at the highest elevations in 1985 declined the most (Fig. 15.5).

Upslope movement of lower-range limits implies that populations are disappearing from the lower elevation part of their range. A compilation of studies across eight countries found that the strongest predictor of population losses was

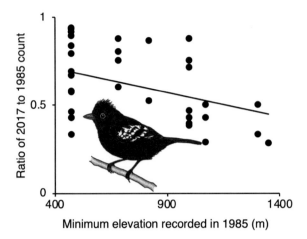

Figure 15.5 Bird density changes between 1985 and 2017 plotted against elevation on a mountain ridge in Peru. For the 38 species that declined in numbers, and at least 3 were caught in each year, the ratio of numbers caught in 2017–1985 is plotted against the lowest elevation an individual was caught in 1985. Others not included increased in numbers. The variable antshrike (illustrated) is one of four species, all of which used to be present at higher elevations, that disappeared altogether and hence are not on the plot. (B. G. Freeman, M. N. Scholer, V. Ruiz-Gutierrez, et al. [2018]. Climate change causes upslope shifts and mountaintop extirpations in a tropical bird community. *Proceedings of the National Academy of Sciences* 115:11982–11987)

the rate at which maximum temperatures are increasing, with rates varying from no increase to 0.08°C annually. The chance of observing at least one population disappearing was about seven times higher in localities with a 0.08°C rise than in those with no increase. Climate likely interacts with other factors to contribute to these and other declines. Chapter 19 describes a possible example of interactions between threats along mountainsides by asking how climate change could have contributed to the virulence of a fungal disease now threatening many frogs. Across Europe and North America, a decrease in the number of bumblebees correlates with conversion to agriculture, but increases in temperature and altered precipitation have likely contributed as well. Coral reef declines are attributed to a warming ocean and ocean acidification, with additional threats from predation and overgrowths of algae, as we consider in the next chapter.

15.4 Conclusions

The link between CO_2 and climate is unequivocal and is generally accepted as the major determinant of temperatures in the past. CO_2 is now at a higher concentration in the atmosphere than it has been for at least 2.5 million years. However, climate in the more distant past was warmer than even the worst predictions for the next century (Figure 15.1). So why should we be so concerned? For humans, some consequences of climate change include sea-level rise, dry places getting drier, wet places getting wetter, hot places getting uninhabitable, fires, and an increase in hurricanes. These changes are happening quickly (for an example, see Color plate 11.). Species are similarly affected, and their ability to deal with climate change is compromised by other threats.

16

Predictions of Future Climate and Its Effects

Prediction of climate is not an easy task, both because it is difficult to measure the many factors involved (value uncertainty) and because these factors can interact in many different ways (structural uncertainty, Chapter 10). We are not only unsure of how quickly temperatures will increase in the face of increasing CO_2 but also of how much more CO_2 will be released. Rather than attach uncertainty to CO_2 release, in its fifth report the Intergovernmental Panel on Climate Change (IPCC) considered four different scenarios that depend on how quickly we reduce fossil fuel emissions (Fig. 16.1). Annual emissions are presently lower than the most extreme scenario, but higher than the others. While emissions have recently decreased in the United States and Europe, they are increasing in India and elsewhere, with the result that 2019 recorded the highest emissions since recordkeeping began (Fig. 16.2) (2020 saw a reduction of 9% over 2019 associated with the pandemic).

The total CO_2 released by humans correlates with the maximum temperature that will be reached in the future. The IPCC estimates that 1 trillion tonnes (a million million tonnes) of carbon released give about a 1 in 3 chance of a 2°C average global temperature rise (about half that amount had been emitted by 2011). Three of the four scenarios in Figure 16.1 predict that we will pass this point around about 2050. The remaining scenario leads to the lowest output and may keep us below 1 trillion tonnes; it assumes not only cuts starting in 2020 but carbon capture.

Given these scenarios, this chapter:

(1) Summarizes expected the consequences of climate change.
(2) Considers what these consequences might mean for species extinctions, with an example drawn from coral reefs.
(3) Evaluates ways to slow and reverse emissions, emphasizing the importance of renewables.

16.1 Consequences of Climate Change: One, Two, and Three Degrees

One degree: This is where we were in 2020. That is, the average temperature across the world in 2020 was 1°C warmer than the average temperature across

Ecology of a Changed World. Trevor Price, Oxford University Press. © Oxford University Press 2022.
DOI: 10.1093/oso/9780197564172.003.0016

PREDICTIONS OF FUTURE CLIMATE AND ITS EFFECTS   167

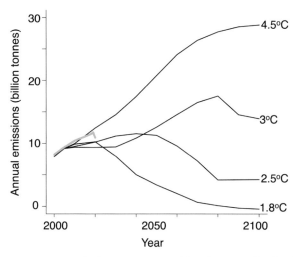

Figure 16.1 Four scenarios for how emissions of CO_2 (tonnes of carbon) due to fossil fuel burning may change over this century, as used by the Intergovernmental Panel on Climate Change in modeling climate. Actual emissions are the thick grey line. The scenarios are based on assumptions about social pressures, technology, population, and so on. The high emissions scenario is now considered extremely unlikely, as it requires a large increase in coal use. The lowest emission curve becomes negative. In this scenario biofuels are burned and the emitted CO_2 is captured underground, so biofuels actually draw CO_2 out of the atmosphere. Temperature increases in 2100 are the predicted rises with 0.5 probability (i.e., there is a 50% chance that the attained temperature will lie beyond this prediction). Based on the SSP database hosted by the IIASA Energy Program at https://tntcat.iiasa.ac.at/SspDb. The actual emissions reference is cited in Figure 16.2.

the world 120 years ago (Fig. 10.2). If fossil fuel emissions were to completely stop tomorrow, CO_2 would decline to 350 ppm by 2100, being absorbed by the oceans, and temperatures at the end of the century might be similar to what they are today (i.e., there would be a 50% chance of a 1°C rise over historical levels). Alternatively, annual 5% cuts in emissions after 2020 are expected to still lead to a temperature increase to midcentury, followed by a slower decline. A 1°C rise is about as warm as it has been over the last 10,000 years. This rise has been considered relatively safe, but is similar to estimated average temperatures for the last inter-glacial, associated with sea levels 6 m–11 m higher than now.

Two degrees: A 2°C rise is problematic for several reasons. First, a 1°C rise is already causing extreme events, as we noted with respect to heat waves in Chapter 15. While it will get wetter globally, we expect redistribution of rainfall to create droughts in some places and floods in others (Fig. 16.3). It takes some time for sea level rise to happen, but under all four emissions scenarios, the

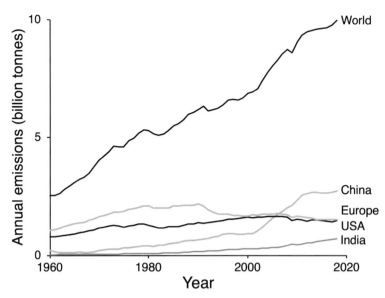

Figure 16.2 Annual emissions from 1960 to 2020 due to fossil fuels and cement production (P. Friedlingstein et al. [2020]. Global carbon budget 2020. *Earth Systems Scientific Data* 12: 3269–3340, 2020. Z. Liu, P. Ciais, Z. Deng, et al. [2020]. Near-real-time monitoring of global CO_2 emissions reveals the effects of the COVID-19 pandemic. *Nature Communications* 11:5172. J. Tollefson. [2021]. COVID curbed 2020 carbon emissions—but not by much. *Nature* 589: 343). Note that the world summary includes international shipping and airplane emissions, which are excluded from the country emissions. https://data.icos-cp.eu.

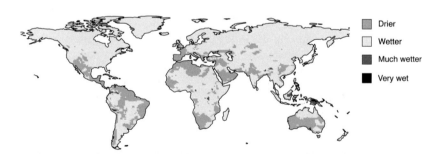

Figure 16.3 (Color plate 12) Predicted changes in precipitation averaged over the years 2041–2060, with respect to the average in the period 1986–2005. Changes are modeled on the scenario leading to 2.5°C, with probability 0.5 by the year 2100 of Figure 16.1. In this case, wet, wetter, and very much wetter are annual increases of <20 cm, 20–40 cm, and >40 cm a year, respectively. In most locations experiencing a decrease, the decrease would be less than 5 cm/year, but these are often relatively dry already. (National Center for Atmospheric Research NCAR GIS Program. 2012. Climate Change Scenarios. https://gisclimatechange.ucar.edu/gis-data, accessed August 23, 2020)

PREDICTIONS OF FUTURE CLIMATE AND ITS EFFECTS   169

end-of-the-century prediction is that of a further 0.3 m increase in sea level, with a probability of 0.95. Even if we stopped all emissions tomorrow, a greater than 1.5 m rise is expected over the next 2,000 years. Third, movements of species as they track the changing climate bring undesirable side effects, such as an increase in the geographical scope of malaria. Fourth, all sorts of positive feedbacks are possible over the longer term, especially as CO_2 levels are much higher than they were during the previous interglacial periods (Fig. 16.3). The impact of positive feedbacks is difficult to assess. As ice melts, the sun's radiation is reflected less, resulting in increased warming. A very big question is whether the terrestrial environment will turn from an absorber of carbon to an emitter of carbon, as the northern regions warm up and permafrost melts. With a 2°C rise, positive feedbacks may take us on the route to 3°C.

Three degrees: A 3°C or greater warming is seen to be disastrous, even without considering potential feedbacks. Mark Lynas (2008) writes about the Amazon rainforest at 3°C:

> A new unrecognizable landscape is born. In the deepest parts of the basin, where once the only sound was the howling of monkeys and the rustling of leaves, a moaning wind has arisen. Dust gathers in the lee of burned out tree stumps. Nearer to the ground, a gentle hissing sound is heard. Sand dunes are rising. The desert has come.

This scenario seems realistic, given the recent droughts in the Amazon we described in Chapter 15. Predictions of how precipitation patterns will change (Fig. 16.3) include a drying out of the Amazon, which would be amplified by the confounding effects of deforestation, which itself reduces local precipitation (Chapter 13). In the most extreme emissions scenario considered by the IPCC, there is a 0.5 probability of a 4.5°C rise or more by 2100. Lynas notes that the tremendous rate we are emitting carbon into the atmosphere takes us into uncharted territory, with possible feedbacks driving us on to 6°C and beyond. Then, the only habitable places on earth would be at the poles, with even the possibility of the earth losing all its water. While we are not on course for this dramatic outcome, even the other, less dramatic scenarios lead to rises of at least 3°C, 2.5°C, and 1.8°C, respectively, with 0.5 probability by the end of the century (Fig. 16.1).

16.2 Species Responses

One way to assess effects on species has been to consider the present range of temperature or other climatic variables a species experiences (termed its climatic envelope). The next step is to ask how quickly a certain temperature contour (e.g.,

170 ECOLOGY OF A CHANGED WORLD

the point of regular freezing) will move over time. This rate is called the climate velocity and is measured in kilometers per year. Climate velocity is more or less equivalent to the rate at which a species would have to move to stay within its climatic envelope. For example, trees moved north at the rate of about 1 km per year in North America during recovery from the last ice age, but under the intermediate scenarios of Figure 16.1, across about 30% of the world they would have to move more quickly. Mitigating the need to move, species might adapt and persist in small geographical areas before expanding their range again when conditions improve. In fact, land extinctions in the fossil record over the entire period of the ice ages were not exceptional (Chapter 25). Temperature fluctuations were large (Fig. 15.2), but still many species appear to have been able to track climate over large distances, persist in isolated pockets, and/or deal with novel climatic regimes.

Nevertheless, extinctions induced by the current warming climate will surely be larger than those in previous interglacial periods, for two complementary reasons. First, the way the climate is changing differs from how it changed in the past. Climate is warming more rapidly than it did when earth came out of previous glacial periods. Rates of CO_2 release at present are likely faster than those ever experienced, and the quantity of CO_2 in the atmosphere is far higher than it has been during the entire glacial period. Previous episodes of rapid warming in the more distant past have been associated with extinctions in the ocean. Notably, a period of rapid warming about 55.6 million years ago is thought to be a good analog to what is happening today because it is attributed to release of CO_2. As far as can be ascertained, land animals did not suffer greatly. However, an increase of perhaps 5°C, plus associated CO_2 releases, resulted in the extinction of 30–50% of all bottom-living Foraminifera and the reduction of coral reefs, which were replaced by reefs composed of Foraminifera, as well as other organisms.

The second reason an increase of 2°C will be associated with extinction events is that human impacts generate multiple threats. We consider several examples later in this book, but here we discuss the compounding of threats to coral reefs. Presently, coral reefs occupy less than 1% of the world's ocean surface and cover about 1.5 times the area of California, yet they are home to perhaps 25% of all marine species. Algae living inside corals trap solar energy and nutrients, providing more than 95% of the metabolic requirements for the coral host. When temperatures exceed typical summer maxima by 1°C –2°C for 3 to 4 weeks, the stress results in ejection of the algae and coral whitening (i.e., coral bleaching). The corals may subsequently recover, but growth rates slow, and after multiple episodes, corals die.

The history of coral reefs during the last interglacial suggests that species can track suitable conditions and persist during periods of warming of up to 2°C.

PREDICTIONS OF FUTURE CLIMATE AND ITS EFFECTS   171

During that time reefs declined in the tropics, but they extended their ranges north by up to 800 km. Coral reefs are expanding north again today, with movement of as much as 14 km in one year. Under experimental conditions, corals that experience warm temperatures suffer slower growth rates and increased mortality, but after 2 years they can physiologically recover to some extent. For both of these reasons, we might think that reefs could persist in isolated pockets and adapt further to changes, thereby avoiding extinction, at least up to 2°C. However, corals are one of the most endangered groups of animals, associated with multiple threats. First, warming is happening quickly, and although we have seen rapid range shifts keeping up with recent climate velocities, it appears that conditions at higher latitudes have become increasingly less hospitable, thus slowing movement. Second, removal of fish species has harmful cascading effects (e.g., Fig. 21.4). For example, overfishing has led to increases in sea urchin densities, placing the urchins at greater risk of disease. Disease-related declines of urchin populations then result in algal overgrowth of the reef (Chapter 5). Third when CO_2 concentrations in the ocean are elevated, the subsequent increase in carbonic acid ("ocean acidification") results in higher rates of conversion of calcium carbonate into the soluble bicarbonate form (Chapter 15), making it more difficult for coral to grow.

Rapid warming, overfishing, and high CO_2 levels reflect some of the stark differences between the present episode of climate change and the past interglacial. Together these effects have created extreme stresses. Between 1995 and 2017, for example, coral cover in many places along the Great Barrier Reef declined by 60% and virtually disappeared from some locations. They are attributed to massive bleaching events in 1998, 2002, 2016, and 2017, as well as cyclones and two outbreaks of crown-of-thorns starfish, a predator on corals. Few coral reefs may survive at 500 ppm atmospheric CO_2, just as their volume was substantially decreased 55.6 million years ago during the rapid warming at that time.

16.3 Limiting CO_2 Build-Up

Various conservation measures could possibly reduce CO_2 emissions. For example, we could reduce the number of kilometers driven by cars and increase use of public transport, or the automobile industry could increase car efficiency (kilometers per gallon). A problem with increased efficiency, however, is that the associated price reductions generate increased consumption. These may actually surpass the original efficiency savings, at least in a world where people are getting generally wealthier (Jevons' paradox, Chapter 12).

Alternative routes to reduce emissions despite ongoing consumption include capturing carbon during the generation of energy from coal or biofuel; using

172 ECOLOGY OF A CHANGED WORLD

cleaner sources of energy on a massive scale, including solar, wind, hydroelectric, nuclear, or even natural gas (which is cleaner than coal); increasing forest cover; and relying more heavily on biofuels. Any switch to sources of energy that release less CO_2 should help. All such sources have some undesirable side effects, from dams limiting water flows and destroying forests, to nuclear waste, to bats and birds flying into wind turbines, although none appear to be as serious in the long term as continuing to burn coal. Next we consider two of these options in more detail: biofuels and fracked natural gas. Neither is currently working well, which suggests the importance of turning to renewable sources, especially wind.

Biofuels

Biofuels are plants used for fuel rather than food. At present, crops such as corn and sugarcane are the main biofuels. In theory, all the carbon released when the crop is burned for energy is recaptured when the same crop is grown the following year; hence it is said to be carbon neutral. In practice, however, it takes energy to grow, harvest, and refine the crop. Table 16.1 presents some calculations for corn grown as a source of ethanol in the United States. This table shows that when we account for the emissions associated with growing, harvesting, and refining, net CO_2 emissions from biofuel are only about 20% less than those from oil. The reason is that growing and refining the biofuel emits more than three times the quantity of CO_2 than is emitted when extracting and refining oil. Nevertheless, on the face of it, this is a saving, albeit a relatively small one. Some biofuels are quite efficient, and sugarcane gives about an 85% reduction in CO_2 emissions because it can be refined more easily than other crops.

Table 16.1. Emissions of CO_2 for the Manufacture of Gasoline from Oil or of Ethanol from Corn*

	Gasoline	Ethanol from corn
Extracting/growing	1	7
Refining	4	11
Burning for energy	20	19
Recovered when growing	0	−17
Total	25	20

*Grams of carbon per megajoule energy (T. Searchinger, R. Heimlich, R. A. Houghton, et al. Use of U.S. croplands for biofuels increases greenhouse gases through emissions from land-use change. *Science* (2008) 319:1238–1240).

PREDICTIONS OF FUTURE CLIMATE AND ITS EFFECTS   173

There is, however, a problem with this accounting. Biofuels as used today are made from crops that take up land, land that could be used for something else. Further, if wildland is converted to biofuel, the cost of land conversion in terms of emissions can be large. Vegetation is burned and rots; even harvested timber eventually decays, releasing CO_2. Converting midwestern grassland to corn releases CO_2 into the atmosphere, which would take more than 90 years for the 20% savings in CO_2 to compensate (Fig. 16.4). When tropical forest on peat (compressed vegetation) is converted to palm oil plantations in Indonesia (e.g., around Gunung Palung National Park [Fig. 13.7]), the small savings in CO_2 from their use as biofuel takes several hundreds of years to compensate for the CO_2 released (Fig. 16.4). One reason is that the peat continues to emit CO_2 for many years as it dries out. Given that we are concerned about CO_2 release in the next decades, use of these kinds of biofuels is clearly making things worse with respect to emissions, not better.

Growing crops for biofuels places pressures on the world's food supplies and natural habitats. Moreover, with current methods, crops grown on land can provide only a small proportion of all energy requirements. In 2019, 40% of corn grain in the United States was used for biofuels (Chapter 7), implying that 10% of all agricultural land in the United States was used to provide corn grain for biofuel. This number has held steady for 10 years, but it provides only 7% of total liquid fuel consumed in the United States. New biofuels (e.g., algae cultured in tanks) may be more successful.

Fracking

Over the past 20 years, extraction of natural gas from underground deposits has become technologically feasible. This extraction uses a technique known as hydraulic fracturing (fracking), whereby water and chemicals, many of which are themselves pollutants that have affected the health of local communities, are forced underground under pressure to eject the gas. The procedure has sometimes contaminated water supplies, and in one case, one could turn a tap on and set the emerging fluid alight. In 2005, the U.S. Congress passed a provision in the Energy Policy Act that exempted gas companies from the Clean Air and Clean Water Acts, thereby stimulating a great surge in natural gas exploration across the nation. The advantage of natural gas is that, for the same amount of energy, the burning of gas releases just over one-half the CO_2 compared to burning coal, which is one reason U.S. emissions have recently declined (Fig. 16.2). However, natural gas consists mainly of methane, with high greenhouse gas warming potential over the short term, some of which is released during extraction and

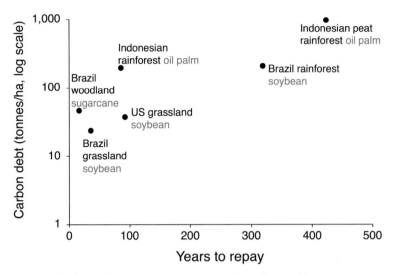

Figure 16.4 Carbon released as a result of conversion of natural land to that growing biofuels, and time to repay the generally small gain in CO_2 savings, which depends on the type of biofuel (in gray). Estimates are based on carbon released over 50 years, but especially Indonesian peatlands are expected to release carbon beyond that time horizon. (J. Fargione, J, Hill, D. Tilman, et al. 2008. Land clearing and the biofuel carbon debt. *Science* 319: 1235–1238)

processing. Given the serious problems associated with biofuels and natural gas, it appears that the only viable way forward is renewable energy.

16.4 Conclusions

It is essential to move away from fossil fuels quickly. The cleanest sources of energy are wind, solar, hydroelectric, and geothermal power. Various studies have shown how these sources can quite rapidly replace gas, oil, and coal. For example, a Stanford University analysis showed how New York State could in principle become entirely carbon neutral in less than 15 years. In the analysis, 17,000 offshore wind turbines would provide 50% of New York's energy requirements, with the rest being provided by other sources, including solar panels on houses and increased efficiency of electricity transmission. In this case, vehicles would run directly on electricity or would be powered by liquid hydrogen fuel, which is itself manufactured from water using electricity. In the case of New York, which does not have its own oil or coal resources, the transition would come at little economic cost. Two externalities also favor New York's transition to renewables: less pollution, hence better health and reduced damage from climate change. Between 2004 and 2009, the state of New York contributed an estimated 0.636% of world fossil fuel CO_2. With renewable energy, this release would be removed from the equation. But returning to the main theme of this chapter we note that even if establishment of a carbon-neutral economy happens rapidly, one should be aware that temperatures will rise before they fall.

17

Pollution

Pollution is the contamination of the environment by human activities. Air pollutants include CO_2 and methane, which pollute by virtue of contributing to global warming, and particulate matter, ozone, and other gases, which directly harm us. Water and soil pollutants include runoff from mining activities, nuclear waste, pesticides, herbicides, and fertilizers. The ocean contains *trillions* of plastic particles. Light pollution kills 100 billion insects annually in Germany. In the United States, more than 60 million pigs produce a quantity of sewage that is greater than half the total of human sewage, but rather than being treated, it is primarily held in storage tanks, allowed to settle, and then sprayed on surrounding fields. Even dead bodies have become pollutants. In India, under the Parsi religion, after death, the human body is left for wild vultures to consume. Vultures have declined by more than 99% in the past 20 years (see Fig. 24.1), poisoned by the pollutant diclofenac, a painkiller taken by humans and given to cattle. Because the vulture population has declined, bodies take longer to decompose.

In a cost-benefit analysis, many pollutants would be banned if we were to account for externalities, that is, sum the costs to everyone (Chapter 14). These externalities are pollutants that are released by individuals and companies but that generate costs shared by many, including future generations. However, some externalities are so apparent that government intervention has successfully reduced their impact, as described for emissions from power plants in Chapter 12. In this chapter, we present a short introduction to the effects of some several major pollutants, including fertilizers, herbicides, and pesticides.

The chapter considers:

(1) Why fertilizers pollute at all, when they are designed to promote vegetation growth.
(2) Air pollution and acid rain.
(3) Herbicides and pesticides.
(5) The principle of biomagnification whereby species at the top of the food chain are especially at risk from poisonous chemicals.
(6) How legislation and other government interventions have helped to reduce pollution.

Ecology of a Changed World. Trevor Price, Oxford University Press. © Oxford University Press 2022.
DOI: 10.1093/oso/9780197564172.003.0017

17.1 Pollution by Fertilizers

Humans now add a similar amount of reactive nitrogen to the world each year as natural processes do (Fig. 17.1). Human contributions come from three different sources: the major one is fixation of atmospheric nitrogen for use as fertilizer, with secondary inputs from fossil fuels and crops, notably soybeans, that harbor symbiotic nitrogen-fixing bacteria in their root nodules (Chapter 4). Ultimately, most of the excess reactive nitrogen we have generated will be returned to the atmosphere as nitrogen gas through denitrification. This process is primarily undertaken by bacteria that use nitrates as an energy source in conditions of low oxygen, on both the land and in the ocean. On land, denitrification is sufficiently rapid that it would take less than a century for the system to recover if inputs ceased.

One might expect reactive nitrogen to be good for the environment, given that it stimulates plant growth. However, as is generally the case, unusual conditions create systems that are out of balance. Accordingly, only a few plant species are good at growing in conditions with high levels of reactive nitrogen, and these outcompete other species. The addition of reactive nitrogen to experimental

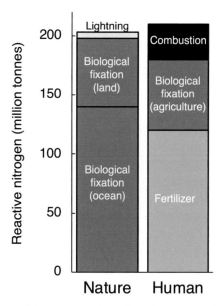

Figure 17.1 Estimates of the amount of reactive nitrogen contributed to the world in 2010. Human biological fixation refers to planting nitrogen-fixing crops, such as soybeans. (D. Fowler, M. Coyle, U. Skiba, et al. [2013]. The global nitrogen cycle in the twenty-first century. *Philosophical Transactions of the Royal Society B* 368: 20130164)

178 ECOLOGY OF A CHANGED WORLD

plots led to a 15% decrease in the number of plant species persisting on average, according to a survey of more than 130 experiments. As always, impacts have ramifying effects on other species. For example, butterfly populations in Holland have decreased by at least half over the past 25 years on heathland, but much less so in woodlands and grasslands. Reactive nitrogen should have particularly strong effects on heathlands, which are historically nitrogen poor; it appears the plant species that butterflies prefer have consequently decreased.

In water, competitive dominance of a few algae and bacteria under high-nitrogen conditions is even more striking than on land. Extreme algal growth produces eutrophic water bodies (eutrophic comes from a Greek word meaning "healthy growth"). These overabundant algae clog waterways and kill plants via a variety of effects, including competition for light. Death of the algae and plants followed by their decomposition by bacteria uses up oxygen, resulting in hypoxic (low oxygen) and anoxic (no oxygen) conditions, which in turn result in fish die-offs and dead zones, where little animal and plant life of any form can survive.

In the oceans, less than five dead zones were thought to be present in 1910. Contrast that figure with today's count of more than 400 dead zones reported around the coasts of the world. Most occur seasonally and are followed by annual recoveries. For example, off the Louisiana coast in the Gulf of Mexico, a low oxygen dead zone appears at the end of the summer each year, as a result of agricultural runoff from the Mississippi River. The size of the zone has varied more than fourfold over the past 25 years and sometimes extends over 20,000 km^2. Years of high river flow have resulted in a larger zone, associated with a larger fertilizer flush.

Finally, reactive nitrogen directly affects animal growth and development, especially in the water, where aquatic animals are continuously exposed to dissolved nitrates. A major debilitating effect is to convert hemoglobin to an inactive form. Laboratory experiments on frogs, toads, and salamanders have found, on average, that when placed in nitrate at concentrations present in some places in nature, mortality is increased by 15% and developmental defects (such as extra toes or loss of gonads) doubled (Fig. 17.2).

17.2 Air Pollution and Acid Rain

Air Pollution

In the years leading up to the pandemic, air pollution was the cause of more premature deaths than communicable diseases. The lockdown in Europe in spring

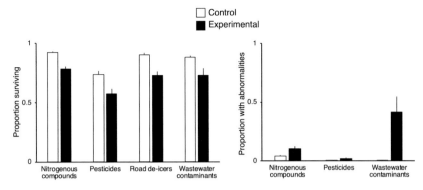

Figure 17.2 Amphibians (frogs, toads, salamanders) have lower survival (*Left*) and more developmental abnormalities (*Right*) when exposed to various pollutants. Shown are means and standard errors from 248 experiments that measured survival and 34 experiments that measured developmental abnormalities. (A. Egea-Serrano, R. A. Relyea, M. Tejedo, et al. 2012. Understanding of the impact of chemicals on amphibians: A meta-analytic review. *Ecology and Evolution* 2: 1382–1397)

2020 resulted in a 30% reduction in the use of coal and oil. A single month of lower emissions is thought to have resulted in 11,000 fewer European deaths in 2020, which would otherwise have occurred as a result of air pollution. In the United States, the Environmental Protection Agency (EPA) publishes a daily map of air quality across the United States. The EPA calculates an air quality index for five major air pollutants: ground-level ozone, particulate matter, carbon monoxide, sulfur dioxide, and nitrogen dioxide. In calculating this index, the EPA weights each pollutant according to its human health hazards. Of the five, particulate matter and ozone are considered the more important threats.

Small particulate matter is a major health hazard. It is also the most easily measured and compared across the world. New York's small particulate matter is, on average, about one-third below the level considered acceptable by the EPA. On the other hand, in 2016 New Delhi averaged almost 12 times this level, and on some days in November 2019, more than 30 times, partly a result of the many fires being set by agriculturalists around the city.

Reactive nitrogen in the atmosphere combines with a cocktail of other compounds to produce ozone. Ozone is harmful to human health, causing respiratory diseases that have been responsible for thousands of premature deaths. Ozone also inhibits plant growth by damaging and killing cells and reducing photosynthesis. Experiments in Europe imply that both crops and forests have suffered substantial losses due to elevated atmospheric ozone, with growth rates slowed by this factor by about 5% compared to preindustrial conditions.

180 ECOLOGY OF A CHANGED WORLD

Acid Rain

Sulfuric acid is derived from sulfur dioxide and is released primarily from refineries and power plants. Nitric acid is derived from nitrogen gases released by industry and as a side-product of denitrification. The consequence is acid rain. The United States has recorded rainfall with a pH as low as 2.1. Acid rain releases bound trace metals from the soil, especially aluminum, which interferes with nutrient uptake by plants. Acid also debilitates aquatic animals. For example, a reduction in pH of just 0.5 units can kill trout. Acid rain has led to large areas of dead forest and the poisoning of lakes, particularly in the northeastern United States and northern Europe. As a result of government initiatives, sulfur dioxide concentrations in the United States decreased by 80% between 1980 and 2014, and nitrogen dioxide concentrations decreased by 60% (Chapter 12). Many lakes have now recovered, but acid rain is likely becoming an important problem in industrializing countries.

17.3 Herbicides and Pesticides

Humans have developed many synthetic chemicals such as plastics, pesticides, herbicides, and medicines, and we are continually manufacturing new ones. These new inventions generally improve on old products. For example, pesticides are useful because they kill organisms that harm humans or their crops. However, pesticides are inevitably pollutants as well because they kill organisms other than the primary targets, including animals that parasitize or prey upon the pest. Pesticides that target insects, as well as herbicides that target weeds, also directly affect other animals (e.g., Fig. 17.2) including us. In the United States, more than 30 million kg of the herbicide atrazine is added to crops, mostly to corn in the midwestern United States (Chapter 8). At ecologically relevant concentrations, this chemical causes developmental abnormalities in humans, frogs (in frogs it can turn males into hermaphrodites by promoting the synthesis of estrogen), and other animals. Because of its adverse effects, atrazine was banned in the European Union in 2004.

Unsurprisingly, the impacts of pollutants on one species have ramifying effects throughout the food web. For example, in Minnesota wetlands, frogs in ponds with high atrazine concentrations are infected with more flatworm larvae than those in ponds with low atrazine concentrations. Chapter 6 describes the life cycle of one species of flatworm in California, in which the adult stage infects large fish-eating birds, and juvenile stages pass through two intermediate hosts, a snail and a fish. The life cycle of the flatworm species studied in Minnesota is very similar, except that frogs rather than fish are a common intermediate host.

POLLUTION   181

Infected frogs may carry hundreds of flatworms. Infection levels are high for two reasons. First, atrazine weakens the frogs, making them less able to mount an effective immune response (the higher the atrazine concentration, the fewer white blood cell aggregates are seen in the liver). Second, atrazine, as an herbicide, kills plankton in the water. Increased water clarity apparently stimulates the growth of algae on stones and other surfaces at greater depths in the body of water. Snails feed on the algae. In an experiment, snail abundance increased by up to four times in atrazine-treated water, and this increase correlated with increased numbers of flatworms in the frogs.

17.4   Biomagnification

Species that live in water are constantly exposed to the atrazine present in the water. However, ingested atrazine is not long retained in animals because it is water soluble. Unlike atrazine, other manufactured chemicals are fat soluble and remain in the body. Polychlorinated biphenyls (PCBs) are one such chemical. The electrical industry used them for insulation and in cooling fluids. PCBs cause cancer. They are listed by the Environmental Protection Agency as one of the more dangerous persistent chemicals in the environment.

Because they are retained in the body, fat-soluble chemicals become a special problem for predators high up in the food chain, due to a process known as biomagnification (Fig. 17.3). To understand biomagnification, consider that, conservatively, a wolf eats about 100 times its own weight in meat during its lifetime. Therefore, if all the PCB of a wolf's prey were to be retained in its body, by the end of its life it would have 100 times the PCB concentration of its prey. In fact, in a 1992 study from around Hudson Bay, Canada, killed wolves had an average concentration almost 20 times that of their main prey, the caribou (Fig. 17.3). At the time, PCB concentrations in the breast milk of indigenous Inuit populations of northern Canada were almost 10 times higher than that of wolves, resulting from concentration in their food (whales, fish, and seals) and in milk. Note that PCBs were manufactured far from where the Inuits live and were banned from production in the United States in 1979. By 2012, concentrations in the breast milk of Inuit populations had declined to 20% of their 1992 levels.

Dichloro diphenyl trichloroethane (DDT) classically exemplifies biomagnification. DDT was developed as an important pesticide in the 1940s. It is carcinogenic and causes eggshells of birds to thin, so that the eggs break and regularly fail to hatch. In the 1950s, the United States' national bird, the bald eagle, declined to less than 500 pairs breeding across the lower 48 states. DDT was implicated in its decline: a typical bald eagle carried about 10 times the DDT concentration of the fish it ate. DDT was banned in the United States in 1979, and the number of

182 ECOLOGY OF A CHANGED WORLD

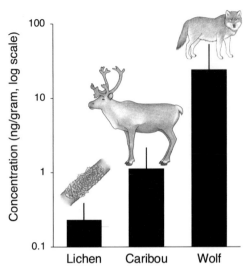

Figure 17.3 Concentrations of the most prominent form of PCB at three different levels of a food chain from a study near Hudson Bay, Canada, in 1992 (lichen is eaten by caribou, which are eaten by wolves). Error bars are standard errors. (B. C. Kelly, M. G. Ikonomou, J. D. Blair, et al. [2007]. Food web-specific biomagnification of persistent organic pollutants. *Science* 317:236–239; B. T. Elkin. [1994]. Organochlorine, heavy metal, and radionuclide contaminant transfer through the lichen-caribou-wolf food chain. pp. 356–361 in J. L. Murray and R. G. Shearer (eds.). *Synopsis of research conducted under the Northern Contaminants Program*. Indian and Northern Affairs Canada, Ottawa, Environmental Studies 72)

bald eagles has now increased to over 5,000 pairs, to the point that it is no longer classified as an endangered species. In 2006, a global agreement set the stage for a general, albeit voluntary, ban of DDT. However, the World Health Organization continues to endorse the use of DDT in Africa remains because of its effectiveness against mosquitoes and consequently malaria.

17.5 Remedial Action

The 1962 publication of Rachel Carson's book *Silent Spring* is viewed as the pivotal moment in understanding the dangers of pesticides, especially DDT. An environmental movement sprang from the concerns she raised, ranging from events such as a fire burning on Ohio's polluted Cuyahoga River in 1967 to a growing awareness of poor air quality and other hazards resulting from human activities. The EPA, introduced in Chapter 12, now regulates pesticides, emissions, and other hazards, and evaluates the safety of hundreds of products. The Clean Water Act (1972) limited discharges into rivers and lakes and has done much to improve water quality, although enforcement remains an issue. Similar efforts are being made in other countries. For example, Beijing has more than halved its levels of air pollution over the past 10 years. However, unlike carbon emissions, fertilizers, and nuclear waste, the production of new synthetic chemicals creates a moving target for regulation, with effects sometimes known only many years after application.

17.6 Conclusions

In many countries, preventing pollution is a lower priority than the economy. When the economy is booming (i.e., there is high demand), air pollution increases, and so does the human death rate. This is an example of Lauderdale's paradox that quality of life does not always increase with wealth (Chapter 12). Pollution not only comes from industry, mining, automobiles, and agriculture, but also from local activities. Anecdotes from India indicate the scale of the problem. In one example, people grind silver ornaments in their homes, flushing the by-products down the sink; silver is very toxic to aquatic organisms. In another example, fish were collected from a stream by using poison and then sold at the local market, resulting in the loss of fish-eating birds, such as kingfishers. At present, pollution seems to be increasing at levels that are commensurate with other threats, but causing long-term effects that are less well understood.

18

Invasive Species

Habitat conversion, climate change, and pollution, covered in Chapters 13–17, are the first three of the big six threats to the environment. The next three threats (invasive species, introduced disease, and harvesting) involve the addition or removal of species from the food web. They have more clearly caused species extinctions than the first three. The next two chapters, Chapters 18 and 19, discuss the addition of species, including pathogens, to places where they were formerly absent, and the following three, Chapters 20, 21, and 22, discuss the removal of species by harvesting.

In any given location, alien species are those species which have become established beyond their former pre-human range. The U.S. Department of Agriculture defines invasive species as "alien species whose presence does or is likely to cause economic or environmental harm or harm to animal or human health." This definition implicitly excludes humans, which would otherwise be the most dramatic example of all invasive species. Invasive species are a small proportion of all alien species but impacts of even one can be very large. Several examples have already been described, including the fayatree in Hawaii, which outcompetes native plants (Chapter 4), rabbits in Australia which outcompete sheep (Chapter 7), and cottony cushion scale insects in California which devastated citrus crops when they first arrived from Australia (Chapter 8).

Alien species are most common in human-disturbed environments, but up to 5,000 plant species may be established in native habitats in the United States, and some of these are considered invasive. Although it has been difficult to get an accurate measure of the number of alien species, when studied the number arriving has been found to be increasing, perhaps exponentially (Figs. 18.1, 18.2). A Global Invasive Species Database maintains a list of "100 of the World's Worst Invasive Alien Species" (the list includes pathogens). In 2013, following a ballot among experts, rinderpest, now declared extinct, was replaced with a water fern native to southeast Brazil, which clogs waterways in the southern United States and elsewhere.

The route to becoming an invasive species can be conveniently partitioned into three stages. First, the organism must be transported and released. Second, it must become established in the new location; many releases fail. Finally, in order to have substantial impacts, the species must expand beyond its point of introduction and become common.

Ecology of a Changed World. Trevor Price, Oxford University Press. © Oxford University Press 2022.
DOI: 10.1093/oso/9780197564172.003.0018

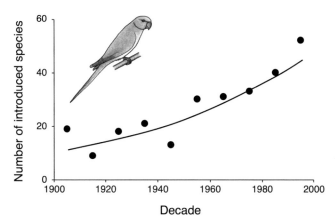

Figure 18.1 Number of alien bird species established in Europe, grouped by preceding decade. Illustration is of the rose-ringed parakeet, from India, established in several European countries and now common in London. (T. M. Blackburn, E. Dyer, S. Su, et al. [2015]. Long after the event, or four things we (should) know about bird invasions. *Journal of Ornithology* 156:S15–S25. The original data for this figure comes from S. Kark, W. Solarz, F. Chiron, et al. [2009] Alien birds, amphibians and reptiles of Europe. In: DAISIE (ed.), *Handbook of alien species in Europe*. Dordrecht: Springer, pp. 105–118)

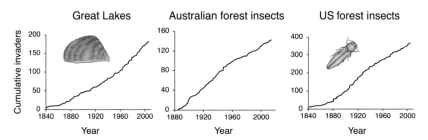

Figure 18.2 *Left:* Rise in the number of alien species, including algae, fish, mollusks, and plants, in the Great lakes. Sixty-five percent of these invasions have been attributed to ballast water from ships. *Center:* Cumulative number of alien insect species detected in Australian forests. *Right:* Cumulative number of alien insect species detected in U.S. forests. Illustrations are of the zebra mussel and the emerald ash borer. (A. Ricciardi. [2006]. Patterns of invasion in the Laurentian Great Lakes in relation to changes in vector activity. *Diversity and Distributions* 12: 425–433; H. F. Nahrung and A. J. Carnegie. [2020]. Non-native forest insects and pathogens in Australia: establishment, spread, and impact. *Frontiers in Forests and Global Change* 3:37; J. E. Aukema, D. G. McCullough, B. von Holle, et al. 2010. Historical accumulation of nonindigenous forest pests in the continental United States. *Bioscience* 60:886–897)

186   ECOLOGY OF A CHANGED WORLD

This chapter considers each stage in turn before addressing potential control mechanisms. The goals are to:

(1) Describe how species are moved from one location to another.
(2) Describe factors that lead to successful establishment.
(3) Consider why some predators become invasive.
(4) Describe costs and control measures.

The next chapter considers the effects of introduced pathogens, which have caused some of the more dramatic recent declines in population sizes.

18.1 Transport and Release

Alien species arrive in a new location in three ways: deliberately, accidentally, or following the breakdown of barriers, making it easier for the species to move. Deliberate introductions used to be common. For example, in 1905 one hundred starlings were released in New York as part of a program to establish all the bird species mentioned in Shakespeare's plays in Central Park. Now, about 100 million starlings are found over all North America, and they are a major pest, especially in vineyards. Some species of animals have been transported and released for biological control. Some of these introductions have gone spectacularly wrong, of which the cane toad in Australia is the classic example. This species is native to South America and was being used to control the abundance of sugar cane beetles in Puerto Rico, apparently successfully. It was introduced to Australia in 1935, where it failed to control the beetle. It has spread through much of the wetter parts of the country, causing declines in native animals through competition and predation. Furthermore, the toads themselves are poisonous to some predators when eaten.

Although the rate of deliberate release is declining, deliberate movement of species is increasing. For example, the pet trade is booming. This has resulted in more bird species escaping and becoming established (e.g., Fig. 18.1). Likewise, live plants for ornamental gardens are increasingly imported, and they regularly become established in the wild. Finally, aquaculture involves the movement of fish and shellfish, some of which have escaped and become invasive.

Many more species arrive by accidental transport and continue to do so (Fig. 18.2). Pets, live plants, and organisms used for aquaculture all carry unwanted hitchhikers along with them. Among the more serious of these are insects whose larvae bore into wood, which have arrived in packing cases, as well as on imported plants. For example, the wood-boring Asian longhorned beetle, which attacks elms, maples, and willows, probably arrived in the United States from China in a shipping crate in the 1990s. Many aquatic species have arrived in the ballast water of ships (water carried to balance the ship when it is not carrying a full cargo) or attached to the hulls of boats (Fig. 18.2).

While transport by humans is the most common means of alien species movement, it is not the only one. Climate change is leading to the opening of a warm water seaway north of Canada, which is thought likely to lead to the movement of marine species between the Pacific and Atlantic. The opening of the Suez Canal led to many marine species from the Red Sea moving into the Mediterranean, mostly on ships. In this case, both the loss of a land barrier and human transport were required for movement.

18.2 Establishment: Becoming Alien

Successful establishment is correlated with frequency of arrival, conditions in the native range, and the absence of natural enemies. Various societies that released British birds across the world kept detailed records of their efforts, and these records show that the number of times a species is introduced is an important predictor of whether it persisted. This principle applies to other species as well. The entomologist Bryan Beirne reviewed 159 different species of insects introduced in Canada between 1880 and 1970 for various biological control projects. About one-quarter actually became established, and those that did so were released in large numbers (more than 30,000 individuals were released for half of these species, whereas a typical number of released individuals was 5,000). Species that became established were also released more often than those that did not: 70% of insect species released on more than twenty occasions became established, whereas only 10% of those released on fewer than ten occasions did so. Many chance factors cause small populations to go extinct, so the more attempts and the larger the numbers, the greater the possibilities that these factors will be overcome. Therefore, a major reason for the increase in introduced species across the world (as illustrated in Figures 18.1 and 18.2) is simply that animals, plants, and fungi are transported around the world much more often than they were in the past through activities such as international trade. For example, Figure 12.4 illustrates just how much trade has increased from China to the United States since 1985.

Conditions in the new location also contribute to successful establishment of alien species. Establishment requires the per capita growth rate, r, to be greater than 0, at least for a short time, so the population can build up its numbers. The per capita growth rate depends on characteristics of both the new environment

188 ECOLOGY OF A CHANGED WORLD

and the introduced species. In fact, alien species are common in urban environments, with familiar examples being the house sparrow, rock pigeon, house mouse, and brown rat. First, human-disturbed habitats have few well-adapted native species, freeing up resources. Second, the more successful alien species have long persisted in urban environments in their native range and are well adapted to these environments. Many nuances can be added to these basic principles. The fayatree, described in Chapter 4, has become established in Hawaii because it colonizes naturally disturbed habitats where competition is low anyway (recent lava flows), but it also had a competitive advantage over natives in these habitats because it is a nitrogen fixer.

A third cause of successful establishment is escape from enemies present in the native range but absent in the new environment, which results in both a high initial population growth rate and ultimately large population sizes. The cottony cushion scale insect exemplifies this principle (Chapter 7). This species was accidentally introduced from Australia, probably on an acacia plant, and became a major pest on citrus. Two predators of this insect in Australia, a fly and lady beetle, were not transported when the cottony cushion scale insect arrived, but they were subsequently introduced and now successfully keep the scale insect at low levels. The lady beetle is a predator, but parasites and pathogens also get left behind. Parasites may be lost because they were simply not present on the few founding individuals. They may also be lost after introduction, either by chance or because they require intermediate hosts or habitats that are not available. For example, 473 plant species from Europe that are now growing in the wild in North America carry on average only one-sixth the number of fungal species that they had in Europe. The starling in the United States carries one-fifth of the forty-four species of ectoparasites found in its native range in Europe. All the same, in birds such as the starling, it has been difficult to demonstrate that loss of enemies is a major contributor to successful establishment. Instead, the number of introduction attempts, climate similarity between introduced and native range, and the presence of already established alien species, are important predictors.

In the next section, we move from factors that affect establishment of an alien species to factors that make it become invasive—that is, it causes economic or environmental harm. Introduced species, such as rats and starlings, are important pests, eating crops and processed food. Here it is the large population sizes and close associations with humans that cause harm. These species are often pests in their native range as well, and they simply continue being pests in their introduced range. Other species only become invasive in places where they have been introduced. We consider effects of introduced species on native species, including the biological principles that explain why they should sometimes be so devastating.

18.3 Invasive Pests and Predators

A major reason for the harm caused by invasive species on native fauna and flora is that natives have had no previous experience of the invader. In a few cases, we understand the underlying mechanism. The plant genus *Viburnum*, for instance, contains about 160 shrubs and trees found in both North America and Eurasia. Many species are ornamental and planted in gardens. The viburnum leaf beetle is native to Europe and Asia, and was first noted breeding in North America in the 1970s. Females excavate small cavities at the end of twigs into which they deposit their eggs at the end of the summer. The larvae hatch the following spring and eat leaves, before leaving the shrub to form a pupa in the soil. They emerge as adults later in the summer, where they consume more leaves. Viburnum has defenses against the beetle, including a wound response whereby it grows tissue across the twig cavities. The tissue crushes and expels the beetle's eggs.

Unlike the case in Europe and Asia, heavy infestations have defoliated and killed viburnum plants in North America. Researchers created twig cavities with a pair of scissors and found that species from Europe and Asia have a stronger wound response, with more tissue growth, than in North America (Fig. 18.3). Consequently, viburnum species from Europe and Asia more effectively expel and crush the eggs. Two viburnum species from the beetle's native range had low wound responses, but in these two cases the majority of the larvae died, presumably because of a chemical defense. Hence, viburnum species in Europe and Asia reduce the impacts of the leaf beetle in at least two different ways, neither of which is as efficient as viburnum species in North America.

North American viburnum species provide an interesting experimental demonstration of poor defenses by species that have not previously experienced the introduced species. In many other cases, the presence of poor defenses has been inferred rather than demonstrated. The Burmese python escaped via the pet trade into the Florida Everglades late last century. It increased dramatically after 2000, associated with declines of over 85% in bobcats and white-tailed deer, and the near disappearance of foxes, rabbits, and raccoons (see Fig. 24.1). The snakes even occasionally eat an alligator. The losses are likely due not only to predation, but also to competition, because snakes reduce the food supply of other predators. For example, both the python and native foxes eat rabbits. The snake has presumably been successful because Everglades animals are not adapted to defend against constrictors. Large snakes that kill by constriction, as opposed to venom, were last present in North America more than 16 million years ago.

Islands used to have few ferocious predators, and consequently island species were notoriously tame. On his trip to the Galápagos Islands in 1835, Darwin found it easy to catch finches by hand. The unsuspecting nature of island species

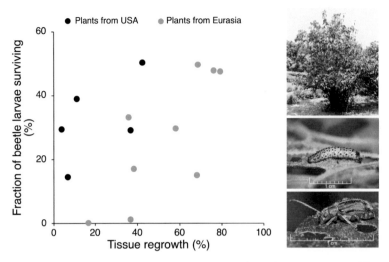

Figure 18.3 In Ithaca, New York, *Viburnum* species from the native range of the viburnum leaf beetle (Europe, Asia) and its introduced range (North America) were grown in the same environment. The proportion of larvae surviving when fed leaves of different species is plotted against the average fraction of a wound that was covered in scar tissue after several weeks. The wound was created by puncturing twigs from five plants. Wound recovery averaged higher in plants from the native range. Note that in the two species from the native range with particularly low wound recovery, larval survivorship was low. (G. A. Desurmont, M. J. Donoghue, W. L. Clement, et al. [2011]. Evolutionary history predicts plant defense against an invasive pest. *Proceedings of the National Academy of Sciences* 108:7070–7074). (Photos: Tea viburnum from China, Gregory Kohs (CC 3.0), adult leaf beetle (Wikipedia CC 3.0), larval leaf beetle, Donald Hobern (CC 2.0))

has therefore led to many extinctions, exemplified by the infamous "snake that ate Guam." Native to New Guinea, eastern Australia, and nearby regions, the brown tree snake was accidentally introduced to the Pacific island of Guam in the late 1940s. Genetic studies show that the arrivals came from the Island of Manus in Papua New Guinea, and perhaps only ten individuals in total arrived. At the time, New Guinea was used as a military transport staging post, and the snakes may have ridden over in the wheels of airplanes. Today 2 million snakes occupy the island. The native fauna of Guam was completely unprepared for the arrival of the snakes, having no antipredator defenses. Of the 13 species of native forest birds originally found on Guam, only the Guam swiftlet persists. It manages to do so because it nests on the roof of three caves, which are inaccessible to the snake, despite the snakes' regular attempts to get to the nests. Half of the original twelve species of lizards have been lost. In a good example of cascading effects through the food chain, spiders seem to have become much more common because of the reduction in the number of birds and lizards that prey on them and compete with them for food.

If an introduced species and former prey or host become established together, the introduced prey or host may help maintain the introduced species, thereby increasing its impact on natives. On Guam, the curious skink and other introduced species are present at high densities. Like the snake, the curious skink comes from the New Guinea region and is likely to be much better at avoiding predation than native species. However, every curious skink has to die eventually, providing a meal for the snake. Thus, the snake is present at high densities, even as the native fauna declines, and may continue to consume individuals from native species. Hence, we do not see cycles in which the predator becomes rare when the prey gets rare, as may happen when predators specialize (Chapter 4).

18.4 Costs

The economic costs of invasive species run into the billions of dollars. They include the costs of control as well as direct financial losses, such as crop and timber loss. Idiosyncratic examples include invasive zebra mussels in Lake Michigan fouling water pipes, brown tree snakes in Guam on power lines causing outages, and the Japanese knotweed growing into walls and concrete in the United Kingdom. One can directly measure such economic costs. To that one could add an assessment of the economic value of wild species by asking people what they would be prepared to pay to preserve a threatened species, which is analogous to "no use" assessments for national parks (Chapter 13). Putting all these different costs together, a 2010 government commission in Great Britain estimated the annual cost of invasive species to be $2.5 billion.

In the United States, the cost of the more than 400 alien insect forest pests alone has been put at $5 billion. These pests fall into three main categories: those that bore into the wood, those that suck sap, and those that eat leaves. Of these, borer costs are estimated at $3.5 billion, and one species, the emerald ash borer, accounts for more than half of this sum. Native to Asia, this beetle has killed hundreds of millions of ash trees in the eastern United States since its discovery in Michigan in 2002. A breakdown of invasive wood-borer costs is shown in Figure 18.4. Most costs result from tree removal and replacement and property losses. Direct costs in terms of lumber are relatively small because ash is not an important commercial tree. These findings suggest that economic and environmental interests can be aligned because people want to preserve some species for aesthetic reasons.

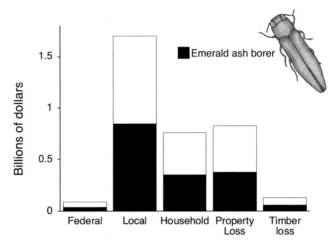

Figure 18.4 Estimates of annual costs of 71 species of alien wood-boring insect species combined, and the subset due to the most invasive, the emerald ash borer (illustrated). Cost categories include (1) federal government expenditures (survey, research, regulation, management, and outreach), (2) local government expenditures (tree removal, replacement, and treatment), (3) household expenditures (also tree removal, replacement, and treatment), (4) residential property value losses and (5) timber value losses to forest landowners. (J. E. Aukema, B. Leung, K. Kovacs, et al. [2011]. Economic impacts of non-native forest insects in the continental United States. *PLoS ONE* 6: e24587)

Given the great costs of invasive species once they have arrived, legislation is directed toward preventing arrival. For example, the impact of the zebra mussel (a native of Central Asia) in the Great Lakes, led the U.S. Congress to pass the 1990 Invasive Species Act, requiring ballast water management for ships. The most common method of doing this is via a mid-ocean ballast water exchange. A second bill, passed in 2007, was specifically directed toward protecting the Great Lakes. This bill increased water purification requirements and funded the Army Corps of Engineers plan to construct a permanent electric dispersal barrier to keep invasive fish species from migrating up the Mississippi, through the Chicago River, and into the Lakes. "Protecting the Great Lakes from invasive species (and restoring the locks along the Mississippi) will boost Illinois' economy while protecting our environment," said Senator Barack Obama, when announcing that funding had been approved. "The Asian Carp Barrier," he observed, "will preserve Lake Michigan's ecosystem and prevent foreign species from wiping out other fish and plant populations." In 2019 the construction of a new barrier, with costs approaching $1 billion, was approved.

18.5 Conclusions

While a great deal of effort and expense goes into curbing the impact of invasive species, we will never remove most aliens from most places. Indeed, many alien species were introduced long ago. *Cannabis* for example, was introduced into India about 3,000 years ago. Some of these species are now an integral part of the community. Introduced large mammals, such as the horse in North America, have helped restore some wildlands to conditions that are more akin to what they were before humans caused mammalian extinctions, including those of native horses in North America (Chapter 25). On the island of Oahu, Hawaii, alien bird species disperse seeds from native plants, although they more efficiently disperse seeds from alien plants than from native ones. Despite the positive effects of introduced species, it is critical to prevent, as far as possible, further transport and introduction, for one alien species that turns invasive can cause great havoc and further simplification of natural ecosystems.

19

Introduced Disease

Some of the steepest population declines, and even extinctions, have resulted from disease outbreaks. Communicable diseases have caused the extinctions of amphibians, the occasional complete failures of wheat crops, and the devastation of human populations. Sometimes epidemics result because host populations have increased in density, raising possibilities for contact and transmission. For example, the bacterial and fungal diseases of sea urchins on coral reefs may have always been present in these populations, but they became an especially serious problem when sea urchin densities increased. Often, however, virulent diseases result from the movement of a pathogen between species. Most of the virulent diseases in humans have been contracted from keeping domestic animals and consuming wild ones. Movements may be between populations as well, such as transmission from Europeans to Native Americans.

Diseases that are passed on from one species to another are known as spillovers, with the subset that passes into humans called zoonoses. Diseases that have become more prevalent in humans over the past 20 years are given the special name, emerging infectious diseases, or EIDs. More than 150 zoonoses, including ebola, are not self-sustaining, and so they require repeated reinfections from their main host. Other pathogens that make the jump into a new species become self-sustaining in the new species. In humans, over one hundred diseases fall into this category, including AIDS (Chapters 2 and 7) and measles (Chapter 6). The HIV strain presently infecting 30 million people probably entered about 100 years ago from chimpanzees, whereas measles evolved from rinderpest about 900 years ago. Finally, about forty diseases are known to freely move between humans and other animals. Leishmaniasis, due to a protist transmitted by sandflies, falls into this class.

The virus SARS-CoV-2, which causes the disease COVID-19, emerged in December 2019 in Wuhan, China. In 2020, SARS-CoV-2 infected more than 84 million people worldwide. By the end of 2021, COVID-19 had killed at least 5 million people, and likely millions more, as causes of death have not always been recorded in some countries. COVID-19 stands for coronavirus disease (2019) and SARS-CoV-2 for severe acute respiratory syndrome-coronavirus-2. The number 2 is added because it follows the SARS outbreak in 2003, which was more lethal (it had a mortality rate of about 10%, whereas the current one was initially about 2%). In the 2003 outbreak, people developed symptoms

Ecology of a Changed World. Trevor Price, Oxford University Press. © Oxford University Press 2022.
DOI: 10.1093/oso/9780197564172.003.0019

coincidentally with becoming infectious so the disease was more easily controlled than COVID-19 and resulted in only 840 SARS deaths in total. Both of these coronaviruses are present in bats, which tend to have strong immune systems. Four other coronaviruses are self-sustaining in humans and cause the common cold. After contracting a cold, short-term immunity develops, as well as some cross-immunity between the two strains. The possibilities for at least some cross-immunity from the cold viruses and between different SARS-CoV-2 strains, the effects of seasons, and the length of time immunity lasts after contracting COVID-19 or after vaccination are all active areas of research. In a fully susceptible population, the original strain of COVID-19 had a reproductive value (R, number of infections per infected person) of between two and three under normal social practices ($R_0 = 2$–3). To lower the reproductive value below one, then, more than 65% of the population would need to become immune (recall that the proportion is given by the expression $1 - 1/R_0$; see Chapter 6), or employ social practices (masks, social distancing) to reduce transmission. The δ variant of the virus emerged in India in 2020, and was more contagious than the initial COVID strain, so in a naïve population of hosts it would have a higher R_0. In a good example of natural selection at work, the δ variant rapidly became the dominant strain in many countries, to be followed by the even more transmissible omicron (O) variant. The omicron strain is more infectious but less virulent than δ, suggesting the possibility of evolution towards cold rather than flu like symptoms. However, cross-immunity between strains is incomplete (as shown also for the common cold), vaccine efficiency against one strain is often less effective against others, and immunity wanes with time. It is evident that COVID-19 will be self-sustaining in humans and vaccination will be regularly required to prevent serious infection.

In this chapter, we consider more generally the causes and consequences of disease outbreaks in both humans and wild species. We ask how diseases arrive, spread, and may be a threat to both rare and common species. We describe:

(1) The main reason invasive pathogens debilitate host populations, which is that hosts are not adapted to the pathogen nor is the pathogen adapted to the host.
(2) An example of how an additional threat—in this case climate change—may accentuate the impact of a pathogen.
(3) Fungal diseases in crops, which illustrate many of the principles underlying the spread of EIDs.
(4) The ramifying impacts of introduced diseases on the food web.
(5) The reasons common species may be particularly threatened by invasive pathogens.

19.1 Undefended Hosts

Just as invasive predators and pests can devastate native species (Chapter 18), some introduced pathogens can be exceptionally virulent. Across the Americas, more than 50 million people are thought to have died from diseases introduced by Europeans. By some estimates, central Mexico had a population of more than 22 million in 1500, which was reduced to less than 2 million by 1630 (Fig. 19.1). Different pathogens sequentially caused multiple epidemics, including bacteria that caused the bubonic plague and viruses that caused measles, smallpox, and influenza.

Why were diseases so much more virulent for Native Americans than they were for the Europeans that brought them over in the 1500s? At least four reasons have been suggested:

(1) Europeans had evolved to be more resistant to these diseases because they had been long exposed to them. Presumably, the ancestors of these Europeans suffered much mortality when the pathogens were first contracted. Accordingly, a high death rate of Native Americans would be expected because their ancestors had not been exposed to these diseases.

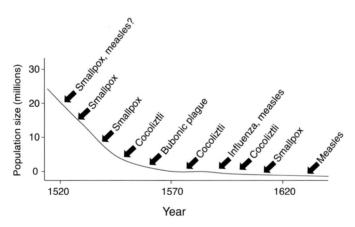

Figure 19.1 Estimates of human population sizes in central Mexico and possible associated epidemics. Cocoliztli refers to a disease whose underlying causes are still disputed, but most evidence indicates it is due to a bacterium, related to typhoid. (W. Borah and S. F. Cook. [1969]. Conquest and population—demographic approach to Mexican history. *Proceedings of the American Philosophical Society* 113:177–183)

INTRODUCED DISEASE 197

(2) Given that diseases should have greater control over virulence levels than hosts (Chapter 7), it is possible that pathogens had also evolved to combat resistance in Europeans, which would have made them especially virulent when entering a naïve population.

(3) Most people catch most diseases, such as measles, when they are young. Consequently, we evolve a strong immune response at that age. Subsequently, because of the acquired memory of past diseases, adults can more easily mount a response and clear the infection before it is debilitating. Adults not exposed to the disease in childhood are particularly susceptible to debilitation, to which neither the virus nor the host is adapted. Thus, when a disease is introduced into a naive population full of susceptible adults, we might expect it to be particularly devastating.

(4) If the disease is maintained in Europeans, who can tolerate it, it will be regularly reintroduced into native populations and maintain high virulence. Resistance, where hosts clear the disease, differs from tolerance, where individuals carry the disease. Tolerant individuals can be an important reason why a disease will spread (see the next section of this chapter).

When a disease first enters a naive population, we expect a high level of virulence because of the reasons listed above and because such populations typically harbor a high density of susceptible hosts, which facilitates transmission.

Pathogens causing devastation to humans have mostly been bacteria and viruses (Fig. 19.1), but fungal infections have been particularly destructive of other species. At least two-thirds of recorded wild population declines due to diseases are attributed to fungi. One recent infection comes from the aptly named fungus *Pseudogymnoascus destructans* likely introduced by a European tourist, which infects bats. *Pseudogymnoascus destructans* first appeared in New York in 2006 and has now spread across North America (Fig. 19.2). It has killed over 6 million bats (the associated disease is white nose syndrome). The fungus causes dehydration and irritation during hibernation. Consequently, the bats come out of hibernation more frequently, depleting their fat reserves. Reflecting a long period of coevolution with the fungus, European bats do not suffer in the same way. (One suggestion is that a particularly susceptible North American bat species, the little brown bat, is small and carries relatively few reserves anyway.) The fungal disease caused by *Pseudogymnoascus destructans* has a very high kill rate, but it does not disappear when host numbers are reduced, as happens in many other pathogens (Chapter 6, Fig. 6.2). First, in common with other fungi, it can survive for long periods of time outside of the host (*Pseudogymnoascus destructans* persists in the soil of bat roosts). Second, it infects multiple species, some of which are tolerant and act as a reservoir. Vaccination, by spraying an edible gel on the colony, has been seriously

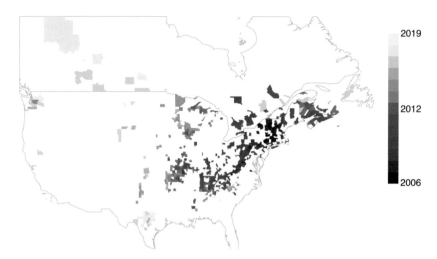

Figure 19.2 (Color plate 13) Dates of first occurrence of confirmed or suspected white-nose disease, or else, in 15% of the cases, the detected presence of the fungus, in counties of the US and administrative districts of Canada. (North American Bat Monitoring Program, Fort Collins Science Center, U.S. Geological Survey)

considered, and although no vaccine has been developed, several susceptible-exposed-infected-resistance (SEIR, Chapter 6) models have been proposed to determine what fraction of the population should be immunized to eliminate the disease. As with human influenza, modelers conclude that about 50% of the population would need to be vaccinated, but because spores can persist in the soil, vaccination would need to be long continued.

Amphibians (frogs, toads, salamanders) are particularly endangered at present (Fig. 19.3), with most of the threat coming from one invasive fungal species, *Batrachochytrium dendrobatidis*. DNA evidence indicates that *Batrachochytrium dendrobatidis* was likely confined to East Asia until about 100 years ago. It was first detected as a killer of frogs in the late 1970s in both Australia and Central America. A reasonable hypothesis is that transport of amphibians for use as experimental lab animals and as pets carried the fungus from one part of its range to another. Out of 110 species of harlequin toads in the region between Costa Rica and Peru, two-thirds appear to have been lost between 1980 and 2000 (a few have been rediscovered). As in the case of white nose syndrome in bats, the fungus can persist as spores and infects many species. It has a high kill rate of some species, whereas others show greater tolerance and act as a reservoir and carrier. In Costa Rica, tolerant species have increased their geographical range, even as susceptible species have declined.

INTRODUCED DISEASE 199

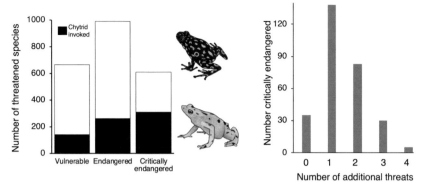

Figure 19.3 *Left*: The number of amphibian species in each of the three threatened IUCN categories, with the filled-in portion representing those where *Batrachochytrium dendrobatidis* has been invoked as a major threat. *Right*: Partitioning of the 299 critically endangered species where *Batrachochytrium dendrobatidis* has been invoked as a threat, according to the number of additional threats listed in the IUCN red list evaluations (out of the six we consider: Climate change, Harvesting, Pollution, Habitat loss, Invasive species, Disease). The Harlequin Poison Frog from Colombia (above) and the Panamanian golden toad are both critically endangered (the Panamanian golden toad may be extinct in the wild but is maintained in captivity). For both species, the pet trade, habitat loss, and pollution have been attributed as reasons underlying the decline. *Batrachochytrium dendrobatidis* is considered the major threat to the Panamanian golden toad but has so far not been found in the Harlequin Poison Frog. (Derived from the *IUCN Red List of Threatened Species*. www.iucnredlist.org. downloaded on August 1, 2020)

19.2 Reservoirs

If pathogens are not self-sustaining in the host but instead are continuously reintroduced into the susceptible population, any selection pressures that favor the pathogen to evolve reduced virulence are absent, so the pathogen may remain virulent. Particularly problematic is the widespread presence of species that are tolerant of the pathogen because then the pathogen can continually leak in to susceptible populations. In this way, tolerance differs from resistance, which refers to individuals that do not carry the pathogen.

In the previous section, we described possible examples of reservoir populations and species, including some species of bats and frogs where tolerance has likely contributed to population declines in species that are not tolerant. A further example of devastation by leakage from a tolerant species is that of rinderpest. The disease used to kill up to 90% of cattle during an outbreak and perhaps 200 million cattle in total died from rinderpest in Europe in the 1700s.

200 ECOLOGY OF A CHANGED WORLD

The disease originated in Asia, where it was less debilitating. Rinderpest appears to have been the first biological weapon:

> The secret weapons of the invaders [from the east into Europe] were the Grey Steppe oxen. Their value was a strong innate resistance manifested by a slow spread of the virus and by the absence of clinical signs. A troop of Grey Steppe cattle could shed rinderpest virus for months, provoking epidemics that devastated buffalo and cattle populations of invaded countries. The sequelae were no transport, untilled fields, starving peasants, and overthrown governments. (Scott, 2000, p. 14)

This passage refers to leakage from oxen into cattle, but rinderpest also leaked into wild populations with devastating effect. Rinderpest arrived on the African continent in 1889. The last big outbreak from 1959 to 1961 killed about 80% of the wildebeest in the famous Serengeti Reserve in Tanzania. The fact that rinderpest disappeared when cattle were vaccinated (and the wildebeest recovered; Fig. 3.4) implies that it could not be maintained in wild populations. As we noted in Chapter 18, rinderpest is now thought to have been eliminated completely after cattle vaccination.

19.3 Multiple Factors

Many factors together increase the possibility of extinction. In the case of harlequin toads affected by the fungus *Batrachochytrium dendrobatidis,* those species living at mid-elevations seem particularly prone to extinction, and many of the early disappearances came after a relatively warm year, suggesting an interaction with climate (Fig. 19.4). For example, the Jambato toad of Ecuador and the Monteverde harlequin toad from Costa Rica were both last seen in 1988, right after a warm year in 1987. Fungal infections and climate change may interact as follows: In montane habitats such as Monteverde, higher overall cloudiness resulting from higher evaporation increases nighttime temperatures (by insulation) and reduces maximum daytime temperatures (by shading). The net result is a predicted increase in the optimal range for *Batrachochytrium dendrobatidis.* Although the role for climate change is disputed, the example indicates how multiple threats generally interact.

19.4 Fungal Diseases in Crops

Monocultures of wheat, rice, and other crops are planted at high density, making them especially susceptible to disease. They illustrate many of the main principles

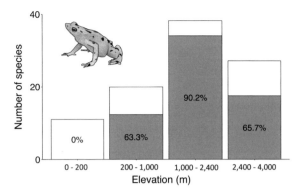

Figure 19.4 The number of species of Harlequin toads (including the Panamanian golden toad, see Fig. 19.3) that used to be found in various elevational belts in Central America and the northern Andes. Gray shading is of those species not seen since 1998, based on a paper published in 2006. (J. A. Pounds, M. R. Bustamante, L. A. Coloma, et al. [2006]. Widespread amphibian extinctions from epidemic disease driven by global warming. *Nature* 439:161–167)

we have covered so far, including causes of high virulence, and interacting threats:

(1) Pathogens can move from the natural host to the crop, where they may be more virulent. An example is that of the fungus causing wheat blast, which jumped from a native grass into a nonresistant wheat in Brazil in the 1980s. Wheat blast grows on the seeds. It can cause 100% loss of the crop. Consequently, in some parts of Brazil, the quantity of wheat now planted has declined by 95% over the past 30 years. The fungus may be continually reintroduced into the wheat from several grass species, including one commonly planted for cattle pasture. This creates a large population that has so far overcome attempts to breed resistant strains of wheat, or effectively use fungicides.

(2) As in the case of invasive organisms in general, increased trade has introduced pathogens from one part of the world to the other. In 2016, wheat blast got to Bangladesh, probably in an import (the blast's dormant stage is within the seed head, making it hard to detect).

(3) Weather, including altered patterns driven by climate change, has affected dispersal. In 2004, Hurricane Ivan is thought to have carried spores of a fungus that causes rust of soybean (another fungal disease that grows on leaves, causing brown patches) from South America into the United States. It is now found throughout the southeastern United States and

202   ECOLOGY OF A CHANGED WORLD

Mexico. As is true for species more generally, various plant pathogens are moving northward in the United States in response to climate change. The estimated average rate is about 6 km per year.

(4) Hybridization between strains creates novel forms that may make them especially able to exploit hosts (just as plant breeders use hybridization to improve their varieties). For example, a native fungus may be adapted to the local climate and bequeath some genes through hybridization to the introduced form. While we have no clear example where this has aided adaptation to local conditions, hybridization with residents is a common feature of invading organisms. A rust that affects barley in Australia resulted from a hybridization event between a wheat rust and a native grass rust.

19.5 Impacts on common species

A feature of the spread of infectious diseases is that common species can be at risk. Figures 6.2 and 6.6 in Chapter 6 imply that when only a few susceptible host individuals are present, the disease should be lost from the population, preventing extinction. However, we have seen examples in this chapter where (1) some species act as reservoirs (they are carriers of the disease and show only mild or no symptoms), (2) some pathogens can exist off the host, and (3) additional threats combine to affect populations when they are reduced to low numbers. These factors imply that once-common species can be driven to extinction by disease. In this section, we consider another consequence of being common that can make common species at greater risk of extinction than rare ones, even without these additional factors. This happens if individuals can be infectious for a relatively long time before they die.

Suppose an infected individual carrying a pathogen enters a population of 200 susceptible individuals. It encounters many others, thereby generating an epidemic that kills many individuals sufficient to cause a rapid population decline. Now suppose the same infected individual enters a population of just 50 individuals and, because of the lower densities, meets one-quarter the number of individuals than it would in a population of 200. Given that the large population passes through a population size of 50 susceptible individuals on the way to extinction, one may wonder why it too was not protected from extinction when it was at that size. The answer is that, by the time it is reduced to 50 susceptible individuals, this initially large population also contains many infected ones. Figure 19.5 shows a model of this process—with parameters for the death rate at $d = 0.02$ deaths/infected individual/unit time, and among the surviving

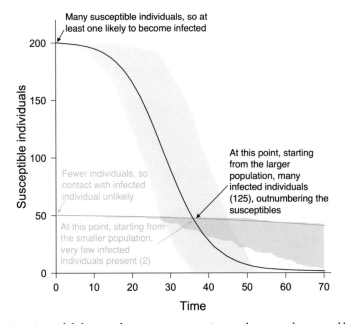

Figure 19.5 A model showing how common species can be more threatened by a new disease than rare ones. The model is a simplified version of the SEIR model (Chapter 6), but only allows for the Susceptible and Infected classes and does not consider births. The model introduces a single infected individual into a population at time $t = 0$. The death rate of infected individuals is set to $d = 0.02$ infected individuals/ infected individual/day and an infected individual has a probability of $p = 0.001$ of infecting every susceptible individual in the population. The solid lines are the average rate of spread of the infection, and the shadings give the range for 90% of the outcomes (determined by simulation). This model assumes that individuals are encountered in proportion to their density, so the infection is passed on to more individuals in a denser population. (M. C. Fisher, D. A. Henk, C. J. Briggs,. [2012]. Emerging fungal threats to animal, plant and ecosystem health. *Nature* 484: 186–194)

individuals a transmission rate with probability $p = .001$ of passing the disease on to every susceptible individual in the population in that time interval. In the model, starting from a population of 200, by the time the number of susceptible individuals has been reduced to 50, 125 infected individuals are also present, making it likely that the remaining susceptible individuals will become infected. On the other hand, if a single infected individual enters a population of 50 susceptible individuals, the infection takes much longer to take off, and sometimes, it does not become established at all because the individual dies before it meets another one.

19.6 Ramifying Effects on the Food Web

Just as removal or addition of predators can have enormous effects on entire communities, diseases send similar ripples through the food web. The elimination of rinderpest in Africa has had ramifying effects on the communities: (1) a tripling of wildebeest in 10 years (see Fig. 2.4 in Chapter 2), (2) large increases in the numbers of lions and hyenas, and (3) a decrease and loss of wild dogs, probably because lions and hyenas compete with them for food, as well as prey on them.

A remarkable example that affects our own food supply and health comes from the spread of white nose syndrome in the United States and the consequent decline in bats (Fig. 19.2). Apparently, because fewer insects were being consumed by the bats, soy and corn yields went down by 5% and 10%, respectively in counties affected by white nose syndrome, both because more insects were eating the plants (Chapter 8) and because insects carry plant fungal diseases. Consequently, farmers in areas where white nose disease was present used 40% more insecticides and 20% more fungicides than the average use.

19.7 Conclusions

When considering predation, we noted that top predators, though often rare, have important ramifying effects on the food web, and that they are often the first to be removed by humans. Besides top predators, common species also play a large functional role in communities. Here we add an additional twist, which is that common species can be especially at risk from pathogens. For example, bats are under threat from white nose disease. Threats to common species from disease make predictions of the future even more uncertain.

20

Harvesting on Land

Harvesting has major impacts on wild populations on land, and the threats it poses are much greater than many appreciate. Although hunting is rife across the world, conservation issues associated with the capturing or killing of animals on land are largely tropical. All three continents spanning the tropics—Africa, Asia, and South America—as well as many tropical islands, are experiencing unsustainable losses of animals, associated with heavy hunting pressures. Many forests have lost most of their large mammals and birds. They are aptly named "empty forests."

The greatest take of animals and birds is in Africa, estimated to be between 4 and 20 times more than what it is in South America. In Asia, extensive trapping continues, so that in many locations, populations have been depleted and catch is low. For example, the eight species of pangolin (scaly anteaters) are small nocturnal mammals eaten locally, and they were formerly widespread across Africa and Asia. They are sought after for their meat and, in Asia, for their scales, which are made into a powder and used in various treatments, despite any evidence that they have medicinal value. All species have declined, but in Africa somewhere between 400,000 and 2.7 million continue to be harvested each year. Both the export and the import of pangolins is illegal. However, in April 2019, Singapore customs agents impounded bags of scales derived from 38,000 pangolins in transit from Africa (along with a large quantity of elephant ivory), which must be only a fraction of the trade. The global demand for products such as these is partly connected to rising wealth in China; ivory has recently sold at $3,000/kg.

The pangolin situation highlights not only the extent to which wild animals are harvested, but also how difficult it is to estimate the true level of harvest, given that much happens at the village level and is often illegal. Further, impacts on nature, though clearly extensive, are essentially impossible to quantify because exploitation on land has been going on so long. Indeed, many large and important terrestrial species went extinct when humans first started hunting in earnest (Chapter 25). Therefore, in this chapter we concentrate on what is known about current exploitation rates and on how impacts have changed over the past 50 years or so. The specific goals are to:

(1) Describe current patterns of exploitation as far as they are understood.
(2) Assess declines of species attributable to hunting.

Ecology of a Changed World. Trevor Price, Oxford University Press. © Oxford University Press 2022.
DOI: 10.1093/oso/9780197564172.003.0020

206 ECOLOGY OF A CHANGED WORLD

(3) Consider some economic issues.

(4) Note the effects of harvesting on other species.

20.1 Patterns of Exploitation

Harvesting can be ranked from (1) subsistence to (2) hunting for commercial in-country markets up to (3) hunting for export. Most harvesters are the rural poor, sometimes employed by organized crime. Captures include not only for meat, but also for pets; release in religious ceremonies; medical research; "medicinal" parts (e.g., pangolin scales, rhino horn); ivory; and furs. Information regarding captures for subsistence comes mainly from interviews, captures for in-country sales by direct observation at markets, and captures for international trade from documented legal trade, plus occasional seizures of contraband.

Subsistence and Local Markets

Across the world, at least several hundred million people eat bushmeat. The proportion of the catch eaten by the hunter's family and friends, rather than being sold in the market, varies from 100% (some Amazon tribes) to less than 10% (some West African countries). In Gabon, West Africa, in one village surveyed in 2002, each person typically ate 0.25 kg of bushmeat a day and little fish or chicken. By contrast in the capital city, people ate almost no wild meat, but consumed 0.45 kg of fish and chicken. Across rural communities of Africa, the poorest people eat more bushmeat than those who are better off, whereas in towns the richest people eat the most bushmeat. Economics helps explain these contrasting patterns. First, assuming bushmeat is something urban dwellers want to eat at least occasionally, only the richer members can afford to do so because transport and marketing result in bushmeat costs exceeding that of other protein sources. Second, hunting is labor/time intensive, so that in rural communities, those without other sources of income are the primary hunters. The need for substantial investment of effort in hunting is clear because in many studies, the amount of bushmeat consumed correlates with the distance hunters need to travel to get it (Fig. 20.1).

Many wild animals are killed for meat, but just how many has been hard to quantify. According to a 2002 study of carcasses in markets along the river border between the West African countries of Cameroon and Nigeria almost 1.5 million animals were traded each year in this region (covering 56,000 km^2),

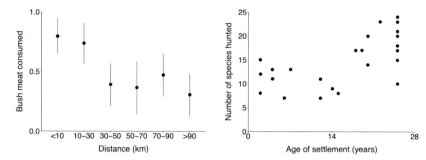

Figure 20.1 *Left:* Pooled across Ghana, Tanzania, Madagascar, and Cameroon, villages less than 30 km from areas with wild animals consumed more bush meat than villages further away. Surveys were conducted between 2004 and 2009. Within each country every household was ranked according to grams/meat/day consumed and then divided by the maximum rank, so that in each country the household with the highest consumption had a value of 1. The Y axis is approximately the average rank (standard errors are based on village mean values, $N = 96$ villages). *Right:* Across 31 different settlements in seven countries of the Amazon Basin, relatively few prey species are harvested in settlements younger than about 15 years. (J. S. Brashares, C. D. Golden, K. Z. Weinbaum, et al. [2011]. Economic and geographic drivers of wildlife consumption in rural Africa. *Proceedings of the National Academy of Sciences* 108:13931–13936; A. Jerozolimski and C. A. Peres. [2003]. Bringing home the biggest bacon: A cross-site analysis of the structure of hunter-kill profiles in Neotropical forests. *Biological Conservation* 111:415–425)

of which 95% were mammals (an undetermined number may have come from outside the region). Small forest antelopes make up more than 25% of the harvest, but other carcasses in the market include gorilla, chimpanzee, leopard, and elephant. This study and others have led to estimates of more than 10 million antelopes alone being annually killed in the tropical forests of Africa in the early 2000s. Although exploitation rates are generally unsustainable, as we consider later in this chapter, it is not clear that catch is decreasing. More people, easier accessibility to more distant locations (roads), more efficient kill methods, and rising prices associated with increased demand, all lead to increased catch. In remote locations, including the Amazon Basin and the Congo (Figure 13.4), areas with good numbers of mammals persist. Across the Amazon, when new settlements are established, hunters target additional species over time, as the more valuable and more easily hunted species become rarer (Fig. 20.1).

In Africa, increased harvest rates have been best documented on the island of Bioko off the west coast. The island is about 70 km by 25 km wide and is still well

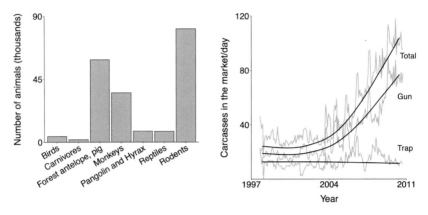

Figure 20.2 Exploitation on the island of Bioko off west Africa. *Left:* Summed numbers recorded in the market across 14 years (1997–2010) split by kind of animal. *Right:* Daily numbers over time in gray, split according to catch method. Smooth lines are drawn by eye. (D. T. Cronin, S. Woloszynek, W. A. Morra, et al. [2015]. Long-term urban market dynamics reveal increased bushmeat carcass volume despite economic growth and proactive environmental legislation on Bioko Island, Equatorial Guinea. *PLoS ONE* 10: e0134464)

forested. Commercial hunters from the mainland do most of the hunting and sell their catch at a market in the capital. Between 1997 and 2010, researchers found a total of 200,000 wild animals in the market, about 95% of which were dead on arrival. The composition is shown in Figure 20.3. Mammals formed the vast majority, with birds and reptiles making up the rest. Numbers recorded increased at an exponential rate from 1997–2010 (Fig. 20.2), for at least two reasons. First, new technology (guns) led to increased harvests, especially of the seven species of monkey present on the island; and second, greater wealth on the island likely led to greater demand for meat. In 2007, the killing of monkeys was banned, leading to a drop-off in captures for a short while, before hunters realized that the law was not being enforced, or perhaps could be circumvented using bribes.

Besides demand for meat, the pet trade and various cultural traditions have major effects on species. In Cambodia, more than 600,000 wild-caught birds are sold each year for release at two temples. Many probably die after release, given the conditions they are held in. Indonesia's markets contain more than 10,000 birds daily available for pets. Few owls were recorded in the 1990s, but almost 2,000 in surveys between 2012 and 2016. The increase appears to have been driven by culture: owls are marketed as "Harry Potter birds," reflecting their widespread appearance in both the Harry Potter books and films (a large owl went for about $60).

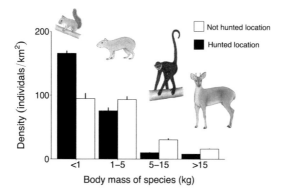

Figure 20.3 Density of small (<1 kg, primarily squirrels and monkeys), medium (1–5 kg, tortoises, birds, some monkeys), large (5–15 kg, monkeys) and very large (>15 kg tapir, peccary, and deer) animals in six heavily hunted sites near villages and six remote (not hunted) sites in the Amazon. Vertical lines are standard error bars. Illustrated: squirrel species, agouti, black spider monkey, red-brocket deer. (C. Peres. [2000]. Effects of subsistence hunting on vertebrate community structure in Amazonian forests. *Conservation Biology* 14:240–253)

Trade Databases

The previous assessments were based on observations at the market. Trade databases are an alternative means of assessing harvesting rates. International trade of wild species is thought to be annually worth billions of dollars and to include hundreds of millions of plants and animals. In an effort to regulate trade in 1973, at the initiative of the International Union for the Conservation of Nature (IUCN), the Convention on International Trade in Endangered Species of Wild Fauna and Flora (CITES) came into force. CITES requires that both the exporting and importing country give permits for species seriously threatened (CITES "Appendix 1 species"), export permits for less seriously threatened species or sometimes populations (CITES "Appendix 2 species"), and species proposed by a country where the species is found in order to instigate co-operation from other countries (CITES "Appendix 3 species"). For example, the elephant is an Appendix 1 species, except in southern Africa, as are all eight species of pangolin everywhere. This makes it extremely difficult to legally trade these species or their products. CITES requires annual documentation of wildlife trade from each country, including discoveries of illicit trade, which can be used to assess wildlife harvesting.

Beyond CITES, which is concerned with international trade, the IUCN has evaluated more than 100,000 species and has listed those known to have been

210 ECOLOGY OF A CHANGED WORLD

traded either internationally and/or within country. Putting these sources together, more than 7,500 bird, mammal and reptile species are recorded as subject to trade; this amounts to more than 25% of all mammal and bird species and 12% of all reptiles. About 1,200 of these species have been traded internationally as pets. While much has been made of these large figures, it should be clear that many records are idiosyncratic. For example, the snowy owl is considered traded because people were caught with two such owls that they had hand-reared and sold between countries within Europe, and a frog species in Chile is recorded as being used for fishing bait in one town. A relatively small proportion of all species traded are heavily exploited. Parrots and falcons are especially important internationally traded birds.

20.2 Effects on Species

While increasingly empty forests are apparent to field researchers, and some species declines are obviously a consequence of hunting, quantitative data are hard to come by. One can assess differences between areas subject to different levels of hunting or use the few available time series, where the same site has been monitored.

Comparisons across Sites

According to a review of 170 studies, areas considered to be subject to low or no hunting pressures have on average about five times as many animals as areas subject to hunting. Figures 20.3 and 20.4 show examples from two case studies. In Gabon, population sizes are about three times lower near villages than they are when one goes more than 15 km away from any village (Fig. 20.4). As of 2014, about 50% of Gabon's countryside was more than 10 km from any village, suggesting plenty of opportunities for increased hunting impacts. Across the Amazon region, large species are rare in heavily hunted areas, but many areas remain to be exploited (Fig. 20.3). Preferential harvesting of large species is a common theme throughout the history of human exploitation, as will be emphasized in the next Chapter, which describes exploitation of the oceans. The targeting of large species is especially the case when hunters use bows, guns, or (in South America) blowpipes, but even snares are first directed at large species.

We can ask which species have been completely lost from a hunted location. In this case, estimates of what species were originally present can be made from knowledge of historical geographical ranges and habitat preferences. Twenty-nine out of forty-five mammal and bird species (60%) have been eliminated

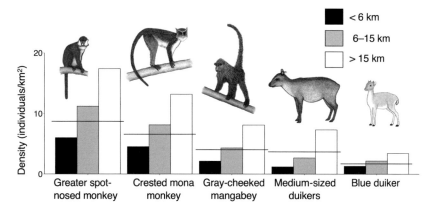

Figure 20.4 Densities of three species of monkeys and four species of antelopes at various distances from villages in rural Gabon. The monkey estimates are based on visual sightings, whereas the antelope estimates are based on dung. Three medium-sized antelope species are grouped together because their dung cannot be distinguished. The horizontal black bars delineate 50% of the maximum for each. (S. E. Koerner, J. R. Poulsen, E. J. Blanchard, et al. [2017]. Vertebrate community composition and diversity declines along a defaunation gradient radiating from rural villages in Gabon. *Journal of Applied Ecology* 54:805–814)

from heavily hunted locations in southern Laos (mainland Southeast Asia), where easily made metal snares are now deployed on an industrial scale (more than 200,000 were confiscated between 2010 and 2015). The species lost include the three largest (elephant, rhinoceros, and tiger). Indeed, the last rhinoceros on mainland Southeast Asia, in Vietnam, was shot for its horn in 2010, and only 400 wild elephants remain in Laos. Such depletion contrasts with a largely nonhunted site in Borneo where only four out of thirty-six species (11%) have disappeared, one of which was the much-persecuted rhinoceros.

Time Series

For some species and locations, censuses across time are available, enabling direct quantification of recent changes. The forest elephant of Africa is secretive, but based on surveys of dung, its population size more than halved between 2002 and 2013. An aerial "Great Elephant Survey" of savannah elephants revealed a decline of 8% a year between 2010 and 2014 (Fig. 20.6). We know this is largely a consequence of poaching for ivory because carcasses can be counted, even from the air, and are many more than expected under death from natural causes.

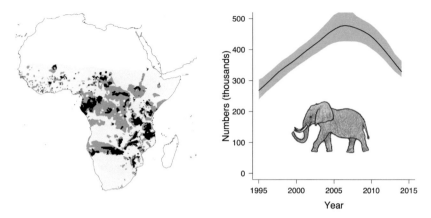

Figure 20.5 *Left:* Range of the two African elephant species combined in the 1600s (grey) compared with 1987 (dark grey) and 2008 (black). The area occupied in 2008 was 34% that of 1987. *Right:* Population changes of savannah elephants in aerial censuses of certain areas between 1995 and 2014. The population recovered from poaching in the 1980s but is again declining. The shading gives the 95% confidence limits. In total it is thought that somewhere between 375,000 and 500,000 savannah elephants remained. (W. J. Ripple, T. M. Newsome, C. Wolf, et al. 2015. Collapse of the world's largest herbivores. *Science Advances* 1:e1400103; M. J. Chase, S. Schlossberg, C. R. Griffin, et al. 2016. Continent-wide survey reveals massive decline in African savannah elephants. *PEERJ* 4: e2354)

20.3 Economics

One can use models to ask first what is the sustainable yield at current population sizes, and second what is the maximum sustainable yield possible? The answers can be used to (1) determine if populations are being harvested sustainably, (2) learn what would have to be done to maximize yield to hunters (if that was the goal), and (3) estimate how many people can be supported at sustainable yields.

The logistic equation is obviously a great simplification, but it provides a guideline, which we illustrate by considering harvesting in Gabon, as reported in Figure 20.4. From Chapter 2, the change in population size given logistic growth is:

$$\frac{dN}{dT} = rN\left(1 - \frac{N}{K}\right)$$

where N is the population size at the beginning of the year, r the intrinsic growth rate, and K the carrying capacity. Intrinsic growth rates are estimated from the age at which females first breed, the number of offspring they have, and their potential lifespan, sometimes using information from related species. For example, the growth rate of the blue duiker, a small antelope (see the rightmost point in Fig. 20.4), is inferred to be high ($r = 0.49$ individuals/individual/year) because females reproduce at 1 year of age, have an average of 1.3 offspring each year, and live for 7 years. In the absence of other estimates, K is taken to be the density far from villages, which for blue duiker is 3.43 animals/km^2 (of course, it may be more if harvesting has happened in these locations as well). With logistic growth, recall that the maximum sustainable yield occurs when the population is at half the carrying capacity, (i.e., for the blue duiker, 1.715), which generates a maximum sustainable yield of 0.45 animals/km^2/year. For the blue duiker the density near villages of 1.31 is less than that inferred to maximize sustainable yield, and sustainably harvesting from this level requires harvest of 0.4 animals per km^2 per year.

According to this simple model, for all the species combined in Figure 20.4, maximum yields are about 20% higher than present sustainable yields. Allowing populations near villages to increase to the maximum sustainable yield by not harvesting at all would take a few years, which would temporarily reduce family incomes and nutrition. At present levels, a sustainable harvest across all species might be 1.2 animals/km^2/year. Harvesting estimates from Equatorial Guinea are slightly above this estimate of sustainability. However, as described above, about six antelopes/km^2/year were thought to be harvested on the Cameroon/Nigeria border in the early 2000s, which would be clearly unsustainable.

Is harvesting in general worthwhile? While bushmeat demonstrably improves human health for those without other sources of protein, further calculations can be used to show that sustainable harvesting would provide sustenance for less than one person/km^2. A rough estimate is that in the Congo Basin we would require 250,000 km^2 of land to replace harvested bushmeat by locally produced beef (recall that this is the least efficient source of farmed protein, Chapter 9). While large, this is just 7% of the area of the Congo Basin. That can be contrasted with the large effects we anticipate by harvesting many species to 50% of their natural level, or in many cases, below 50%.

20.4 Effects on the Food Web

The targeting of large species (e.g., Fig. 20.3), is quite different from what happens in nature, where large species are often top predators or are subject to low predation rates because they can defend themselves (e.g., elephants). The loss of large animals due to human exploitation has ramifying effects through food webs (Chapter 4). The woolly and spider monkeys are the largest primates

214   ECOLOGY OF A CHANGED WORLD

in the Amazonian region. They are common in nonhunted regions, where together they can account for about 70% of the total primate biomass, but they are completely absent in many hunted areas. Along with the ground-dwelling tapir, which is also hunted, these large monkeys are the only seed dispersers for trees that have large fruits. One elegant study showed that trees with large seeds have especially dense wood that stores much carbon. The replacement of these trees by other species that don't require large monkeys for seed dispersal might reduce 5% or more of the carbon storage facilities of the local forest.

20.5 Conclusions

The threat from the harvesting of wildlife on land is large. Clearly, the past 30 years or so have seen a serious increase in hunting, leading to exploitation of ever more remote areas, with many places now devoid of large species. Besides the sheer number of animals being hunted, protected areas, the cornerstone of current conservation efforts, are themselves subject to hunting pressures. For example, in the early 1990s, Lambir Hills National Park in Borneo had a more or less intact fauna, but establishment of a local bushmeat market precipitated large declines. Hornbills have largely disappeared, and the number of primate species has been halved. Instead of 25 species of birds and mammals eating fruit at trees in 1997–1998, there were just 10 remaining in 2005–2006, with large fruits not being eaten at all (see Fig. 24.1). Possible solutions to the threats from harvesting are the subject of Chapter 22, after a consideration of what is happening in the oceans.

21

Harvesting in the Ocean

Harvesting is by far the most important reason for the decline of marine species. Indonesian archaeological sites indicate that people were eating sharks and tuna as long as 40,000 years ago. Human waste sites on California's Channel Islands contain otter bones, many abalone shells, and sea urchins from about 8,000 years ago. But it is only at the end of the nineteenth century that it became apparent that fish stocks could be exhausted. As fish catches have declined, technology has enabled more efficient exploitation of old fisheries and the opening up of new ones. The invention of the trawl in the 1300s was replaced by the steam trawl in the late 1800s, followed by improved designs that enable trawl fishing over rough seabeds, trawls 60 m across, and, most recently, the ability to trawl as deep as 1,300 m. Other inventions include radar to locate fish shoals; long lines with over 2,500 hand-baited hooks on lines many kilometers in length dragged behind a ship; and floating logs in the sea, monitored using GPS, which attract fish (called fish-aggregating devices).

In 1995, Daniel Pauly coined the term *shifting baselines*. He noted that the starting point against which we evaluate the world is often when we first become aware of it, with one consequence being that older people infer greater deterioration. Therefore, to assess the true effects of humans, we need to know what the state of the world was before exploitation began. This study is difficult, but it is not nearly as difficult for the ocean as it is for land, which has been heavily affected by humans for thousands of years (Chapter 25). Hence, in this chapter we consider:

(1) The status of marine mammals, fish, and turtles.
(2) The effects of harvesting on other species in marine food webs.
(3) The present state of fisheries.

The next chapter deals with economic analyses and routes toward remedial action.

Ecology of a Changed World. Trevor Price, Oxford University Press. © Oxford University Press 2022.
DOI: 10.1093/oso/9780197564172.003.0021

21.1 Assessment of Abundances of Marine Species

As in the case of land, the two routes to evaluate how populations have changed is to compare sites that are differently impacted by human activities and historical time series.

Comparisons across Sites

Some areas of the world still remain relatively pristine and can be used as a baseline. The chain of small northern islands in Hawaii is reasonably well protected, but the large islands to the south have been more heavily fished. Off Pearl and Hermes Atoll, 80% of all fish biomass consists of large predatory fish, about one-quarter of which are sharks (Fig. 21.1). By contrast, sharks are almost never seen in the main island chain, even though small fish are of comparable abundance. As we observed earlier regarding land (Fig. 20.3), it is the large species and large individuals that disappear first. These individuals and species are often targeted because they are the most profitable, but even if individuals are randomly caught, average size declines because all fish start off as a small egg and need time to grow to large size. Large species tend to have low per capita growth rates, making recovery difficult. Because the

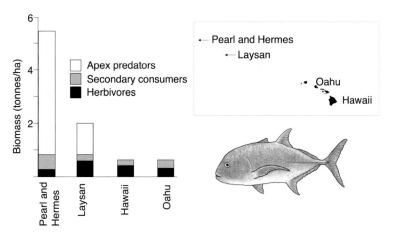

Figure 21.1 Abundance of herbivorous fishes, small predators, and large predatory fish, predominantly the giant trevally (illustrated), and sharks, off four islands in Hawaii in 2000 and 2001. Compare the relatively protected and isolated northern islands with those in the south. (A. M. Friedlander and E. E. DeMartini. [2002]. Contrasts in density, size, and biomass of reef fishes between the northwestern and the main Hawaiian Islands: The effects of fishing down apex predators. *Marine Ecology Progress Series* 230:253–264)

top predators are large species, we fish down the food chain, removing secondary and tertiary consumers first and eventually ending up with species such as jellyfish.

Comparisons across Time

Besides comparisons across space, historical records of population sizes can be used to infer human effects. For example, human waste sites on California's Channel Islands show that 8,000 to 3,000 years ago the size of shellfish fossils decreased. Given the recent heavy impacts of fishing, we can also use records of catch over the past 400 years, as well as people's recollections and their photographs over the past 100 years. Such data can be combined with models, such as the logistic growth model, to estimate past populations; one such model used for a whale species is described in the next section. Use of these methods suggests that many large marine species are close to 10% of their former abundance, meaning they have suffered huge declines (Fig. 21.2). Some groups (whales, turtles) have declined over the past 200 to 300 years and others (sharks, deep sea fish) during the past few decades. The next sections describe estimates of past population sizes based on time series records for marine mammals, cod, and sharks.

Whales and Seals

A drawing in sandstone in Korea indicates that whales were captured as far back as 8,000 years ago, but exploitation did not begin in earnest until the eighteenth century. Whales and seals were decimated across all oceans over the next 200 years. The major product was oil for lamps, at least up to the end of the nineteenth century. A moratorium on whaling, which came into force in 1966, has led to some recovery despite incomplete compliance. In fact, whales and otters/seals are the only groups thought to be substantially higher at present than their historic minimum (Fig. 21.2).

Historical population sizes have been modeled by combining catch records, which cause population declines, with estimates of species' reproductive rates, which lead to population increases. Consider the southern right whale (Fig. 21.3). Three pieces of information can be used to estimate the population size of this species before harvesting started in 1770: (1) recent population size (estimated to be 7,500 in 1997); (2) the number of whales killed per year, estimated from ship records; and (3) the intrinsic growth rate for the species (r), estimated to be about 0.075 individuals/individual/year, based on the recent population increase, as well as from birth rates and longevity. Given this information, it is possible to incorporate a harvesting term into the logistic equation to estimate the number

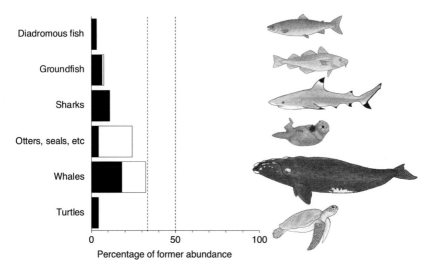

Figure 21.2 Abundance of large marine animals in the year 2000 as a fraction of inferred abundance before human exploitation. The black shading is the historic low. Several groups are at their lowest today, but whales, seals, and groundfish (marginally) have recovered from historic lows. Diadramous fish are those that spend part of their life in freshwater and part in the ocean, such as salmon. Groundfish include cod. Dashed lines indicate abundance that would lead to maximum sustainable yields under the logistic model (50%) and a model for biomass (not numbers) of species on Georges Bank, primarily groundfish (from Fig. 3.5). (H. K. Lotze and B. Wörm. [2009]. Historical baselines for large marine animals. *Trends in Ecology and Evolution* 24: 254–262)

of whales in 1770. According to this model, the change in the number of individuals from one year to the next (dN/dT) is:

$$\frac{dN}{dT} = rN\left(1 - \frac{N}{K}\right) - \text{number of individuals killed per year}$$

The population before exploitation is assumed to be at carrying capacity, K, so our goal is to estimate this value based on the three pieces of information listed above—that is r, the present population size, and mortality due to whaling. First, we guess a reasonable value for K, the population size in 1770, and use observed harvesting rates followed by population growth, to project the population size to 1772. We then project to 1773 and so on. For example, say we assume $K = 50,000$ whales. An estimated 1,100 whales were harvested in the first year (Fig. 21.3),

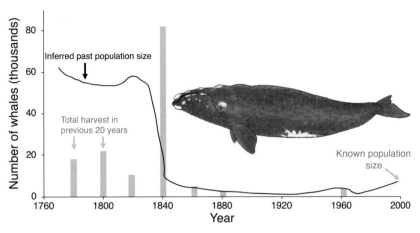

Figure 21.3 The gray bars estimate the total number of southern right whales killed from whaling in the 20 years prior to the indicated date, based on reviews of ship logbooks. In total, about 150,000 were harvested. The black line is the modeled population numbers, back-calculating from a size of 7,500 in 1997, and with an assumption that the intrinsic growth rate $r = 0.075$ individuals/individual/year. This gives an estimate of 60,000 animals present before whaling. In the early 20th century, almost no whales were caught because almost none were available to catch. (C. S. Baker and P. J. Clapham. [2004]. Modeling the past and future of whales and whaling. *Trends in Ecology and Evolution* 19:365-371)

reducing the population to 48,900. The logistic growth formula with $K = 50,000$ implies that the population of 48,900 should now increase by 80 individuals, but another 1,100 whales were thought to have been killed, so after two years we have 47,880 whales. This process can be repeated all the way to the present. We then compare the number we get from the model to the present-day population size. If the match is poor, we go back and try different values of K until the model predicts a present-day population size that is close to the actual value. Using a similar but more sophisticated model along these lines, the original southern right whale population was estimated to be about 60,000, eight times larger than it is at present. Also, the population was reduced to perhaps 60 females, at its lowest point in the 1920s.

Cod

Cod were fished on Georges Bank off the northeast coast of the United States at least 5,000 years ago. Fish longer than the average human were caught into the early twentieth century. Historical population sizes have been estimated directly from a ship's log from 1852 and separately from a mathematical model of carrying capacity similar to the one outlined above. The two estimates are in rough agreement and indicate that the biomass of cod was about 12 times it is currently. The introduction of mechanized trawls precipitated the collapse, especially from 1950 onward when European countries began to heavily exploit the fisheries. In 1992, the Canadian fisheries minister declared a 2-year moratorium on cod fishing, which has since been extended indefinitely because cod have shown few signs of increase. Cod eat Atlantic herring, which increased ninefold in the absence of cod predation. The large numbers of herring compete directly with and prey upon young cod, contributing to the cod's failure to recover, a phenomenon known as predator–prey reversal. Some indicators, however, suggest that cod are starting to come back. Once enough cod get beyond a certain size, it is possible that a recovery, if it happens, will proceed relatively quickly.

Sharks

Shark populations have been particularly hard hit recently and have undergone a worldwide decline, largely due to the Asian market in shark fin soup. From catch reports and fin numbers, 100 million sharks appear to have been harvested in the year 2000, and 97 million in the year 2010, which is an annual catch of about 7% of the world's shark population. With a recovery rate estimated at 5%, this level of fishing translates into a global decline of about 2% per year. The numbers in Figure 21.2 are for the year 2000, so present population sizes are lower than indicated in that graph.

21.2 Effects of Fishing beyond the Target Species

Fishing has many adverse effects, beyiond those on species that are targeted. These include (1) plowing the ocean floor by trawlers, sometimes as much as seven times a year over the same area; (2) the capture of many species as bycatch, not only fish, but also birds, turtles, and dolphins; and (3) "ghost fishing" in which discarded or lost gear continues to capture fish.

Such direct effects on nontarget species are complemented by indirect effects, whereby exploitation of one species affects another. As on land, large predatory species are particularly important to the maintenance of intact communities, and their removal leads to various trophic cascades. We have already considered the killer whale, sea otter, sea urchins, and kelp forests along the Pacific coast (Chapter 4). The loss of sharks in the Atlantic in the 1990s led to an increase in the cownose rays upon which they feed. Because rays are the major predators on scallops, this resulted in the collapse of a 100-year-old scallop fishing industry (now being replaced by aquaculture). Other examples come from reefs and sea grasses:

Coral Reefs

In Chapter 16 we considered how climate change and ocean acidification might lead to the demise of corals. Nevertheless, some reefs that have not been heavily fished are in relatively good shape (e.g., Fig. 21.4). Thus, although warming oceans and increased atmospheric CO_2 contribute to reef deterioration, these factors do not seem to be the biggest problem. Instead, it appears that the loss of large fish has made corals particularly susceptible, through a trophic cascade. This effect is most apparent in the Caribbean, where species such as the goliath grouper, which grows to be larger than a human, used to be common, but now only small fish are caught. The fossil record of the coral community of the Caribbean is dominated in many places by a branching coral, *Acropora*, which goes back at least 125,000 years. The corals started dying off in the 1980s, and less than 1–2% of the reef remains alive at many sites. A possible sequence of events is as follows. Overfishing resulted in a sea urchin becoming extremely abundant because predatory fish used to eat it, and herbivorous fish compete with it for food. Like the fish before them, sea urchins graze on algae on the reef, keeping the algae down. An unidentified pathogen caused mass mortality of the sea urchin in the years 1983 and 1984. As we have seen, very high abundances make a species particularly susceptible to virulent diseases. Then the algae overgrew the coral.

Seagrass

Also in the Caribbean, seagrass enhances sediment stability and is home for many species of fish and invertebrates. In the 1980s, along the Florida coast, seagrasses suffered mass mortality. A clear predisposing factor was the absence of large herbivores. Three hundred years ago, turtles and manatees were extremely common, but now they are at less than 5% of their former abundance (Fig. 21.2). In the absence of these grazers, turtle grass beds grow tall. The tall beds slow currents, shade the bottom, and decompose in place, thereby reducing oxygen in the water. These new conditions have led to the spread of plant diseases and to die-offs. Similar effects are present elsewhere in the world, such as in the Great Barrier Reef in Australia.

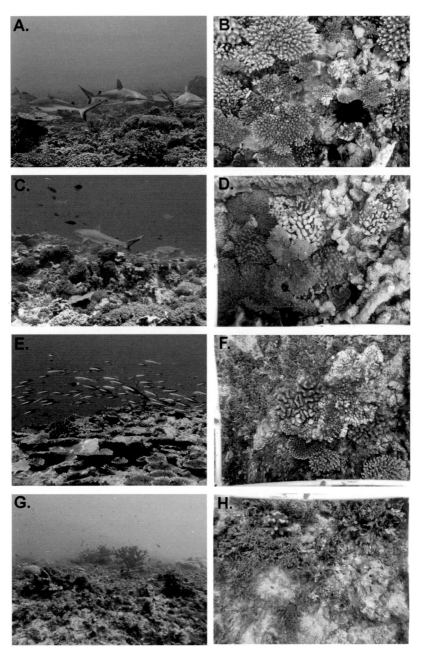

Figure 21.4 (Color plate 14) Photographs of the state of the reef from the Line Islands, central Pacific, on a gradient from heavily fished (G, H, bottom) to protected (A, B, top). Corals in heavily fished regions suffer from algal overgrowth. Note the number of sharks in protected areas. (S. A. Sandin, J. E. Smith, E. E. DeMartini, et al. [2008]. Baselines and degradation of coral reefs in the Northern Line Islands. *PLoS ONE* 3:11, figure provided by Stuart Sandin)

21.3 Present State of Fisheries

We conclude this chapter with a general assessment of the state of the fishing industry. The United Nations Food and Agriculture Organization (FAO) collates annual fish catches as reported by national governments, and reports that catch remained more or less steady between 1990 and 2018, albeit with the largest catch ever in 2018. These statistics underestimate total catch because of both illegal catches (e.g., catching more fish than government regulations allow) and underreporting (e.g., ignoring artisanal fishing—that is, inshore fishing done by local fishermen). Together these sources of underestimates are known as IUU (illegal, unreported, unregulated) fishing. Another estimate of fish take worldwide up to 2010 is reported in Figure 21.5. This estimate came from reading local publications on fish catch, using estimates of the numbers of fishermen from the size of fishing villages, and employing various other kinds of detective work. For example, the number of fishing docks (weirs) in the Persian Gulf was determined using images from Google Earth. In this case, scaling up from the number of fish brought in at monitored weirs, catch was determined to be about six times higher than catch reported from the countries around the Gulf.

According to these reconstructions, global catch peaked in 1996 and is in decline (Fig. 21.5). Harvest from areas that have been long exploited, such as

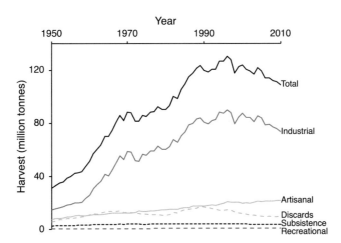

Figure 21.5 Partitioning of total seafood catch into four categories. Artisanal refers to fish caught by local fishermen and sold in local markets. The gray line running along the X axis represents the recreational catch, which is always <1% of the total. (D. Pauly and D. Zeller. [2016]. Catch reconstructions reveal that global marine fisheries catches are higher than reported and declining. *Nature Communications* 7:10244)

224 ECOLOGY OF A CHANGED WORLD

1958   1980-1985   2007

Figure 21.6 Trophy fish catch and size in Key West, Florida (for the middle photo the exact year is uncertain). The average weight of a fish in 2007 is estimated from the photos to be one-eighth that in 1958. The decline in size is largely due to large species not being caught, but the average length of shark species decreased by 50%. One might compare this trajectory to that of wheat, which has gone the same way for different reasons (Fig. 9.6). (L. McClenachan. [2009]. Documenting loss of large trophy fish from the Florida Keys with historical photographs. *Conservation Biology* 23:636–643; Photos *Left and Middle*: the Monroe County Library, *Right:* L. McClenachan, all provided by Loren McClenachan)

the northwest Atlantic (Chapter 3) decreased substantially. In more recently exploited areas, such as the Indian Ocean, harvest leveled off. Greater regulation and enforcement have made a small contribution to the decline in catch. Notably, in some countries a strict limit has been established on catch, and greater restrictions have been placed on discards (whereby unwanted or illegally caught bycatch is thrown back into the water; see Fig. 21.5). But the main reason for the decline in catch is overexploitation. In its 2020 report, the FAO considered 34% of all assessed fish stocks to be currently fished beyond sustainable levels and only 6% to be underexploited; the remainder may be fished sustainably, but most are unlikely to be at the maximally sustainable yield.

21.4 Conclusions

A promising sign is that despite large declines in populations of marine species, relatively few species are thought to have gone extinct. Only two marine mammals are known to have disappeared as a result of human exploitation: the Caribbean monk seal, last seen in 1952, and the Steller's sea cow, which was only discovered in 1742 on a few islands off north Russia and was extinct by 1769. The Steller's sea cow was a slow-moving grazer of seaweed. Fossil records indicate that it was previously more widespread across the Pacific and may have already been greatly reduced by native hunters. Otter hunting appears to have contributed to its final extinction by leading to an increase in urchins and other grazers, and to the associated depletion of the seaweeds on which the sea cow depended. Again, we see the importance of two general themes of this book: multiple causes contribute to extinctions, and ramifying effects through the food web have been driven by removing the top predators. Multiple causes, and impacts of the loss of large animals, are features that likely applied much less in past mass extinction events, emphasizing the unique period of earth's history we are living through.

22

Harvesting

Prospects

Among the six threats we have considered—climate change, overharvesting, pollution, habitat loss, invasive species, and disease—overharvesting is particularly egregious because resources are often reduced below a level that would give sustainably higher catches. With more restraint, more fish could be harvested and more animals could be hunted on land. Restoration of overfished stocks could increase fishery production by millions of tonnes and increase the contribution of marine fisheries to food security. In 2013, it was estimated that allowing fish stocks to recover to levels that would lead to maximum sustainable yields would generate more than $30 billion a year extra in fish yields, but to succeed, more than 12 million fisherman would have to give up their livelihoods. In this chapter, we:

(1) Ask why restraint usually does not happen, so that fisheries are often harvested to low abundances.
(2) Determine what incentives could be used to improve conservation efforts.
(3) Consider the role of government legislation.
(4) Consider possibilities for technology to lower impacts.
(5) Evaluate the promise of protected areas (parks and sanctuaries).

Examples are mainly from the oceans, but the principles apply on land as well.

22.1 Overharvesting

The issues associated with overharvesting are exactly those considered in the economics chapter (Chapter 14). Many people target a common resource, whereby benefits accrue to a single individual, but costs fall on many individuals. The person who catches the fish gets the value of the fish, but the costs (i.e., the removal of an individual fish from the population) are shared by the larger group, which is other fishermen, all of whom could have harvested that fish. The benefit-to-cost ratio for an individual fisherman is high, so they continue to harvest.

Ecology of a Changed World. Trevor Price, Oxford University Press. © Oxford University Press 2022.
DOI: 10.1093/oso/9780197564172.003.0022

This externality (Chapter 14) is known colloquially as the tragedy of the commons. which refers to depletion of stocks when everyone works in their own interest, rather than the collective showing restraint. Table 22.1 shows a made-up example applied to the use of land for grazing. Suppose five herdsmen tend 50 sheep each and half the sheep die each year (death rate, $d = 0.5$ individuals/individual/year). This means that 125 of the 250 sheep survive and each of the 5 herdsmen expect to have 25 sheep by the end of the year. Now suppose that one herdsman adds 50 more sheep to his flock. This means that instead of the usual 250 sheep, 300 are introduced in total. Because of increased competition for food, let's say the death rate goes up to 60%. The result is that only 120 sheep survive, and each herdsman has on average 24 sheep, so the wealth of the community has declined. But the herdsman who introduced 100 sheep has $100 \times 0.4 = 40$ sheep surviving, whereas before he had 25, so it paid for him to introduce them.

Benefits accrue to one individual, but costs are shared by many. The shepherd introducing the extra sheep has a high ratio of benefits to costs, but value to the community decreases. The end result of a process such as this will be harvesting beyond the maximum sustainable yield, overexploitation of the resource, and everyone ultimately ending up with fewer than 25 sheep. So-called goatscapes— barren lands with a few herdsmen on them—are a common feature in many parts of the world.

The tragedy of the commons principle equally applies to fishing, where fish are being removed from a system rather than sheep being added to pasture. Consider five boats catching ten fish each. Let's suppose the total caught (fifty) is at the maximum sustainable yield and everyone is doing well. However, if one more boat goes out, perhaps ten fish are caught by this boat as well, but it places us above the maximum sustainable yield, and fish stocks decline. In spite of this decline, still more boats go out because even a catch of two fish is profitable. Finally, perhaps twenty boats will eventually be out there, catching two fish each for a total catch of less than what the original five boats were bringing. The result of such a scenario is the "bionomic" equilibrium: overfishing, low stocks,

Table 22.1. Worked Example of the Tragedy of the Commons

Sheep	Sheep for Each of 4 Herdsmen	Sheep for Fred	Death rate	Survivors	Fred's Survivors
250	50	50	0.5	125	25
300	50	100	0.6	120	40

In the first line, five herdsmen graze 50 sheep each. In the second line, one of those herdsmen, Fred, decides to double the number he grazes, increasing the overall death rate, yet he still benefits.

228   ECOLOGY OF A CHANGED WORLD

almost no profit, but many employed (Chapter 3, Fig. 3.5). Beyond this point, it becomes economically unprofitable for any further intensification of effort. Fish species may persist at low densities at this equilibrium, and that may be part of the reason why no oceanic fish or whale is known to have gone extinct. However, the basic principle of economics is that given demand, scarcity often leads to price increases. Speculators that stockpile may actually benefit from extinction, as it pushes prices still higher. In 2020, a single tuna fetched $1.8 million on the Tokyo market. In cases like this, fishermen may be prepared to increase their effort to search for increasingly rare fish.

Beyond an increase in value in response to scarcity, a second reason why a species may not persist at a bionomic equilibrium is that it continues to be harvested as bycatch of other more common species that remain targeted. This is similar to the principle of subsidies introduced in the invasive species chapter (Chapter 18). There we noted that the population size of an invasive predator can remain high even when it has consumed much of the native fauna because it is accompanied by a prey animal that can escape from it when young and maintain a high population. In the fishing example, relatively common fish species subsidize the costs to the fisherman of catching rare fish species.

22.2 Incentives to Conserve

How could we conserve populations, and incidentally raise yields, if that was an additional goal? Possibly, benefits could be manipulated by the consumer, as noted for coffee on land, when people began to demand organic and/or coffee grown under forest cover ("shade coffee") (Chapter 14). For example, the Marine Stewardship Council certifies fisheries as sustainable and attempts to persuade consumers to buy from such fisheries. Another route to conservation is through local enforcement. Interested parties could mutually monitor each other to make sure no one takes more than their allotted catch, so that the community as a whole does not overexploit the resource. While this strategy may work sometimes, a community acting in its own interest often does not follow such restrictions. First, fishing may not be the main source of income, but it acts as a supplement, with little incentive for control. Second, heavy discounting is always an issue, whereby the people who make up the community have little concern for the future.

Given motivations to exploit, government legislation and reserves are two ways that have been used to raise fish stocks and sometimes increase yield. These two routes to the preservation of species lie at the heart of conservation biology and form the basis of much of the last five chapters in this book. In the next sections, we explore examples of their application to marine conservation.

22.3 Legislation and Quotas

More than 10% of fish biomass harvested and even more in terms of value comes from oceans beyond national jurisdiction. Fishing in these waters comes with few regulations. Closer to land, the Exclusive Economic Zone regulation, established in 1976, gives countries sole jurisdiction over waters 370 km from their coastline. Especially in the more economically developed countries, for example, the United States, Australia, and New Zealand, fishing is increasingly regulated within these zones. Consequently, according to the Food and Agriculture Organization, in 2016, 74% of U.S. marine fish stocks were being maintained at a level at or above that required for maximum sustainable yields, up from 53% in 2005. However, globally an increased intensity of fishing in developing countries has led to net declines, with at least one-third of all fisheries now thought to be overexploited (Chapter 21).

Quotas are commonly used to regulate fishing. In this case, government agencies tell ship captains how much they can catch. The problems are twofold: working out what the quota should be and enforcing it.

Setting a Quota

Quotas can be set based on different types of data, such as counts of the number of salmon entering rivers, but they are often based on predictions of total population size in an upcoming season, based on the previous year's data. As in climate change and weather predictions (Chapter 10), methods to predict population sizes are becoming increasingly sophisticated, but they will always be uncertain. Such methods are applied mainly to well-managed fisheries in developed countries because of the costs and expertise required for the modeling itself and the need for good data and historical time series. Here, we draw on an example from the herring fishery in southeast Alaska.

From the age of 3 years, millions of herring come inshore to deposit their eggs on marine vegetation in early spring, usually the end of March. The bulk of the fish are harvested commercially for the eggs taken from females (sac roe), much of which is exported to Japan. Later, local people collect the deposited eggs from fish that survived to spawn. The herring are important food for other marine species, including mammals, birds, and other fish. A quota for the commercial harvest is set with the goal of maintaining sufficient stock for all three purposes. Since 1980, the Alaska Department of Fish and Game has used the following data to predict the biomass of fish that would spawn in the absence of harvest: (1) biomass of spawning fish in all previous years, used to predict annual mortality due to natural causes; (2) the proportion of each age class that spawns; (3) the proportion of each age class that is harvested; (4) the average weight at each age; (5) the total number of age-3 fish recruiting each year. For example, the proportion of fish that spawn in a certain year might be 10% for age-3 herring, and 100% for age-8 herring. A logistic equation that relates the proportion spawning to age is fit to all years in order to predict numbers spawning in an upcoming year from the age structure of sampled fish. Recently, sea temperature has been added to the models, which helps largely because it improves the prediction of natural mortality (warmer oceans result in higher mortality). These data are then used to predict the year's spawning biomass.

The top left panel in Figure 22.1 shows the annual estimated pre-fishery spawning biomass in Sitka Sound, the largest herring stock in southeast Alaska. The righthand panel plots predictions against this estimate. One can see that predicted and observed are well correlated, but still a prediction of 60,000 tonnes can be associated with observed numbers of 40,000 tonnes or 100,000 tonnes (One can see how much variation is present in nature from age structure patterns

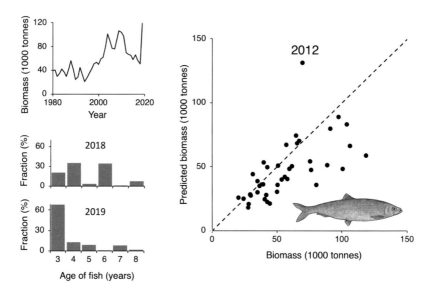

Figure 22.1 *Top left:* Estimated biomass of herring present at Sitka Sound, Alaska, 1980 to 2019, prior to harvest. *Lower left:* Age class distributions of mature spawning herring (based on numbers, not biomass) in two different years. *Right:* Predictions of biomass plotted against observed biomass. Predictions are based on an updated (2020) modeling approach. The line is that of perfect correspondence. (Alaska Department of Fish and Game, www.adfg.alaska.gov/index.cfm?adfg= commercialbyareasoutheast.herring#harvest, and S. Dressel and K. Hebert, *personal communications*)

across years, Figure 22.1). One year, 2012, was an outlier, with the prediction much higher than estimated pre-fishing biomass. This occurred when spawning biomass had been increasing steadily for at least the 14 previous years: the prediction was for a continued increase, but the population decreased.

Harvest levels for Sitka Sound herring are set according to the following rule: if predicted biomass falls below 22,700 tonnes, then no fishing is allowed, but once this threshold has been reached, allowed harvest increases from 12% of the predicted biomass to a maximum of 20% at a biomass of 41,000 tonnes. The maximum 20% harvest rate in Sitka has been allowed since 1983, the time when this sliding scale was instituted. Note that the occasional overprediction (as happened in 2012) is counterbalanced by the 14 prior years of under forecasts. As in cases we have considered throughout this volume (e.g., average temperatures in the future, and temperature on a particular day in the future), the average is more easily predicted than single events. Modelers at the Alaska Department of Fish and Game have shown that it is the average that matters for long-term maintenance of stocks, and that occasional overpredictions do not create high risk. Finally, because fishermen sometimes do not fill the quota, the long-term average exploitation rate is about 15%. For example, in 2019 and 2020 there was no harvest because fish were too small to be commercially viable (see Fig. 22.1 for 2019 age classes among spawning fish).

Enforcement

Besides the difficulty of predicting population sizes, quotas involve the difficulty of regulation. In Alaska, ships' landings are monitored. Elsewhere, when national governments assume jurisdiction over local areas, a "tragedy of tragedy of the commons" sometimes takes effect, whereby incentives for local communities to manage their own lands is reduced, and instead they harvest more than they otherwise would.

Many political issues affect the setting of fishing quotas, especially given competing interests. Consider the Eastern Atlantic bluefin tuna, which is fished by people from multiple countries both in the Mediterranean, where it breeds, and the east Atlantic, to which it migrates annually. In 2006, scientists advised that the annual quota should be 15,000 tonnes. The International Commission for the Conservation of Atlantic Tunas (ICCAT) set a quota at 22,000 tonnes, but even more fish were harvested, both as part of independent national policies and illegally (illegal harvest was estimated at more than 60,000 tonnes in 2008). As a result of efforts from concerned parties—including fishermen, merchants, governments, and conservationists—ICCAT adopted a recovery plan, setting quotas and rules on season, fish size, and bycatch allowed, and increased investment in monitoring

232 ECOLOGY OF A CHANGED WORLD

and reporting. This reduced the estimated catch to less than 10,000 tonnes in 2010, followed by a gradual increase. In 2018 ICCAT switched to emphasize management of stocks. The quota was set at 32,240 tonnes in 2019, and 36,000 tonnes for each year 2020–2022, considered by some scientists to be too much too soon, especially as illegal fishing is continuing, and its extent unclear.

In summary, legislation presents issues associated with the setting of quotas and enforcement, often resulting from competition between different interests. A possible resolution comes from the establishment of reserves, as described in the next section.

22.4 Reserves

Harvesting from a reserve may be much easier to limit or halt altogether than the imposition of quotas. Reserves can maintain populations and also replenish stocks outside of them. An excellent example comes from the island of Saint Lucia in the West Indies. In the early 1990s, the tourist industry was beginning to suffer because of a lack of fish on the reefs. In 1995, the government established a series of fully protected reserves, covering 35% of the 11 km coastline around the town of Soufrière (Fig. 22.2). Fishermen objected. The government stepped in with some financial help for the first year of the restrictions, when catches declined. After a few years, catches increased beyond those before the reserves were established. Fishermen set their nets along the edge of the reserve: as fish become crowded in the reserve, they swim out. Unfortunately, in this age of multiple threats, two species of predatory lion fish, accidentally introduced from the Indian Ocean and tropical west Pacific as a result of the aquarium trade have become established throughout the Caribbean and were first reported from St. Lucia in 2012. They have reached densities of more than 2,000 per ha elsewhere, and spear fishing in the reserves is now allowed to control their numbers.

In Florida in 1962, the government established two no-fishing zones (totaling 40 km^2) around the Apollo launch site. Virtually all world record sport fish from Florida and the Gulf coast are within 100 km of this no-fish zone. For example, an evaluation in 2001 found that thirty-one of thirty-five different world records associated with the sport fish "red drum" come from this area. You may wonder how a single species of fish can have multiple world records and why these records weren't set when fish were plentiful and large. First, the records are based on both the sex of the angler and the strength of the line, which comes in multiple classes. Second, many new classes were introduced in 1980.

In 2014, between 2% and 3% of the whole ocean was under some sort of protection, but fishing was completely prohibited in less than 0.1% of the ocean. A survey of eighty-seven marine-protected areas found that strong benefits to

HARVESTING: PROSPECTS   233

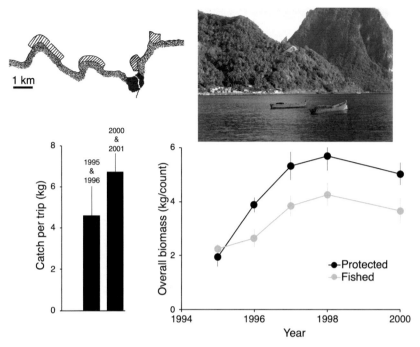

Figure 22.2 *Upper Left:* Location of marine reserves (shaded) established near Soufrière, St. Lucia, in 1995. *Bottom Left:* Catch per fishing trip. The number of trips was similar both before and after the establishment of the reserve. *Bottom Right:* Increase in fish biomass inside and outside the reserves, based on visual censuses. Error bars are standard errors. (C. M. Roberts, J. A. Bohnsack, F. Gell, et al. [2001]. Effects of marine reserves on adjacent fisheries. *Science* 294:1920–1923; Photo provided by Erik Gauger)

nature are present if (1) fishing is banned; (2) regulations are well enforced; (3) the reserve is large (>100 km²); (4) it has been long established (allowing recovery); and (5) it is isolated (isolation appears to aid in enforcement, making fishing boats less likely to visit illegally). All these conditions were met for Kingman Island in the Line Islands (Fig. 21.4), which has some of the highest fish biomass recorded from anywhere in the world. Findings such as these, plus depletion of stocks, have led to a general increase in marine protected areas. In November 2021, Panama, Ecuador, Colombia and Costa Rica announced the creation of the Eastern Tropical Pacific Marine Corridor that unites and expands their current protected areas up to a total of 500,000 km², some of which includes complete no-take zones. Protection from international fleets, most notably China's huge number of ships, is likely to continue to be an issue. Conservation efforts can be reversed as well: in November 2021, the president of the island state of Kiribati

234 ECOLOGY OF A CHANGED WORLD

in the Pacific announced his intention to relax fishing restrictions in its marine
protected area.

Land

Contributions to conservation from incentives and legislation on land are similar to those considered for the ocean, but issues associated with reserves are somewhat different. Government-sponsored terrestrial reserves are rarely designed with sustainable harvesting in mind, and many are surrounded by people, who regularly encroach. An assessment of sixty tropical reserves indicated large losses of biodiversity from about half of the reserves and found that these losses are especially likely to occur if the surrounding area is being developed. In the tropics at least, local support, coupled with strong enforcement against poaching, is essential for protection. In India, the more developed regions contain well-managed reserves, associated with a rapid growth in tourism as the economy grows. On the other hand, tourists rarely visit protected areas far from the main cities, and hunting is common. General issues with reserves and reserve design are considered further in Chapter 27.

22.5 Technology

Technological advances have been one of the drivers of overexploitation on both sea and land. Examples we have considered include locating fish by radar and using guns. However, as noted in Chapter 12, technology can also be applied to reduce environmental impacts, which is likely to be most successful when it is economically advantageous.

One of the more exciting recent developments in conservation biology has been the rapid rate at which new technologies are being devised and used. For example, a study of twenty-three shark species tracked 1,681 individuals by satellite and also tracked fishing boats that were fitted with satellite gear for safety reasons. The study showed that boats and sharks travel together (both follow fish they want to catch and eat, respectively, over thousands of kilometers). Consequently, the sharks themselves have little respite from fishing. Advances in molecular biology have also contributed. In the late 1990s, researchers used DNA sequences to show that meat from protected whales were appearing in Japan, marketed as whale meat derived from species with quotas. More recently, DNA analyses have shown that fish in restaurants are often not the species advertised; a survey of eighteen sushi restaurants along the east coast of the USA found that none of them was serving the advertised red snapper. Advancing technology is producing the first handheld DNA assessments that can be used to quickly identify species, for example, by customs agents.

Prizes and competitions have stimulated innovative ideas that should lead to more sustainable fisheries. New ways of reducing the use of wild-caught fish for fish feed in aquaculture (Chapter 9) have emerged from the Australian government's initiative, the Blue Economy Challenge. Lobsters are harvested along the East Coast of the United States, with lines from the lobster pot on the ocean floor connected to a buoy on the surface, enabling the pot to be located and hauled up. Whales get entangled in these lines, and the north Atlantic right whale (a close relative of the southern right whale whose numbers were documented in Fig. 21.3) is suffering under this new threat. About 400 individuals are left in the north Atlantic, and the death rate presently exceeds the birth rate. A novel idea has been to attach coiled lines on the pot to an inflatable device which can be activated from a smart phone app. That idea is now being extended to lineless inflatable pots. Such creative approaches are in the development stage, sometimes requiring raising funds through crowdsourcing.

Finally, citizen science is increasingly contributing to conservation efforts. The monitoring of populations requires a large effort, which can be achieved when volunteers are involved, as in the North American Breeding Bird Survey (Chapter 24). Photographs and video now monitor charismatic species. For example, tourists love to watch whale sharks, the filter-feeding species that grows to the length of a typical city bus and is by far the largest fish. Each shark has a distinctive spot pattern, from which individuals can be identified from photographs. Photographs by tourists have provided important information on movements between different countries, with different levels of protection (one shark moved 2700 km between Australia and Indonesia), age, population size, and site fidelity. At first, people had to upload a photograph of a part of the body to a specified website. Now, the ongoing computing revolution means that individuals can be identified from videos taken by the public (e.g., through YouTube).

22.6 Conclusions

As a source of protein, the harvesting of wild animals from both the ocean and land provides important nutrition to millions of people, especially in the tropics. As populations and development have expanded, so harvesting pressures have increased. In tropical forests in South America, circles of depleted areas have arisen around new settlements, with one major driver of such settlements being increased mining activities. Ultimately, we hope advances in farming and aquaculture, plus increased costs of harvesting from the wild, lead to a better balance between the forces of destruction and conservation.

236 ECOLOGY OF A CHANGED WORLD

This chapter completes the middle section of the book, which has considered threats to nature. It appears that the preservation of species through the next 50 years or so will depend heavily on protected areas, as well as on those protected areas retaining their species. In the final section of the book, we focus on the conservation of species.

PART 3
AVERTING EXTINCTIONS

23

Species

We have used the words "biodiversity" and "species" widely throughout this book. Biodiversity encompasses the variety of life. It includes the genetic, physiological, behavioral, and morphological differences among species and individuals. Biodiversity also includes functional diversity—that is, variation in what individuals and species do (e.g., fix nitrogen or eat certain prey). Species are the fundamental unit of biodiversity because once a species is extinct, all of its genetic diversity and distinct functions are lost. Loss of a population is considered less critical than loss of a species because populations of the same species are similar in both their genes and in their function in the ecosystem. Hence, the general focus is on species and their conservation. This chapter:

(1) Defines what a species is. This is essential if species are to be the focus of conservation efforts.
(2) Describes how species are classified.
(3) Asks how many species there are in the world.

The following chapters consider issues of how to conserve species.

23.1 What Is a Species?

It is not always straightforward to define a species. People identify an animal or plant as belonging to a named species because it shares certain characteristics with other members of the same species (e.g., size, shape, color, songs), and it differs from individuals of similar species. Such characteristics are called diagnostic traits, and those traits are used, often in combination, to assign an individual to its species. For example, we identify a robin based on its size, color, and song. If we study a group of species that all occur in the same place, it is often relatively easy to classify individuals to species. Experienced birders have little trouble figuring out the species of birds in their garden. This way of defining species is termed the morphological species concept.

This definition often works well, but the formation of new species is a continuous process, as two populations evolutionarily diverge from one another (Fig. 23.1). It can be difficult to decide when they should be considered different

Ecology of a Changed World. Trevor Price, Oxford University Press. © Oxford University Press 2022.
DOI: 10.1093/oso/9780197564172.003.0023

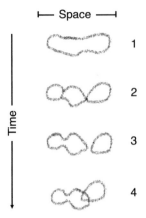

Figure 23.1 A basic model of the formation of two species from one. The outline is of a species geographical range. Once a geographical barrier forms somewhere in the range (e.g., a mountain rises), the two separated populations diverge along different trajectories. If they diverge sufficiently that they do not interbreed when they come back into contact, they are biological species. Divergence may also be initiated if a new population becomes isolated after crossing a barrier (e.g., some individuals fly to an island). (Redrawn from E. Mayr. 1942. *Systematics and the Origin of Species from the Viewpoint of a Zoologist*. New York: Columbia University Press)

enough to be called species. Thus, a second useful definition of a species is "a group of interbreeding populations reproductively isolated from other such groups." When we say that two species are reproductively isolated, we mean that an individual of one species will not mate with an individual of another in the wild, or if a mating does occur, the offspring are either not produced or are infertile (e.g., a mule is the infertile product of a horse and a donkey). This way of defining species is termed the biological species concept. It reflects the fact that if individuals from one form were to mate with individuals from another and produce offspring that could themselves reproduce, the two different forms would collapse into one. Thus, the establishment of reproductive isolation is the crucial step in enabling two species to coexist in the same place.

Most animal species, and probably a large fraction of plant species, form when populations are separated by a geographical barrier of some sort and subsequently evolve differences (Fig. 23.1). When the differences are significant enough that the different populations can be distinguished (i.e., all individuals in one population share some diagnostic characteristics that are not exhibited by individuals in the other population), they are termed subspecies (Fig. 23.2). When the differences are even larger, they are considered different biological species because it is thought that the two forms would not interbreed if they came

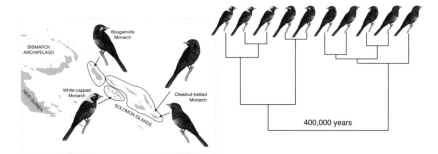

Figure 23.2 (Color plate 15) Monarch flycatchers and species concepts. The right panel shows relationships between 10 individual birds based on differences in DNA sequence: individuals separated by shorter branches are more similar in sequence than those that are separated by longer branches.

(1) *Biological species concept.* The black individuals (upper right on map) from Ugi Island also occur on the nearby island, where they breed with the chestnut form of the chestnut-bellied monarch; hence, they are definitively the same species. The white-capped monarch is geographically separated from the chestnut-bellied monarch, and individuals never come into contact. In this case, experiments have shown that individuals from each respond somewhat to prepared mounts of the other species, as well as its songs, suggesting they may well interbreed if they were to come into contact. Further, genetic differences among all the forms are small, with the last common ancestor estimated (from DNA differences) to be about 400,000 years. This implies that if they were to interbreed, their offspring would likely be fertile. Hence, under the biological species concept, all the illustrated forms are best considered subspecies of one species.

(2) *Phylogenetic species concept.* Based on the DNA sequences, some black and chestnut-bellied individuals are more similar than are two individuals of the same color, implying that these two color forms are one phylogenetic species. On the other hand, 12 individuals (3 are shown) of the white-capped monarch are very similar to each other in their DNA, as are 2 of the Bougainville monarch, and there is no intermixing, implying that they are different phylogenetic species (see Fig. 23.3).

(3) *Morphological species concept.* Distinctive differences suggest the presence of three species (the black and chestnut forms are still considered variants of the same species because they are known to interbreed).

(4) *Resolution.* The IUCN red list now places the white-capped monarch as one species and the Bougainville monarch as a subspecies of the chestnut-bellied monarch. A few years previously, all three forms were classified as different species. (J. A. C. Uy, R. G. Moyle and C. E. Filardi. [2009]. Plumage and song differences mediate species recognition between incipient flycatcher species of the Solomon Islands. *Evolution* 63:153–164). Picture drawn by Emiko Paul.

into contact. Such criteria can be difficult to apply, and the distinction between subspecies and species is often unclear. Therefore, in asking critical questions about the global number of species and strategies for focusing conservation efforts, we are stuck with the dispute about where the boundary between subspecies and species should be drawn (Fig. 23.2).

The advent of DNA sequencing has introduced yet a third definition of species: the phylogenetic species concept (see the Notes and References section for a brief history). Populations are classified as different species if all the individuals in one population are more similar in DNA sequence than they are to individuals from another population (Figs. 23.2, 23.3). This concept is very useful, although it remains difficult to know if such forms would interbreed and merge if they occurred in the same place (i.e., are phylogenetic species also biological species?). One guideline we might use to decide if phylogenetic species are indeed biological species is the length of time the populations have been separated from each other. If differences in DNA sequence are large, we infer that the time is long, and it is unlikely the two populations would successfully interbreed. For birds, subspecies that are separated by about 2 million years according to their DNA sequence differences are considered biological species by some authors.

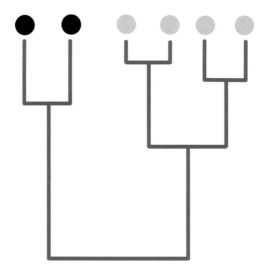

Figure 23.3 The circles indicate six different individuals, four drawn from one population and two drawn from another. The vertical lines measure the distance between individuals based on a sequence of DNA taken from each individual; individuals closer together are more similar in their DNA sequences. When all individuals from one population cluster together (that is, they are more like each other than they are to individuals from another population), the two populations are considered different phylogenetic species.

This approach remains controversial. From the point of view of conservation, if we are concerned about maximizing genetic diversity remaining on earth, using a metric based on DNA difference makes sense, whether or not we wish to consider them species.

Currently, about 11,000 bird species are recognized. However, an application of the morphological species concept, which is often based on color, suggests about two times as many species could be recognized because many subspecies are quite different from each other. Furthermore, studies of the genetics of 437 recognized bird species implied that, on average, they contain just over two phylogenetic species. Therefore, the number of bird species by either of these measures may be 20,000 or more. Species recognized by color or DNA sequence clusters are unlikely to be separable in the fossil record, and such methods have yet to be applied widely. We return to the difficulties of species identification later in this chapter, in consideration of how to estimate the number of species present in the world.

23.2 Classification and Phylogeny

All species on earth are derived from a single ancestral species. It should be possible to describe their relationships and construct a complete tree of life, starting from the present day and connecting back through ancestors to the base of the tree (check out the Tree of Life project at http://tolweb.org/tree). For example, Figure 23.4 shows phylogenetic relationships between humans and our closest relatives based on DNA sequence similarities. Recent estimates, based on both DNA differences and fossils, indicate that we last shared a common ancestor with the orangutans about 16 million years ago, with the gorillas about 7 million years ago, and with the chimpanzees about 6 million years ago (Chapter 11).

Each species is given a two-part name, as we have been using throughout this book. We are members of the species *Homo sapiens*. The first word is the genus name and includes a group of closely related, similar species. The chimpanzee is *Pan troglodytes*, and its close relative, the bonobo, is *Pan paniscus*. Genera are grouped together into families. For example, humans, chimpanzees, bonobos, and gorillas are all in the family Hominidae (the great apes), with relationships depicted in Figure 23.4. This family belongs to the order Primates, along with monkeys and lemurs. The Primates belong to class Mammalia, along with whales, tigers, mice, etc. The Mammalia belong to the phylum Chordata along with fish, snakes, sea urchins, etc. The Chordata belong to the kingdom Animalia (also known as the Metazoans), along with worms, insects, sponges, etc. Beyond the animals lies more diversity. The Animalia are only a small part of one group, the Opisthokonts, which also includes fungi. Beyond the Opisthokonts lies yet more diversity, including the plants (you are more closely related to a mushroom than

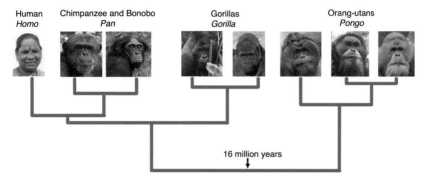

Figure 23.4 Relationships among eight species of the great apes (Hominidae), classified into four genera. The length of the branch approximates the relative time since separation, with that between human (genus *Homo*) and chimpanzees (genus *Pan*) set at 6 million years. Among the different species of orangutan, one species was discovered in 2016. It is a small population of only 800 individuals on a few mountain tops in Sumatra, but it is the one in the figure that is most divergent from the other two more widespread species (A. Scally, J. Y. Dutheil, L.W. Hillier, et al. [2012]. Insights into hominid evolution from the gorilla genome sequence. *Nature* 483:169–175. The new orangutan species is added according to A. Nater, M. P Mattle-Greminger, A. Nurcahyo, et al. [2017] Morphometric, behavioral, and genomic evidence for a new orangutan species. *Current Biology* 27:3487–3498). Photos from left: a friend of the author, Padma (Nitin Jamdar), chimpanzee (Clément Bardot, with permission), bonobo (Kabir Bakie, CC 2.5), eastern gorilla (Wikipedia, CC 2.0), western gorilla (Clément Bardot, with permission), Tapanuli orangutan (Tim Laman, CC 4.0), Borneo orangutan (Eric Kilby, CC 2.0), Sumatran orangutan (Wikipedia, CC 3.0)

a tree). Animals, plants, fungi, algae, and many kinds of single-celled organisms belong to one domain of life, the eukaryotes, defined as species whose cells contain a distinct nucleus. The Archaea and bacteria, whose relationships to each other remain uncertain, form the two other domains. The Archaea occur in the saltiest, hottest, coldest, and most acidic environments but have also been found in less extreme places. The optimal temperature for one species of Archaea that lives near deep-sea vents is 105°C. At 90°C it is too cold for it to reproduce.

23.3 Number of Species

A major gap in our knowledge is that we do not know how many species are present on earth. The main reason lies in the enormous effort required to find and describe new species. So far, slightly over 1.9 million animals and more than 450,000 plant species have been described. Among better-known groups, 64,000 tree species have been described, and based on the rate at which rare species are being discovered with increased sampling, another 9,000 (14%) may still be undocumented. Less conspicuous groups surely contain higher proportions of undescribed species. Starting with the 1.2 million species that had been collated into a database, a research group came up with an elegant way to estimate total species numbers. They computed the ratio of species within families and of families within orders. If one takes a group, such as the birds, in which virtually all species have been described, the ratio represents a complete enumeration. For example, in the most recent classifications of birds, a family contains on average forty-five species. In less well-studied groups (e.g., beetles), the number of families is much better known than the number of species. Assuming the number of species in a family is the same in birds and these groups, one could simply multiply the number of families by 45 to work out how many species are present in these groups. One could also estimate the number of families still to be encountered, based on the rate at which new ones are being described. The analysis is a little more complex than this, and the authors used phylum, class, order, and genus ratios, as well as family, in making their assessments. They estimate that about, 6.5 million eukaryotes currently exist on land and 2.2 million in the oceans. These results imply that about three-quarters of all eukaryotes remain to be described. Even well-known groups are turning up new species. For example, in 1992 a mammal nearly the size of a cow, the saola, was discovered in Vietnam, and a raccoon-like mammal from the Andes, the olinguito (illustrated in Fig. 26.5), was only recognized as a species separate from its closest relative in 2013. More generally, experts guess that 50% of all reptiles may remain to be discovered, but they emphasize that this is just a guess.

Possibly, these estimates of species numbers may turn out to be low. First, whole communities have scarcely been studied. They include parasites of many species and communities in deep-sea hydrothermal vents, which were only discovered in 1977. Second, it is likely that cryptic species, which appear so similar to each other that they have previously been classified as one species, are more common than we once thought. For example, five species of birds present in the same museum drawer were considered the same species for 100 years (two are shown in Fig. 23.5). Then, in the early 1990s, two Swedish ornithologists noticed some differences in songs between birds breeding in the same location. They showed that birds singing one type of song did not respond to the playback of a bird singing a different type, suggesting that they may not mate with other and are thus reproductively isolated biological species. Finally, they determined large differences in the DNA sequence of these birds, implying that the species last shared a common ancestor as long as 6 million years ago (i.e., the same length of

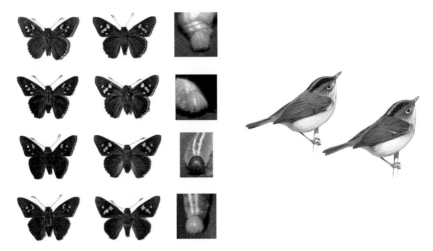

Figure 23.5 (Color plate 16) Cryptic species. *Left:* Four species of skipper butterfly, each represented by the male, female, and caterpillar, not known to be different species until DNA sequencing. Subsequent examination showed the caterpillars to differ visually. *Right:* Two species of flycatcher warbler classified as the same species until 1999. Song and DNA sequence data, plus their presence in the same place with no evidence of interbreeding, unequivocally shows that they are different biological species. (J. M. Burns, D. H. Janzen, M. Hajibabaei, et al. 2008. DNA barcodes and cryptic species of skipper butterflies in the genus *Perichares* in Area de Conservación Guanacaste, Costa Rica. *Proceedings of the National Academy of Sciences* 105:6350–6355, reprinted with permission, Copyright (2008) National Academy of Sciences, U.S.A.; P. Alström and U. Olsson [1999]. The golden-spectacled warbler: a complex of sibling species, including a previously undescribed species. *Ibis* 141:545–568, painted by Ian Lewington)

time separating humans from chimpanzees). The number of recognized species in birds has increased by at least 1,000 in the last 20 years from similar discoveries. Most of these species occur in different locations, so it is not known if they would interbreed if they encountered each other, but they are still considered valid biological species because they are separated by millions of years.

DNA sequence data has been used to test whether a somewhat variable species of skipper butterfly in Central America might in fact consist of cryptic species. Sequences of DNA from 255 individuals, mostly in the Guanacaste forest reserve of Costa Rica, cluster into four phylogenetic species. Although the adults of these four species are virtually indistinguishable, detailed scrutiny of the caterpillars revealed differences in the larval color pattern (Fig. 23.5). The four species were then found to utilize different host plants, with two species each specializing on a different species of palm and the other two feeding as generalists on grasses (i.e., they have clearly different ecological functions). Given that these four species co-occur, we can infer that they are biological species as well because otherwise the DNA sequences would not cluster in the same way as the morphological features.

At the Guanacaste forest reserve, about 1% of all butterfly caterpillars are infected with a wasp. The wasp lays a single egg in the caterpillar, which hatches and then eats the caterpillar from the inside out. Some wasp species were long thought to be generalists (i.e., to infect many species of butterflies). After sequencing the DNA of many individuals, it appears that each wasp only parasitizes one butterfly species. Hence instead of one generalist species, we have multiple specialist species of wasps. DNA sequence differences suggest that these species last shared a common ancestor hundreds of thousands to millions of years ago. On the basis of results from studies like these, many research programs have now instigated what is called DNA barcoding, whereby individuals are classified to species based on similarity of gene sequence. Such classification strategies should provide better estimates of species numbers and thus enhance efforts to conserve biodiversity across the globe.

23.4 Conclusions

The main unit of conservation is the species. In the United States, the powerful Endangered Species Act of 1973 was originally proposed to apply to species only. Once a species is officially classified as endangered, the act prohibits its "killing, harming, harassing, pursuing, or removing from the wild"; a 1978 amendment extended this provision to destruction of the species' habitat. Recognizing the importance of maintaining biodiversity more widely, the act was also amended to allow the listing of distinct population segments, for vertebrates only. These segments are generally identified using DNA sequences (i.e., in our definition, they are therefore phylogenetic species [Fig. 23.3], but they are also sometimes called evolutionary significant units). That approach has been instrumental in maintaining salmon runs in the Pacific northwest, partly because it enables harvesting from populations that are not considered endangered. Overall, conserving segments of species is desirable, but practicalities in most parts of the world mean that the focus of conservation remains on species. In any case, most species remain to be described, implying they are experiencing no targeted conservation efforts.

24

Population Declines

We have already shown that the population sizes of many species are declining, e.g., elephants (Fig. 20.6). In some species, populations are so low that the species is considered functionally extinct—that is, it is not meaningfully affecting other species in the community (e.g., as a predator, prey, or nutrient recycler). Other species have gone completely extinct. In this chapter, we summarize evidence for declines, and in the next, evidence for complete extinction.

Figure 24.1 reproduces examples of declines in population size, chosen because each is attributed to one of the major COPHID threats acting alone: climate change, overharvesting, pollution, habitat conversion, invasive species, and emergent disease. Of these threats, climate change has been the most difficult to isolate as a major sole cause of declines, but it contributes broadly to many cases. Indeed, synergisms between threats are widespread (e.g., see Figs. 19.3 and 19.4). What do we know more generally about how population sizes have changed over time, and what has caused these changes? Here, we evaluate declines in detail, focusing on assessments based on land-use change, causing loss of natural habitat, and monitoring within habitats. Recent changes in wild populations are much more poorly quantified than the rise of humans and their wealth. As introduced in the harvesting chapters and elsewhere, we cover three approaches:

(1) Because most species have not been specifically monitored, one method is to compare species in both natural and human-modified habitats. Here, we ask how known land-use conversion (e.g., from grasslands to croplands) has affected species.

(2) Some locations have been regularly monitored over the past for both what species are present, and in some cases, their densities. One of the largest surveys, designed explicitly to detect population change, is that of breeding birds across North America since 1970.

(3) The International Union for the Conservation of Nature has generated a list of species that are threatened because population sizes are small and/or declining. The list is regularly updated, so one can evaluate changes in individual species.

In this and the next chapter, it is important to bear in mind that data are limited (primarily to plants, mammals, and birds) and geographically biased.

Ecology of a Changed World. Trevor Price, Oxford University Press. © Oxford University Press 2022.
DOI: 10.1093/oso/9780197564172.003.0024

POPULATION DECLINES 249

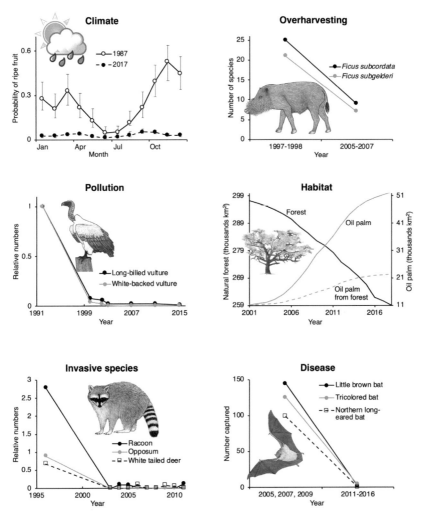

Figure 24.1 Examples of recent declines in populations attributed to a single threat. A. **Climate change** Fruit production in Lopé National Park, Gabon, declined by an estimated 80% between 1987 and 2018, particularly outside the dry season (June–September), attributed to water stress (temperatures have increased by 0.25°C and precipitation decreased at a rate of 0.75 cm every 10 years). Shown is the probability of encountering a tree in fruit in regular surveys, with standard errors. (E. R. Bush, R. C. Whytock, L. Bahaa-el-din, et al. [2020]. Long-term collapse in fruit availability threatens Central African forest megafauna. *Science* 370: 1219–1222). B. **Overharvesting** Total number of bird and mammal species visiting two different species of fig tree in Lambir Hills National Park, Sarawak, Malaysian Borneo. Figs of *Ficus subcordata* are larger than those of *Ficus subgelderi* and attract a different, though overlapping, assemblage of fruit-eaters (R. D. Harrison. [2011]. Emptying the forest: hunting and the extirpation of wildlife from tropical

250 ECOLOGY OF A CHANGED WORLD

Figure 24.1 Continued

nature reserves. *Bioscience* 61:919–924). C. **Pollution** Vulture declines in India resulting from poisoning by the drug diclofenac, based on road transect surveys. Observations of the long-billed vulture were not separated from the slender-billed vulture, which is thought to consist of about 2% of these two species. (V. B. Prakash, T. H. Galligan, S. S. Chakraborty, et al. [2019]. Recent changes in populations of critically endangered *Gyps* vultures in India. *Bird Conservation International* 29:55–70). D. **Habitat conversion and land-use change** Forest loss in Kalimantan (Indonesian Borneo), and rise of oil palm, 2001–2018. Over this time, forest cover declined from 56 to 49%, with most remaining at higher elevations. Oil palm was responsible for 30% of the loss, with fires and conversion to other uses accounting for the rest (see Fig. 13.7). E. **Invasive species** Declines of the three most common mammal species in Everglades National Park, Florida, after the establishment and increase in numbers of the Burmese python from 2000 onward (M. E. Dorcas, J. D. Willson, R. N. Reed, et al. [2012]. Severe mammal declines coincide with proliferation of invasive Burmese pythons in Everglades National Park. *Proceedings of the National Academy of Sciences* 109: 2418–2422). F. **Disease** Numbers of bats caught during standard trapping at a mine entrance in New River Gorge National River, West Virginia, USA, before and after the appearance of white nose disease (Fig. 19.2) (National Park Service; https://www.nps.gov/neri/index.htm). Note that four of these examples (A, B, E, and F) are based on observations in officially protected areas. The other two examples span larger regions, which include parks. For example, in protected areas of Kalimantan (which form about 9% of Kalimantan's total area), forest cover declined from 77% to 73% between 2001 and 2018 (see Fig. 13.7). However, a major stronghold of vultures is in India's protected areas. (Illustrations of affected species in B-F: bearded pig, white-backed vulture, rosewood, racoon, little brown bat; A from Wikipedia (CC3.0)).

For example, although the Living Planet Index compiles more than 1,000 assessments of changes in the number of species, they largely come from monitored sites in North America and Europe and are often in protected areas (20% of seventy-seven marine studies come from protected areas). The comparisons over time from Borneo, Peru, and India detailed in Figures 24.1a–d represent some of the few tropical studies. Many of these studies are based on a census at just two time points. For example, among Indian birds and mammals, only vultures (Fig. 24.1c), the tiger, and a few other large mammals have published data sets extending over more than 5 years. Vultures were specifically assessed because in the early 2000s it was obvious that they had steeply declined, and a roadside survey that enumerated their numbers had fortuitously been conducted before the decline. Tigers are monitored because India's conservation efforts are centered on creating protected areas that contain this species. In this case, censuses suggest that the tiger population has increased over the past 10 years from about

POPULATION DECLINES   251

2,000 to more than 2,800 individuals, albeit still a small fraction of historical levels. In summary, the general status of species in India and many other tropical locations is poorly quantified.

A case in point is the status of insects. Over the past 30 years, many people feel that insect numbers in North America and Europe have decreased (partly based on recollections of large numbers hitting car windshields in the past). Two high-profile European studies indicated declines in the insect abundance of (1) more than 75% over the past 25 years and (2) about 75% over 9 years, associated with a loss of one-third of species. However, both showed a large decline in the first year, possibly because of the destructive sampling method, and in the second shorter study it is hard to rule out that the decline was not by chance. Assessments of multiple studies (which include the aforementioned European studies), find substantially weaker declines on average. Tempered by this lack of clarity, we do have strong evidence of declines in well-monitored species (notably bees and butter-flies) from the USA and Europe, such as (1) the disappearance of bumblebee species from almost 50% of monitored sites in North America and 20% of sites in Europe over the past 50+ years (a detail that is of interest because both cli-mate change in the form of reduced precipitation as well as land-use change are implicated); (2) a decline of about 50% in the number of individual butterflies observed in the western United States compared with 40 years ago, concentrated at sites that have experienced exceptional warming in the fall; (3) the decline of the monarch butterfly in North America this century, attributed to multiple causes, including loss of breeding habitat, pesticides, and migration difficulties; and (4) a complete failure to observe about 10% of all insect species previously monitored at sites in Europe (Chapter 25). Costa Rican biologists have good ev-idence from light traps of a large decline in both moth numbers and moth spe-cies, which they attribute to climate change, through the direct effects of abiotic factors, including a 50% longer dry season and increase in the number of "hot" days, as well as the consequently altered species interactions.

24.1 Land Use

Seventy-five percent of all plant growth may have historically been on land that was used for crops, pasture, forestry, and cities in 2011 (Chapter 13). That was a 10% increase over the 16 years from 1995, based on remote sensing of cropland and night lights. Hence, an increasingly large fraction of land is human-appro-priated. However, native species are present in these landscapes. For example, many bird species live in forest plantations. Therefore the first question that we need to address is the extent to which this habitat conversion affects species.

We start with a consideration of tropical forests as the holders of the largest number of species. A compilation of more than one hundred studies covering all continents and many different groups (plants, insects, birds, mammals, etc.) showed that agricultural land and pasture have less than one-quarter the number of species compared to adjacent tropical forest (Fig. 24.2). Even selective removal of trees reduces species numbers by one-third. Putting values such as these into a global analysis, an assessment of land use over the past 400 years concluded that the number of species in an average location had declined from historical levels by about 15%. Most of this decline took place in the past 100 years and is largely associated with increase in pastureland, cropland, and the associated loss of primary vegetation (Fig. 24.2, right). The number of individuals has also gone down but not as much; around 10%. Developed land may contain few species, but some are common.

A feature of these analyses is that species in altered lands are not simply a subset of the species in the forest. Species associated with humans, such as pests and weeds, do well in agricultural landscapes, and these species are of little conservation concern. On the other hand, a large fraction of species in forests may be unable to persist outside of the forest, or even in selectively logged forest. To evaluate the extent to which many species are dependent on natural habitats, a group of researchers surveyed birds and trees across agricultural and forest landscapes in tropical Ghana and subtropical India, comparing three different locations.

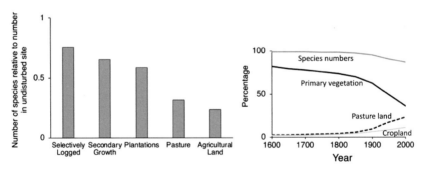

Figure 24.2 *Left:* Proportion of species associated with different uses of land, when compared with adjacent tropical forest. Note that the baseline forest that provides the point of comparison is usually not pristine, but it is always less disturbed than the comparison habitat. *Right:* Estimates of change in land use (dark lines as a proportion: categories not shown are secondary vegetation, urban and forest plantation) plus reduction in species numbers present locally, averaged across the globe for the past 400 years. (L. Gibson, T. M. Lee, L. P. Koh, et al. [2011]. Primary forests are irreplaceable for sustaining tropical biodiversity. *Nature* 478: 378–381; T. Newbold, L. N. Hudson, S. L. L. Hill, et al. [2015]. Global effects of land use on local terrestrial biodiversity. *Nature* 520: 45–50)

First, they surveyed forest, assumed to be the appropriate baseline for the region. Second, they surveyed high-intensity agriculture, perhaps run by agribusinesses, where most tree cover had been removed. Third, they surveyed locations with low-intensity agriculture. These locations are often the smallholdings of villagers, where scattered groves of trees are retained, and where about half the agricultural productivity of the high-intensity sites may be found (Fig. 24.3). In the Indian study of 175 bird species, 76 species were strongly dependent on forests and were rarely seen outside them. By this accounting, almost half of all species in this study depend on little-disturbed forest. Results were similar in Ghana, and for trees. To summarize, a measure of the difference in the number of species present in a location (e.g., forest and field) masks important patterns in turnover as many species are confined to little-disturbed habitats.

We can take our discussion further to assess the value of alternative patterns of land use. Suppose a business enterprise contemplates farming on 100 ha of land but is prepared to sacrifice half the possible income toward the conservation of nature. Land can be converted to obtain equivalent agricultural yields in two ways, either by (1) making 50 ha into high-intensity farming and 50 ha into a forest preserve (termed *land sparing*) or (2) converting the entire area

Figure 24.3 In the plains surrounding the Ganges in Uttarakhand, India, low-intensity subsistence agriculture over the entire region (sharing, *Left*) can result in the same agricultural output as high-intensity agriculture in half the region (*Upper Right*, wheat and poplar trees cultivated for wood), and half is untouched (*Lower Right* (sparing)). With sparing more biodiversity is maintained. (B. Phalan, M. Onial, A. Balmford, et al. [2011]. Reconciling food production and biodiversity conservation: land sharing and land sparing compared. Science 33: 1289–1291; photos by Malvika Onial)

254 ECOLOGY OF A CHANGED WORLD

into low-intensity farming (termed *land sharing*, Fig. 24.3). Assessments of the benefits to nature clearly favor sparing. A few forest species benefit from land sharing because they would then occur throughout the entire 100 ha, whereas they would not be present in high-intensity agricultural areas. However, the large number of forest species not found outside of primary forest are completely lost under the land-sharing option. All these analyses imply that natural areas, in as undisturbed a state as possible, are exceptionally important for the preservation of species, and the reduction of these areas is inexorably associated with the loss of species.

24.2 North American Breeding Bird Survey

Analyses based on land-use change do not directly address changes within habitats. For example, hunting is causing declines in many forest species and is thought to be "overwhelmingly the greatest conservation threat to Afrotropical forest wildlife," but this is not clearly accounted for in the studies considered so far. Assessments of species declines require that we evaluate both the direct effects that people have on biodiversity through habitat modification, as well as effects within habitats. As noted in the introduction to this chapter, published surveys of changes over time come largely from North America and Europe. They are often documentation of the number of species, rather than their densities. Focused on plants and marine vertebrates, these studies have found little change overall in species numbers, with both decreases and increases observed over time. Increases appear to be for many reasons, including the presence of introduced species compensating for the loss of native species, northerly range shifts in response to climate change without corresponding retraction of southerly range limits, and disturbances creating a greater variety of habitats within the study locations. Decreases in species numbers are also common, reflecting loss of species from small natural fragments (Chapter 27) and impacts from the multiple threats we have considered.

Surveys that estimate the population size of species over large geographical areas, encompassing both land use and the other threats, require standardized monitoring. This is exemplified by the North American Breeding Bird Survey, which was initiated in 1966. Public participants stop at 50 standardized locations along a road, each 800 m apart, and count all the birds they see or hear in a 400 m radius for 3 minutes. More than 3,000 such routes across North America are surveyed each May or June (e.g., Fig. 24.4). The number of each species observed can be converted into a density (individuals/ha) and then multiplied by the extent of the species habitat across the geographical range covered to estimate population size. This has been done for almost 450 species of North American land

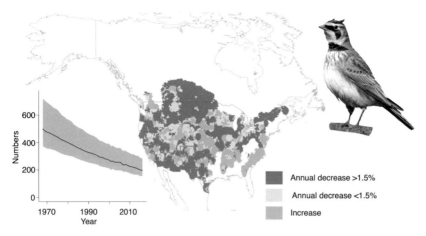

Figure 24.4 (Color plate 17) Counts of horned larks along 108 Breeding Bird Survey routes in the state of Colorado. The black line is the average. The red lines provide a 95% confidence limit for numbers across the state as a whole (i.e., in 95% of the time when we conduct such censuses, the constructed confidence limit should contain the true value) based on differences among the routes, estimates of detectability, and the like. The horned lark shows one of the largest declines of all species, by almost 66%, equivalent to 180 million fewer birds now breeding when compared to 1970. Shades on the map indicate locations showing an annual increase or decrease. (U.S. geological survey, https://www.mbr-pwrc.usgs.gov/bbs. Photo by Brad Imhoff)

birds. Additional survey methods have been used for other species, such as ducks on lakes and ponds, resulting in population size estimates for 529 breeding bird species in North America, which is nearly all of them. The findings are striking (Figs. 24.4, 24.5). Between 1970 and 2010, it appears that the total number of North American birds declined by about 30%, from just over 10 billion to about 7.3 billion (Fig. 24.5). More than 300 species show declines in abundance (e.g., the horned lark, Fig. 24.4), whereas the other 225 show increases (e.g., the red-eyed vireo is estimated to have increased by 40 million birds), but the increases are not nearly enough to compensate for the losses. A complementary data set using a system of weather radar stations across the entire United States found evidence that species are continuing to decline, as the total mass of birds passing overhead during migration decreased by 10% between 2007 and 2017. Grassland species, such as the horned lark, have declined the most.

It is impossible to attribute these declines to one cause, especially as the suffering species are so varied in their habits. A total of about 3.7 billion birds a year are estimated to be killed by the cumulative effect of cats (2.4 billion); buildings, especially windows of buildings (600 million); cars (200 million); and wind

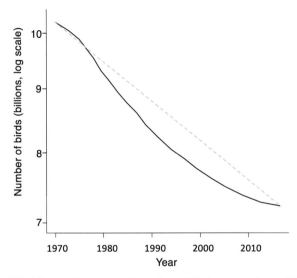

Figure 24.5 Black line: estimated total number of birds summed over 529 species breeding in North America. Between 1970 and 2016 numbers declined from 10.2 billion to 7.3 billion. The dashed gray line is the expected relationship if death and birth rates were constant. (K. V. Rosenberg, A. M. Dokter, P. J. Blancher, et al. [2019]. Decline of the North American avifauna. *Science* 366:120–124)

turbines, cell phone towers, and power lines (62 million). Even if the population size triples after breeding, 3.7 billion comes to 20% of all the North American birds dying each year, so it does not take much to see how additional stresses can result in declines. Conversion of natural land to sterile farmland and cities likely contributes, as does loss of winter habitat.

Disease may also play a part. One identified stressor is the mosquito-transmitted West Nile virus, native to Africa, which was first detected in the United States in 1999 and spread to the West Coast by 2005 (a pattern to be followed by white nose syndrome in bats 7 years later, Fig. 19.2). Because West Nile infects humans and horses, it has been tracked by the Centers for Disease Control. Bird banders work at select locations across the United States, trapping and banding many forest species. The death rate, d, can be estimated from the number of banded birds they recapture the following year. Some species, such as the red-eyed vireo, appear to have been hit hard when the virus first arrived in the region (Fig. 24.6), killing an estimated 30 million birds. Other species, such as the Swainson's thrush have had low survival rates ever since the arrival of West Nile (Fig. 24.6). Swainson's thrush has also shown long-term declines in the Breeding Bird Survey, although any link to the virus has yet to be established.

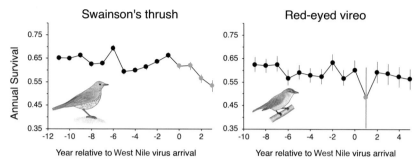

Figure 24.6 West Nile virus was first detected in the eastern United States in 1999, spread to the central plains and some western sites by 2002, and reached all western sites by 2005. These plots show the estimated survival rates of two forest species before and after West Nile reached a location, averaged across all locations. Note the red-eyed vireo appears to have suffered most in the year following the arrival of the virus, whereas the Swainson's thrush has shown continual decline since its arrival. The error bars are 95% confidence intervals and reflect uncertainty in the true survival rate. (T. L. George, R. J. Harrigan, J. A. LaManna, et al. [2015]. Persistent impacts of West Nile virus on North American bird populations. *Proceedings of the National Academy of Sciences* 112:14290–14294)

While these stressors surely contribute to species declines, perhaps the major factor is the introduction of new pesticides—notably a family of chemicals related to nicotine called neonicotinoids, which have strong deleterious effects on insects. Neonicotinoids was first applied extensively in the United States in 2003. Their application increased exponentially from then until 2014, at which point the government agency that collates records (the U.S. Geological Survey) stopped separating them from other pesticides. Neonicotinoids have been implicated in the declines of butterflies and bees. They not only reduce insect numbers as food for birds but affect the birds directly. Experiments show that ingestion of relatively small amounts causes sparrows to lose appetite, lose weight and delay their annual cycle. In some U.S. counties, more than 50% of the decline in grassland birds over the period 2008–2014 has been related to the rise of neonicotinoids, forming a contribution inferred to be at least 10 times greater than changes in land use in these counties (i.e. increased cropland).

A population decline implies that the death rate, d, exceeds the birth rate, b; the population growth $r = b - d$ is negative. If r remains constant over time, then, as we showed in Chapter 2, population size follows the exponential equation:

$$N_t = N_o e^{rt}$$

258 ECOLOGY OF A CHANGED WORLD

When r is less than 0, this equation is termed the negative exponential because it models a decrease in population size down toward zero. As we did for the case of human population growth (Chapter 11), we can test for a constant rate of decline by plotting log of population size against t. With constant birth and death rates, this should be a straight line, with slope r:

$$Log(N_t) = Log(N_o) + rt$$

The dashed line in Figure 24.5 indicates the expectation expected if the decline of North American birds was exponential (i.e., r is constant), which is an average loss of 0.8% per year. The actual pattern suggests that birth and/or death rates have not been constant. Instead, the death rate was higher or the birth rate was lower in the 1970s and 1980s than more recently. One possible explanation is that the fewer birds now present has lowered competition for resources.

24.3 Global Species Assessments

Most species have not been studied as thoroughly as those in the Breeding Bird Survey. However, birds and mammals, as well as most named amphibians and reptiles, have been evaluated for extinction risk. The International Union for the Conservation of Nature is an umbrella organization for about 1,300-member organizations concerned with the preservation of nature. The IUCN sponsors the Red list, which gives a summary of the status of the world's species, based on both unpublished and published data. Each species is classified into one of nine different categories (Figure 24.7). The IUCN has strict criteria for admission into these categories. For example, critically endangered species must either have undergone very rapid decline in recent years or have a very small, fragmented current population size, with high possibility of extinction (Table 24.1). Vulnerable, endangered, and critically endangered are grouped together into "threatened", which often forms the baseline to examine the state of species. For the most thoroughly investigated groups, in 2020, the IUCN considered 13% of all bird species, 22% of all mammal species, and 33% of all amphibian species to be threatened (Fig. 24.7).

The Red list can be used to summarize global trends in populations. First, values are assigned to each species according to their threat status (Fig. 24.7): 1 = least concern, 0.8 = near threatened, 0.6 = vulnerable, 0.4 = endangered, 0.2 = critically endangered, 0 = extinct. For a group of species of interest (e.g., mammals), an index is constructed as

Table 24.1. Simplified Version for Some of the Criteria Required to Classify a Species into One of the Three Classes That Are Together Considered To Indicate Threatened with Extinction*

	Critically endangered	Endangered	Vulnerable
Recent reduction in population size	>90% in 10 years	>70% in 10 years	>50% in 10 years
Small geographic range/occupancy	<100 km^2	<5,000 km^2	<20,000 km^2
Population size	<50 individuals	<250	<1,000
Modeling of extinction probability	50% chance in 10 years	20% in 20 years	10% in 100 years

*Additional criteria to those listed are generally required. For example, small range size alone is not sufficient but must be accompanied by two of the following: severe fragmentation, declining populations, or large population fluctuations in size. The second row lists the criteria based on total range covered by the species, or the area of the range actually occupied (e.g., because of the presence of suitable habitat ["occupancy"]). (*The IUCN Red List of Threatened Species*, iucnredlist.org).

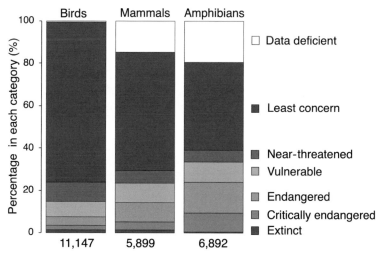

Figure 24.7 (Color plate 18) Fraction of all species in different IUCN categories for the three best known groups, as of September 2020. The total number of species is given at the base. "Extinct" is all species that have gone extinct in the past 500 years but includes two mammal, two amphibian, and five bird species that are extinct in the wild but held in captivity. Data deficient means the species have yet to be assessed. (*IUCN Red List of Threatened Species*. www.iucnredlist.org. downloaded on September 21, 2020)

$$\text{INDEX} = \text{Total score/Number species}$$

If ten species were evaluated and eight were of least concern, one was near threatened and one was critically endangered, the INDEX = $(8 \times 1 + 0.8 + 0.2)/10 = 0.9$. The INDEX has been declining over the years. For example, between 1975 and 2010, for large carnivores, the index went down from 0.83 to 0.81, and for ungulates (deer, horses, etc.) from 0.83 to 0.78. For carnivores and ungulates combined, 114 species went down one or more category, 12 went up, and 370 remained unchanged. The INDEX summarizes species that are most threatened with extinction and presently is largely a summary of vertebrate species, but it corroborates a general picture of widespread declines.

24.4 Conclusions

It is difficult to quantify how much species are declining because we need to monitor within habitats and account for loss of habitat. While some studies have found little change in the number of species in natural habitats, the mere fact that more than 20% of all land vertebrates are considered threatened by the IUCN points to widespread declines. Clearly, some insect species are declining. Even when they don't become extinct, the reduction of formally common species to low levels can have ramifying effects on communities. In this case, the species is considered functionally extinct, but recovery is possible. The extent to which species have been completely lost is as difficult to assess as species declines. We consider the evidence for extinction in the next chapter.

25

Extinction

Extinction, the loss of a species from earth, is a normal part of the history of life. Based on fossils, more than 99% of all species that ever lived are extinct. On some occasions, a large fraction of all species went extinct over a relatively short time period, and the fossil record shows that it took up to 10 million years after those extinctions for diversity to recover. Is a similar scenario being played out now? It is obvious that humans are causing extinctions, but it is important to put this into the context of what has happened in the past, long before we dramatically altered the natural landscape. This chapter therefore considers:

(1) The history of extinction before human influences.
(2) The history of extinctions precipitated by humans up to the present.
(3) An assessment of whether we are in the middle of a sixth mass extinction.

25.1 Extinctions in Earth's History

According to the fossil record of North American mammals, between 65 million years and 1 million years ago (i.e., before known human impacts), on average about 25% of all species present in the fossil record at the beginning of a 700,000-year time interval were not found again after the end of it (Fig. 25.1). In order to translate this into an extinction rate, we make a simplifying assumption that the probability of extinction and of fossilization of a species is constant throughout time, and these probabilities are the same for all species. That means the loss of species from a certain time point follows the negative exponential introduced in Chapter 24, whereby numbers go down by a constant proportion at each time step (Fig. 25.1, right). As discussed in Chapter 2, the death rate, d, is the proportion of individuals that die over a unit time interval. We write:

$$N_t = N_o e^{-dt}$$

Here, N_o is the number of species present at a certain time, and N_t is the number of these species still present at time t after that (that is, it is measuring deaths,

Ecology of a Changed World. Trevor Price, Oxford University Press. © Oxford University Press 2022.
DOI: 10.1093/oso/9780197564172.003.0025

262 ECOLOGY OF A CHANGED WORLD

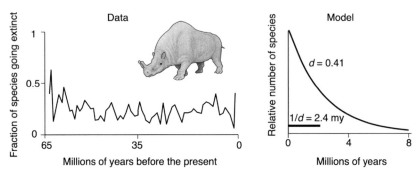

Figure 25.1 *Left:* The proportion of North American mammal fossil species present at the beginning of a 700,000-year time interval that were not recorded in the next one, from 65 million years ago to the present, based on a total of 2,941 species. Illustration is of a species of titanothere, which as a group was present 50–28 million years ago. *Right:* Negative exponential model giving the fraction of all species at the beginning of the time interval present at later times. The line is the average time a species persists (when $d = 0.41$ extinctions/species/my, $1/d = 2.43$ million years). (J. Alroy. [1999]. Putting North America's end-Pleistocene megafaunal extinction in context, pp. 105–143. In R. D. E. MacPhee (ed.), *Extinctions in near time.* Kluwer Academic/Plenum Publishers; M. Foote and D. M. Raup. [1996]. Fossil preservation and the stratigraphic ranges of taxa. *Paleobiology* 22:121–140)

without considering births). One property of the negative exponential is that the average time a species persists is $1/d$ (Fig. 25.1), which is perhaps easier to relate to than the rate itself, d. Substituting the loss of 25% of species into this equation—that is, set $N_t/N_o = 0.75$ and $t = 0.7$ million years—we estimate an extinction rate of $d = 0.41$ extinctions/species/million years, and hence an average duration of a species of 2.5 million years. Even though species persist for longer than estimated from the first and last occurrence of a fossil, it turns out that the extinction rate of fossils that occur at least twice in the fossil record estimates the extinction rate of species, provided the exponential model applies. The reason is that species with particularly short durations often do not leave fossils at all. This cancels the fact that those with longer durations and leave at least two fossils have their duration under-recorded.

The exponential model assumes that all species had the same chances of going extinct and of being fossilized. Extinction rates would be higher if a set of species had both a high chance of never being observed as fossils and a high chance of extinction, which may be expected for species with especially small ranges. In fact, among 4,715 present day mammal species, those with a fossil record have geographical ranges on average five times larger than those without a fossil record. They also have a lower probability of being threatened with extinction (based on the IUCN Red List; see Table 24.1) by more than two times.

We do not know the extent to which the higher threats to species with small geographical ranges is a special feature of the present, reflecting human impacts, or whether it is part of the normal history of life. However, it may be a special feature of human impacts. A study of birds on the Lesser Antilles in the Caribbean used molecular methods to estimate extinction rates by comparing genetic divergence between these birds and related species elsewhere to work out when a particular species arrived in the archipelago. Then the researchers assumed islands where the species was not currently present had lost it through extinction. On islands larger than 1000 km^2, this led to an estimated extinction rate of $d = 0.29$ species/species/million years (duration of 3.4 million years). To be comparable to fossil estimates, we would need to correct for the time it took for a population on the island to evolve morphological differences sufficient to be recognizable as a different species; the island study overestimates species durations and hence underestimates the extinction rate. Consequently, an average species duration of about 2.5 million years may be a reasonable estimate for both North American fossil mammals and Caribbean island birds. Other information could be brought to bear on the question. For example, based on molecular methods, the chimpanzee and bonobo last shared a common ancestor within the past 2 million years, as did two orangutan species (Fig. 23.4). If speciation and extinction are in balance, this is also consistent with an average on the order of 2.5 million years for the duration of species.

What caused extinctions in the past? In specific cases, we usually don't know. However, some large patterns do seem to be linked to displacement by superior competitors associated with changing climates, or because these competitors could enter a new location when a barrier to dispersal disappeared. The songbirds (a suborder of the order Passeriformes, or perching birds) contain about 4,500 species and include many of those most familiar to us (sparrow, crow, starling, finch). But according to the fossil record, songbirds were absent from Europe 25 million years ago. The bird species that were present belonged to different groups, such as the parrots. These other species were replaced by the songbirds, which originated in Australia and diversified in Indonesia, before becoming established in Asia after 34 million years ago, a time when the planet cooled rapidly (Fig. 15.1). They then dispersed and diverged into different species through the rest of the world. One hypothesis for the success of the songbirds is that their relatively large brains gave them an advantage in exploiting new foods, storing food for the winter, or building nests that are better concealed from predators. If intelligence was the songbird's weapon, it was a precursor for the later spread of the supremely intelligent animal, us humans, which is even more dramatically associated with species loss.

264 ECOLOGY OF A CHANGED WORLD

Mass Extinctions

At times in Earth's history, the extinction rate has risen higher than the base-line, generating a mass extinction, which is here defined as the loss of more than 75% of all species over a relatively short time interval, substantially less than a million years. Scientists traditionally recognize five mass extinctions in the last 500 million years. The greatest loss of species happened 252 million years ago and is labeled the "time the earth nearly died." If one adds in extinction events just before this mass extinction, up to 95% of all marine inshore bivalve species were lost. The causes of this extinction are still debated, but the most widely accepted explanation is a massive outpouring of volcanic basalts in the region of present-day Siberia, which led to increased CO_2 and global warming, perhaps by as much as 10°C, associated with ocean acidification. Extinctions on land appear to have happened quickly, over only a few thousand years. This rapid loss of species supports various kill mechanisms, including acid rain from sulfur dioxide emissions, which would have destroyed most vegetation. Volcanism also seems to have caused huge coal fires, resulting in the release of mercury and other poisons.

The most recent mass extinction was still in the distant past, 66 million years ago. Although volcanic eruptions have been invoked as contributing to this extinction, it is now clear that an asteroid drove this event, hitting with the force of several million nuclear bombs. Mammal extinctions across the event have been estimated at 76% (see Fig. 25.1). The dinosaurs went extinct as well.

Extinctions during the Ice Ages

The last few million years have been a time of climate fluctuations. Over the past 2.6 million years, extensive glaciations have been punctuated by relatively short warm interglacial periods, which over the past 400,000 years have occurred at intervals of 70,000 to 110,000 years (Chapter 15, Fig. 15.2, see also Fig. 25.2). It appears that glaciations did not cause many extinctions on land, and they certainly did not cause a mass extinction, in part because many species were pushed towards tropical latitudes without going extinct. In Figure 25.1 note how few mammals present 2 million years ago had gone extinct by 1 million years before the present.

While few land animals may have gone extinct, marine species in shallow-water environments suffered more extinctions, mostly at the beginning of the ice ages. Both cooling and the large drop in sea level associated with tying up water in ice surely contributed to these extinctions (Fig. 25.2). In California, almost two-thirds of all scallop species went extinct between 2 million and 1 million

EXTINCTION

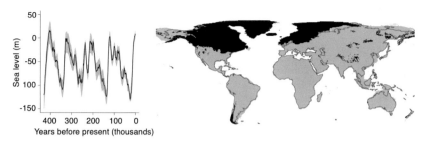

Figure 25.2. *Left:* Sea level, with 5,000 years ago as the baseline, based on an average of seven studies across the world. The gray shading gives 95% confidence limits, reflecting uncertainty in both dates and sea-level estimates. *Right:* Exposed land (gray) and ice (black) at the last glacial maximum, 18,000 years ago. (*Left:* R. M. Spratt and L. E. Lisiecki. [2016]. A late Pleistocene sea level stack. *Climate of the Past* 12: 1079–1092, https://www.ncdc.noaa.gov/; *Right*, Ice: http://crc80 6db.uni-koeln.de/layer/show/6 (J. Ehlers, P. L. Gibbard and P. D. Hughes. [2011]. *Quaternary glaciations—Extent and chronology*, Volume 15. New York: Elsevier); Sea level: E. J. Sbrocco and P. H. Barber. [2013]. MARSPEC: Ocean climate layers for marine spatial ecology. *Ecology* 94: 979, www.marspec.org)

years ago (Fig. 25.3). In the Caribbean, many coral species went extinct before the start of the glaciations (Fig. 25.3). It appears that a gradual closing of the Isthmus of Panama starting about 4 million years ago prevented nutrient-rich Pacific waters from entering the Caribbean, which reduced productivity and generated population declines that culminated in some extinctions. Later, climate change and sea-level fluctuations may have completed extinctions of what were already rare species, providing yet another example of how several factors contribute to population declines.

25.2 Human Impacts

In the popular press, as well as in many articles and books, the question is widely being asked: are we now in the middle of a sixth mass extinction driven by humans? We return to that question after summarizing what is known about extinction since humans became a dominant force.

One million years ago, some elephant species and sabertooth cats went extinct in Africa. These are the first extinctions attributed to humans, but it was the spread of modern humans out of Africa beginning about 70,000 years ago that triggered many more extinctions. Three periods can be distinguished: (1) the loss of large animals (mammals, flightless birds, reptiles) from continents from about 50,000 years to about 10,000 years ago; (2) the loss of many species from

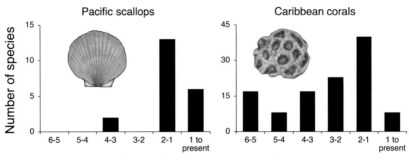

Figure 25.3 Number of fossil species that went extinct in each million-year period. Both of these groups show the most extinctions at the beginning of the Pleistocene, but Caribbean corals show more protracted earlier extinctions. (J. T. Smith and K. Roy. [2006]. Selectivity during background extinction: Plio-Pleistocene scallops in California. *Paleobiology* 32: 408–416; J. B. C. Jackson and K. G. Johnson. [2000]. Life in the last few million years. *Paleobiology* 26 Supplement: 221–235)

islands from 3,000 to 800 years ago; and (3) loss of species over the past 500 years, many from continents.

Large Animals from Continents

At least 200 species of mammals estimated to weigh more than 10 kg went extinct between 50,000 and 10,000 years before the present, and are known only from fossils. That is about 30% of all mammal species greater than 10 kg known to be present at the time, which must be a lower bound on extinctions, given that some extinct species remain to be discovered and some likely left no fossil record. The extinction rate was highest among the very large species: out of sixty-two species weighing more than 1,000 kg (1 tonne), only twelve (20%) are left. Extinctions were particularly severe in Australia and in the Americas (Fig. 25.4).

These extinctions clearly correlate with the arrival of modern humans. Humans first got to Australia about 56,000 years ago. The peak in extinctions of large mammals—including a marsupial "hippopotamus", a giant bird and five species of large reptiles dates to 42,000 years ago. Humans came into North America much later, perhaps just 16,000 years ago (possibly earlier, Chapter 11). Thirty-seven out of fifty-three genera of large mammals were lost from the Americas. In North America extinctions happened between 16,000 and

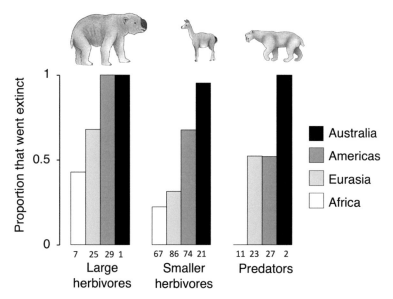

Figure 25.4 Proportion of all large mammals on continents that went extinct between 50,000 and 10,000 years ago, as inferred from fossils. Large herbivores weigh more than 1,000 kg (e.g., elephant), smaller herbivores weigh between 45 kg and 999 kg (e.g. many deer), and predators refers to all species over 21.5 kg (small wolf upward). Numbers refer to the minimum number present in each group before the extinctions, inferred from fossils and unfossilized species that are still present today. Illustrations are of *Diprotodon octatum* (see Fig. 1.1), *Palaeolama mirifica*, a relative of the llama that lived in North America, and a sabre toothed cat, many similar species of which are recorded from 42 mya up to 11,000 years ago. (Y. Malhi, C. E. Doughty, M. Galetti, et al. 2016. Megafauna and ecosystem function from the Pleistocene to the Anthropocene. *Proceedings of the National Academy of Sciences, USA* 113:838–846. The data are at https://megapast2future.github.io/PHYLACINE_1.2/, as described by S. Faurby and J-C Svenning. [2015]. Historic and prehistoric human-driven extinctions have reshaped global mammal diversity patterns. *Diversity and Distributions* 21:1155–1166)

11,000 years ago and included a giant lion (twice as large as the species still present in Africa), sabertooth cat, dire wolf, giant bear, giant armadillo, giant sloth, mammoth, camel, llama, deer, horse, and bison species. The correspondence of extinctions with human arrival holds across other areas, such as Madagascar, the West Indies, and New Zealand. However, rapid climate change coinciding with the end of the last glaciation appears to have contributed to some extinctions especially in the north, perhaps by enforcing reductions in range size.

268 ECOLOGY OF A CHANGED WORLD

Evidence points to direct hunting by humans, plus the trophic cascades subsequently precipitated, as the main reason why large mammals went extinct. As we have seen, large animals mature at old ages and have few offspring and hence a low intrinsic growth rate. This means that even a moderate increase in the death rate would lead to population decline ($r = b - d < 0$). In addition, human hunters likely targeted large animals just as they continue to do so today (e.g., Fig. 20.3). Australia and the Americas experienced higher extinction rates than did Africa and Eurasia (Fig. 25.4), where humans first developed their hunting techniques and where prey presumably had time to adapt to increasingly sophisticated hunting techniques.

The global loss of perhaps 1 billion large-bodied animals had far-reaching consequences. The amount and type of vegetation changed. Fewer plants were consumed and different plant species came to dominate communities (we noted in Chapter 20 how present-day harvesting of large monkeys in the Amazon is affecting dispersal of trees with large seeds). In Australia and North America, fires increased in response to denser vegetation. In these locations, top-down control of vegetation by herbivores was replaced by top down control by fire. In Europe, vegetation apparently became more rampant, resulting in bottom-up control dominating, through limitation by resources such as light and nutrients.

Though speculative, it is interesting to ask if the loss of large animals could affect the climate. From 12,900 to 11,600 years ago, the earth cooled by 0.5°C during a time labeled the Younger Dryas, punctuating a generally warming climate as the earth came out of the last ice age. Suggested contributions to this cooling from the loss of mammals include (1) increased heat reflection from snow (albedo) because mammals no longer trampled snow into mud; (2) less emissions from the animals themselves, resulting in a dip in the greenhouse gas methane (a dip in methane is recorded in ice cores); and (3) increased plant growth resulting in a drawdown of CO_2. However, more trees also contribute to warming because they grow above snow, thereby reducing heat reflection. The net contributions of loss of large mammals to cooling remain uncertain.

Island Extinctions

From about 3,000 to 700 years before present, humans spread across the Pacific. Contemporaneously, between 500 and 1,600 species of flightless rails went extinct. Most of these species were on different islands, but up to five could be found on one island. Easter Island, colonized around the year 900, has lost all its birds, associated with complete deforestation. At one time, 15,000 people may have lived there, but by 1870 just over 100 were left, probably a reflection of

EXTINCTION   269

complete deforestation and habitat degradation. Six extinct land bird species are known from fossils.

In the late thirteenth century, humans reached the last major location to be colonized, New Zealand. Eleven species of large flightless birds (moas) had evolved to eat fruits and other plant materials in the absence of the reptiles and mammals that typically occupy this niche in other geographic regions. Moas ranged in size from 20 kg to 250 kg. Combined, they may have numbered about 160,000 individuals. Within 100 years of human colonization, all moas were extinct. Rapid extinction is expected even under quite moderate hunting pressures. One model assumed that 100 humans arrived and proceeded to increase in population size at a rate of r = 2%/year. Assuming ten people consume one moa a week, simulations like those used for the whale example in Chapter 21 (Fig. 21.3) projected that all moas would be extinct in 80 years, by which time the human population would have increased to about 600. This calculation assumes that eggs were not eaten, but they were. It also assumes that no habitat was lost, but it was. As in the case of mammals on continents, the low intrinsic growth rate of large animals makes extinction especially likely. Based on the reproductive rate of extant large birds and some fossil clutches of moas, the inferred time to maturity averaged 5 years and clutch size was just one or two eggs.

Hawaii was colonized before New Zealand, perhaps about 1,000 years ago. Since then, 80 out of an estimated 140 bird species have been lost (60 of these in the past 150 years, the other 20 known only from fossils). Extinct species include large grazing birds. As in New Zealand, these birds evolved in the absence of mammals and tortoises, but in this case, they were derived from a duck ancestor. Extinctions also included many species belonging to a fantastic radiation of Hawaiian honeycreepers, which evolved to eat seeds, nectar, and insects across the islands. Rats, introduced diseases, and habitat loss appear to be the main causes of Hawaiian bird extinctions.

Island extinctions continue. Two of the three bird species that went extinct between 2000 and 2010 are from Hawaii. On Guam, nearly all native bird species were lost in the past 50 years, largely a consequence of the introduced brown tree snake (Chapter 18).

Extinctions on Continents in the Past 200 Years

The past 200 years have witnessed some spectacular extinctions of formerly widespread and abundant species. In the United States, the passenger pigeon lived in enormous flocks. During migration season in the 1800s, one flock of 3 billion birds took several days to pass. The passenger pigeon went extinct in the wild in the early 1900s, and the last one died in 1914 in the Cincinnati Zoo. One of the two native parrots in the United States, the Carolina parakeet, went extinct in the wild in 1904. Hunting and loss of forest contributed substantially to these extinctions. Australia has lost 30 of its remaining 273 mammals in the past 200 years. These were mostly small and widespread species that appear to have been devastated by introduced cats and foxes, and perhaps also altered fire regimes.

Contrasting with the loss from Australia, only one mammal is known to have been lost from the continental Americas over the past 200 years, a species of opossum from Argentina. In fact, the IUCN lists a total of 278 birds, mammals, and amphibians combined that are known to have gone extinct from anywhere in the past 500 years. That number is about 1% of those present at the beginning, suggesting that, after the removal of large animals from continents and island populations, perhaps the remaining species are more resistant to extinction. However, this figure is surely too low:

First, perhaps four-fifths of all eukaryote species may remain to be described (Chapter 23). Undescribed species are expected to be more often overlooked because they rare and have small ranges. Rarity and small ranges should make the species more likely to be threatened with extinction. For example, mammal species described after 1900 have gone extinct at three times the rate of those described before 1900. Thus, species that remain to be described are probably going extinct at a higher rate than those we know about.

Even among described species, some species have likely gone extinct but are not yet recognized as having done so. In Chapter 19 we considered the extinction of sixty-seven frog species, which is attributed to an invasive fungus, but the IUCN has yet to confirm these as extinct and they are instead considered critically endangered. In fact, a careful assessment of amphibians based on dates of last observation makes it clear that at least 3% of all amphibians have become extinct since 1970. Another study used similar methods, plus expert opinion, to deduce that 7% of all described species of mollusks have now gone extinct (many from islands). Insects may be undergoing undocumented extinctions. A review of studies largely from Europe and North America concluded that about 10% of all species have not been seen for 50 years. In the United Kingdom, twenty-three species of bees and wasps that visit flowers have not been seen since 1990. Three of these species disappeared before 1900. However, we do not know if they are globally extinct.

Correlates of Extinction

The classification of a species as endangered and high on the list of species likely to go extinct depends on recent declines in population size and/or the species' current population size (see Table 24.1). Because all species must pass through a small population size on their way to extinction, small range size and/or small population size is an obvious current-status criterion to list a species as threatened, and the threat to small-range species is often habitat loss. However, species with large ranges and large population sizes have also shown rapid declines. Harvesting may be regularly directed at large, widespread species, as we saw for the Pleistocene megafaunal extinctions and the passenger pigeon, and diseases rapidly spread through dense and widespread populations of animals. One study showed how threats to 96 snail species on a small archipelago off Japan have changed over time. Twenty-five species are extinct. These snails had historical ranges almost 3× smaller than surviving species, and their extinction has been attributed to the clearing of forests. Thirty-nine species are listed as threatened, with predation by invasive flatworm species a major factor, and these do not differ in range size from the remaining thirty-two species. The snail example illustrates how important it is to recognize that there is no single feature, like small geographic range, that pegs a species as high risk. As with so many issues in conservation, identifying threatened species requires consideration of many interacting factors.

25.3 Are We in the Sixth Mass Extinction?

To summarize what we know: among vertebrates we lost more than 30% of large mammal species between 50,000 and 10,000 years ago; 30% of all bird species in the past 2,000 years, mostly from islands; and at least 1% of the remaining continental vertebrates in the past 500 years. How do these losses compare with extinction patterns from the past? A background instantaneous extinction rate of $d = 0.41$ extinctions/species/million years, as we derived earlier in this chapter for mammals from the fossil record, implies that 1 in 5,000 species (0.02%) go extinct every 500 years (using the equation $N_t/N_o = e^{-dt}$). A loss of 1% in the past 500 years is 50 times faster than the background extinction rate, and a loss of 5% is 250 times faster. Going forward, assuming vertebrates are lost at the same rate as in the past 500 years, we will reach a 75% loss (i.e., the level recognized as a mass extinction) in another 14,000 to 70,000 years (depending on whether a 5% or 1% loss has happened in the past 500 years.) Note that this ignores all the extinctions that happened before 500 years ago. This rate is comparable to the estimates of extinctions for the two mass extinctions described earlier, which plausibly took only a few thousand years.

It is quite likely that extinction rates will increase in the near future rather than holding steady, given the perilously low population sizes of many species and the ongoing increase of human impacts, reaching mass extinction criteria sooner. More than 500 species of vertebrates have population sizes of less than 1,000 individuals, and half of these less than 250.

25.4 Conclusions

Habitat loss is most likely to threaten species that have historically had small geographic ranges. But we have also observed many rapid declines and extinctions of widespread, abundant species that have been relatively immune to habitat loss. Driven largely by harvesting, disease, and invasive species, as well as synergistic interactions between different threats, these have proven difficult, if not impossible, to predict. It is possible, that present extinction rates will create a sixth mass extinction perhaps even within this century. This number may need to be qualified for charismatic vertebrates, because of the many directed efforts toward preserving small populations, from zoos to targeted reserves, that have managed to delay or prevent extinctions. Indeed, a 2006 study concluded that at least sixteen bird species would have gone extinct without such efforts. Well-protected reserves may help maintain species even at low population sizes, a possibility addressed in the next two chapters.

26

Species across Space

The number of species varies across space as well as time. Islands tend to have fewer species than areas the same size on continents. Even across continents, the number of species differs greatly from one place to another, as illustrated by tree species in North America (Fig. 26.1). Such an uneven distribution of species numbers has obvious implications for conservation: to conserve many species, we should focus our efforts on species-rich parts of the world, although this needs to be qualified because some areas contain particularly threatened species. This chapter's goals are therefore to:

(1) Ask why species numbers vary across the globe, and especially why so many species are found in tropical regions.
(2) Determine why species in tropical regions tend to have small ranges and are rare, both factors threatening persistence.
(3) Consider how these patterns result in the presence of biodiversity hotspots: locations in the world with an especially large number of threatened species that deserve special attention.

26.1 Why Are More Species Present in One Place Than in Another?

On land, many animal and plant groups show patterns similar to those of the trees (Fig. 26.2). Large numbers of species are found in warm and wet areas, with high net primary productivity (Fig. 26.2; see Fig. 13.1 for primary productivity), whereas deserts and cold regions have relatively few species. Much of the debate about differences in species numbers across the world asks why wet tropical climates have more species than wet temperate climates (temperate climates are defined by the presence of regular freezing, unless temperatures go below –40°C, in which case the climate is considered boreal). Tropical rainforests are particularly rich in species, and mountainous areas in tropical regions have exceptionally many species, associated with variation in climate from tropical at the base to permanent snow. The most species-rich region in the world lies on the equator, straddling the area stretching from the Amazon Basin over the Andes to the coast

Ecology of a Changed World. Trevor Price, Oxford University Press. © Oxford University Press 2022.
DOI: 10.1093/oso/9780197564172.003.0026

274 ECOLOGY OF A CHANGED WORLD

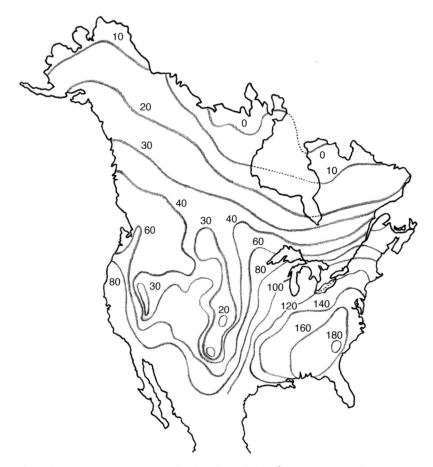

Figure 26.1 Numbers of tree species found in 7,000 km² squares across North America. The lines are contours of tree abundance. (D. J. Currie and V. Paquin. 1987. Large-scale biogeographical patterns of species richness of trees. *Nature* 329: 326–327)

in Ecuador. One in twelve of all the world's bird species can be found breeding in this region. Plants are similarly most diverse here; no other region comes close.

The reasons why the number of species varies so much across the globe are still being debated, but one can contrast two general explanations. Ecological explanations of species diversity posit that tropical climates favor the presence of more species. For example, more plant growth may lead to more food and hence more niches, and lower seasonality means species can specialize on a narrow range of food sources. Historical explanations propose that the tropics have been undisturbed for a long time, during which they spanned a large area, allowing many species to accumulate. In contrast, over earth's history, regions

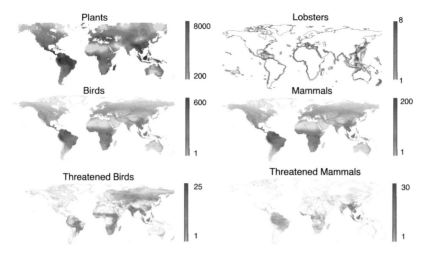

Figure 26.2 (Color plate 19) *Top and middle:* Numbers of species across the globe for four groups. *Below:* Distributions of bird and mammal species in the IUCN threatened category. Resolution for the animal maps is considered to be 10 × 10 km². Resolution for the plant map is considered to be approximately 110 × 110 km² at the equator. The plant map is a result of interpolating between the relatively few sites on the globe where diversity has been measured and is less certain than the others. (Animal maps: C. N. Jenkins, S. L. Pimm, and L. N. Joppa. [2013]. Global patterns of terrestrial vertebrate diversity and conservation. *Proceedings of the National Academy of Sciences* 28: E2602–E2610; C. N. Jenkins and K. van Houtan. 2016. Global and regional priorities for marine biodiversity protection. *Biological Conservation* 204:333–339 (see Figure 26.5 for further details). Plant map: H. Kreft and W. Jetz. [2007]. Global patterns and determinants of vascular plant diversity. *Proceedings of the National Academy of Sciences* 104: 5925–5930)

with temperate climates have sometimes been at very high latitudes, sometimes closer to the equator, and other times quite reduced in size. The argument is that species in temperate regions have had less time than tropical lineages to diversify into many species, and species associated with temperate climates have more often gone extinct.

Ecological Explanations for Patterns of Species Diversity

Very cold and very dry regions do not have the warmth or moisture, respectively, to support plant growth. Hence, it is clear why no species occur in these regions. By extension, it makes sense that only a few species can persist in places with relatively little moisture/warmth. On the other hand, tropical rainforests, with their abundant moisture and high temperatures, may have more species than anywhere else because of the greater variety of available niches. These niches arise through at least two complementary mechanisms: (1) increased abundance and diversity of food and other resources (Fig. 4.6) and (2) an absence of temperature seasonality (Fig. 26.3).

More food may mean more species because some niches in some regions have so little food that they cannot support a species, whereas regions with high primary productivity or a long growing season enable species persistence (see Chapter 4, Fig. 4.6). Among insect-eating birds, twice as many species breed in a Panama forest as they do in a forest in Illinois, but the number of small insect-eating bird species is similar in both places; the difference arises because Panama has many more large species. Hence, more insectivorous bird species occur in Central America because that region can support many large species, and this may reflect the abundance of large insects. Abundant large insects are an expected consequence of both a longer growing season and higher plant productivity.

The tropics are less seasonal than the temperate regions, especially in temperature. In seasonal environments, species must often be generalists to exploit a wide variety of resources that come and go over the year. In the tropics, species can specialize, and hence they partition resources more finely. Birds and bats that specialize on fruit are found in tropical but not temperate climates because their supply of fruit is present year-round in the tropics. Similarly, pollen and nectar are available throughout the year in the tropics. Low seasonality may also result in species evolving to have smaller elevational ranges, and hence more can be packed along a mountainside. A temperate species must adapt to wide-ranging temperatures, allowing it to track a changing climate up and down the mountain more easily and exclude competitors (Fig. 26.3). This argument applies across geographical space as well as up and down mountains: in seasonal environments, each species may evolve to have a broader geographical range, excluding others.

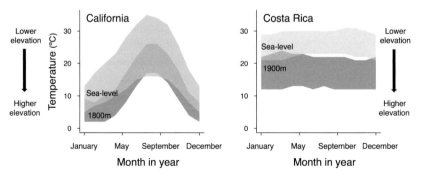

Figure 26.3 Monthly mean temperatures at sites in California and Costa Rica (from *weather-and-climate.com*).

Historical Explanations for Patterns of Species Diversity

Many scientists doubt that ecological hypotheses such as those described above are the major contributor to global biodiversity patterns. Can any of these hypotheses fully explain why there are almost 15 times as many trees in 1 ha of Amazonia than a similar area in Missouri? The complementary explanation goes back to Alfred Russell Wallace, who in the nineteenth century was the first to seriously consider the causes of high numbers of species in the tropics. He argued that the tropics had been more stable and had covered a larger area for a longer time, so that species have been able to steadily accumulate. Wallace focused on the Pleistocene (i.e., the last 2.5 million years): "[I]n the tropics, evolution has had a fair chance while in the temperate zone, with its glacial periods, it has had countless difficulties thrown in its way." Going back further in time, across the past tens of millions of years, temperatures have varied more at higher latitudes than at lower latitudes (Fig. 26.4), presumed to contribute to elevated extinction rates. Such a historical explanation for species richness is also sometimes called the age and area hypothesis. Species in the tropics are not equipped to deal with freezing temperatures or to compete with species that are so equipped, and therefore even as they accumulate in the tropics, they are unable to disperse into temperate regions.

Both present-day ecological conditions and historical factors surely contribute to the high number of species in the tropics, but it has been difficult to apportion their roles because they are highly associated. Warm, wet areas (the tropics), thought to enable more species to persist for ecological reasons, have also remained climatically more stable, thus resulting in greater opportunities for speciation and reduced chances of extinction. Furthermore, many of the more explicit hypotheses predict similar patterns. A lack of seasonality in the tropics may mean that resources can be subdivided more finely (i.e., individual species specialize on narrow portions of the resource spectrum). Alternatively, if more species accumulated for reasons of age and area, then resources might be divided among these species and with each expected to specialize.

26.2 Rarity, Small Ranges, and Distinctiveness in the Tropics

Rarity

The tropics have many species, but these species on average occur at lower densities than temperate species. For example, trees in 0.1 ha plots have been enumerated across the world. A Missouri wood contained 184 trees belonging to 23 species, for an average of 8 individual trees per species. On the other hand, the most species-rich location in the world is in the lowlands of tropical Peru, where one 0.1

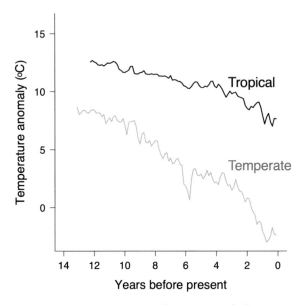

Figure 26.4 Climate through the past 12 million years in shallow marine waters at present-day north temperate regions (between 30°N and 50°N) and in present-day tropical regions (between 23°S and 23°N). Temperature estimates are based on ratios of different alkenones (molecules related to fat storage) in certain algae, calibrated according to conditions at the present day. Temperatures are averaged across 250,000 years, smoothing out effects of climate swings through the Pleistocene. Each curve is the average from multiple places across each region. (T. D. Herbert, K. T. Lawrence, A. Tzanova, et al. [2016]. Late Miocene global cooling and the rise of modern ecosystems. *Nature Geoscience* 9:843–847)

ha plot contained 331 trees belonging to 218 species. The Missouri plot contained one-tenth the number of species, but each is on average 5 times more common than the Peruvian rainforest. Birds and other groups show similar patterns, with many species in the tropics rare. As with the hypotheses to explain species numbers, both ecological and historical explanations are likely to contribute to patterns of rarity. Sorting out their importance has been difficult. For example, specialists should have smaller population sizes than generalists. As noted above, specialization could evolve as a result of lower seasonality, but it could also result from the production of more species, which then compete for resources.

Small Ranges

Vertebrate species in the north temperate zone tend to have larger ranges than those in the tropics (Fig. 26.5). In this case, historical influences clearly

Figure 26.5 *Left*: Locations in the world which have three or more small-range mammal species, defined as being less than the median range size, that is, the range size (182,000 km²) below which 50% of all species lie. *Right*: Range of the racoon (gray; note that the racoon has expanded its range, especially in the west, over the past 80 years) and olinguito (black), a small raccoon-like animal not recognized as a species different from its lowland relatives until 2013. (C. N. Jenkins, S. L. Pimm, and L. N. Joppa. [2013]. Global patterns of terrestrial vertebrate diversity and conservation. *Proceedings of the National Academy of Sciences* 28: E2602–E2610. Available at biodiversitymapping.org, derived from maps at NatureServe and IUCN 2018: The IUCN red list of threatened species. Version 2018-1. http://www.iucnredlist.org. The maps on the right were from the same source)

contribute: recolonization into areas recently vacated by ice sheets is thought to be one reason for the large ranges of species in north temperate regions. If we were to let the system run without any further disturbance, surely barriers would carve up these large ranges, generating species with smaller ranges, each in a different region. For example, the black-capped chickadee is a familiar bird across the whole of the northern part of the United States, but over time it might split into a western and eastern species. Indeed, patterns of geographical replacement like this are common in the Andes. In this case, barriers have divided up continuous ranges, leading to the formation of multiple species, each confined to a small range and geographically separated from each other. Many of these species formed over the last 2 million years as a result of becoming isolated on these different mountains, and they have not gone extinct because even during the ice ages, refugia (i.e., a patch of suitable habitat) remained on each mountain.

Evolutionary distinctiveness

Small ranges and rarity increase the probability of extinction. These factors, plus the sheer numbers of species all point to the tropics as particularly important areas on which to focus conservation efforts. Several researchers have suggested that we should also consider how different species are from each other when assessing the importance of a location to conservation. For

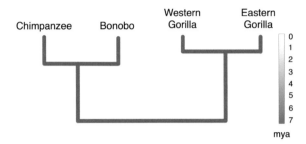

Figure 26.6. A phylogeny of four ape species, taken from Figure 23.4. The scale bar on the right is in millions of years. The bonobo and chimp are separated by about 2 million years, and the two gorillas by about 1 million years. Phylogenetic diversity (Faith's PD) is the total length of all the branches in the tree = 2 + 2 + 5 + 1 + 1 + 6 = 17. For a location in the Democratic Republic of Congo with just bonobo and chimpanzee, PD = 4. For a location in west Africa with just western gorilla and chimpanzee, PD = 14. By this criterion, even though species numbers are the same, west Africa has higher diversity.

example, the bonobo and chimpanzee are similar to each other in genes, ecology, and function, whereas the chimpanzee and the gorilla are more different, so perhaps a protected area that preserves the gorilla and chimpanzee would be favored over one that preserves the bonobo and chimpanzee. An effective way to quantify the distinctiveness of a location is by computing by the total length of the branches connecting all species in a phylogenetic tree. This statistic is known as Faith's Phylogenetic Diversity (or Faith's PD; see Fig. 26.6). Faith's PD has been calculated for birds across the world, and approximately matches patterns of bird species richness. Regions with high phylogenetic diversity include Amazonia, east Africa, and south China, all of which have many species in total, but include Indonesia as well, which has fewer species. On average, however, Indonesian species have been separated from each other for a relatively long period of time. All the regions with high phylogenetic diversity are tropical, reinforcing the age hypothesis, according to which the tropics have been relatively undisturbed.

The presence of many species, their relatively small ranges, and their rarity mean that the tropics contains a large fraction of at-risk species. Threats are accentuated by ongoing land conversion for agriculture and harvesting, which are both especially affecting the tropics. The distribution of threatened species (based on the Red List; Table 24.1) is broadly similar between mammals and birds (Fig. 26.2), and other groups such as amphibians.

26.3 Biodiversity Hotspots

The uneven distribution of threatened species across the globe led Conservation International, a not-for-profit conservation organization (biodiversityhotspots.org), to designate biodiversity hotspots. As described for the year 2000, to qualify as a hotspot, a region must meet two criteria: it should contain at least 1,500 species of named plants (about 0.5% of the world's total) as endemics (i.e., restricted to this region), and it must have lost at least 70% of its original habitat. The combined area of remaining habitat in these areas covers about 2% of the earth's land surface (see Color plate 20). Over 50% of the world's plant species and 42% of all land vertebrate species can be found in thirty-four such hotspots. Note that New Guinea was not considered a hotspot in the year 2000 because it retained substantial forest, but that picture is changing. More recently, areas of special concern have been mapped based on the numbers of threatened species and the level of current protection. The general argument that conservation efforts in small areas could save a large proportion of the world's biodiversity holds. Two examples follow:

The Atlantic Forest of tropical South America contains 20,000 plant species, 40% of which are endemic. Less than 10% of the forest remains, and most of this forest is at higher elevations. Beginning with sugarcane plantations, coffee plantations, and cattle ranching, this region has been losing habitat for hundreds of years, and continues to face severe pressure from urbanization. Three species of lion tamarins (monkeys) are listed as Critically Endangered. Six Critically Endangered bird species are restricted to the small patch of forest near the Murici Ecological Station in northeastern Brazil. One conservation organization that has been active in the region is Saving Species (now reorganized as Saving Nature), which purchased land and passed it on to the park service as a protected area. Their first purchase was of 70 ha of farmland for $333,000 near the city of Rio de Janeiro. This farmland separated two forest remnants, one of which provided a highland refuge for the golden lion tamarin. Within 10 years, forest had regenerated sufficiently for tamarins to increase. An unexpected bonus was colonization by the puma, which has reduced the abundances of predators of the golden lion tamarin. The 70 ha of regenerating tropical forest soaks up carbon used by perhaps thirty-five average Americans for roughly 30 years. Promoting purchases such as these in this way helps attract potential donors.

Madagascar features a unique fauna and flora with many endemic families. Presently only 10% of the 600,000 km^2 island may contain native vegetation. More than 150 birds, mammals, and amphibians are listed as threatened, and 45 are known to have gone extinct in the past 500 years. The central plateau of Madagascar is almost completely deforested. The International Union for the

Conservation of Nature considers Madagascar to be on the frontline of conservation, with threats being both extreme and immediate.

26.4 Conclusions

The question of why many more species occur in the tropics than elsewhere has been difficult to resolve. In the marine realm, the most diverse place in the world is Indonesia, with its many islands promoting speciation, high productivity, and low seasonality. In the terrestrial realm, the most diverse location in the world is tropical South America, again with high productivity and low seasonality, as well as many barriers, including both mountains and rivers, promoting speciation. In both places, stability through time has likely led to relatively low extinction rates and high phylogenetic diversity.

27

Island Biogeography and Reserve Design

One of the major threats to biodiversity is loss of habitat. However, habitat loss is not simply a contraction in area, but separation into patches, between which movement is restricted or absent. Many species are now confined to small fragments in a sea of a human altered landscape. For example, in Kalimantan, Borneo, offically protected lowland forest declined threefold between 1985 and 2001, resulting in the persistence of only sixteen forest fragments larger than 100 km². Small populations are much more susceptible to being lost than large ones, including from chance effects (e.g., a tornado). This chapter considers the effects of fragmentation and how reserves might be designed to minimize species loss. Specifically, it discusses:

(1) The theory of island biogeography, a successful model to explain why small isolated areas maintain relatively few species.
(2) A real example of species loss from habitat fragments.
(3) The underlying causes of species loss from small areas.
(4) Optimal design of nature reserves in light of both theory and empirical studies.

27.1 Island Biogeography

Habitat fragments are islands surrounded by areas inhospitable to the species found in the fragment. We investigate their capacity for retaining species first by modeling species numbers on islands in the sea. Oceanic islands (i.e., those that have never been connected to the mainland) are distinguished by two universal features: (1) small islands have fewer species than large islands and (2) islands far away from the mainland tend to have relatively few species (Fig. 27.1). In Figure 27.1 axes are drawn on logarithmic scales because it is not the absolute increase in island size that matters, but the relative increase. Across oceanic islands, a rule of thumb is that an island 10 times larger than another holds about twice as many species.

Why should small, isolated islands have few species and large islands near continents have many? An elegant model starts with the assumption that the

Ecology of a Changed World. Trevor Price, Oxford University Press. © Oxford University Press 2022.
DOI: 10.1093/oso/9780197564172.003.0027

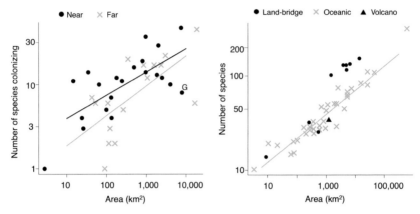

Figure 27.1 *Left:* Number of bird colonization events to islands and archipelagoes, separated into those more than 1,000 km distant from a continent, and those less than 1,000 km distant. The Galápagos Islands (G on the figure, 926 km from South America) has 26 species derived from eight over-water colonization events from South America. The data set consists of smaller species of land birds and is valuable because human-driven extinctions, mostly known from fossils, are included (136 out of 569 species are extinct, or 24% of the total). *Right:* Number of bird species present on islands near New Guinea. Land-bridge islands are those that were connected to New Guinea 18,000 years ago (Fig. 25.2). Note the larger land-bridge islands have more species than similar sized oceanic ones, but the smaller land-bridge islands have similar numbers to equivalent sized oceanic ones. The volcanic island of Long is 50 km offshore. It exploded about 350 years ago, likely killing everything. (L. Valente, A. B. Phillimore, M. Melo, et al. [2020]. A simple dynamic model explains the diversity of island birds worldwide. *Nature* 579:92–96; J. M. Diamond. [1973]. Distributional ecology of New Guinea birds. *Science* 179:759–769)

number of species on an island represents a balance between the number going extinct over a certain time interval and the number of species colonizing the island over the same time interval (Fig. 27.2). When no species are present, of course none can go extinct, but the more species that are present on an island, the more we expect species to go extinct. Hence the extinction curve (i.e., the number of species becoming extinct per unit time) must increase with the number of species present (Fig. 27.2). On the other hand, the more species present on the island, the fewer there are that remain to immigrate from a source. The colonization curve must therefore decrease with the number of species on the island (Fig. 27.2). The intersection of the two curves gives the equilibrium number predicted for the island, where the number immigrating is exactly balanced by the number going extinct.

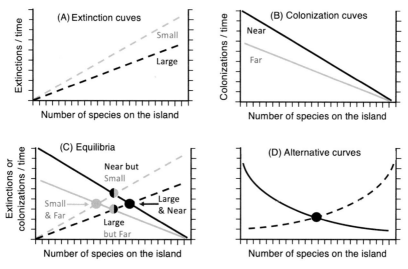

Figure 27.2 *Upper left:* Number of species going extinct in unit time (dashed lines) increases with the number of species on an island and increases relatively steeply on small islands (gray). *Upper Right:* Number of species colonizing in unit time must decrease with more species on an island, as fewer new ones are available in the source. Successful colonization of islands distant from a source occurs less frequently (gray). *Lower Left:* Putting the upper graphs together results in four equilibria. *Lower Right:* More realistic extinction and colonization curves. (R. H. MacArthur and E. O. Wilson. [1967]. *The theory of island biogeography.* Princeton, NJ: Princeton University Press)

This model can be adapted to explain variation in species numbers among islands that differ in size or distance from the mainland. On large islands, the extinction curve is assumed to be less steep than that on small islands: fewer species go extinct in a given time period. This effect of island size may arise because each species on a larger island has a larger population, lowering its chances of being reduced to dangerously low levels. The effect could also occur because larger islands have a greater diversity of habitats that provide refuges for different species. Because large islands have lower extinction rates at equilibrium, more species will be maintained on a larger island than on a smaller one (Fig. 27.2). With respect to distance, fewer species successfully immigrate per unit time on to a distant island than on to a near island. The result is that isolated islands have fewer species at equilibrium than islands closer to a source.

The model has been subject to many extensions and tests. One important modification is the shape of the extinction and immigration curves. The straight lines drawn in Figure 27.2A–C assume that the probability of an immigration or

286 ECOLOGY OF A CHANGED WORLD

extinction event does not depend on the number or kind of species already present. For example when twice as many species are present, twice as many go extinct in a certain time interval. More reasonably, the probability of a species' extinction should increase with the number already on the island because once more species are present, each should on average have a smaller population size, for example, due to competition for limited resources. In turn, this means that as the number of species present on an island increases, the proportion of all species that go extinct during each time interval increases (i.e., the tangent to the curve becomes steeper; see Fig. 27.2D). In addition, the number colonizing per unit time as a fraction of the ones already there should decline because the best colonists (e.g., species that are strong fliers) get to the island first and poor colonists only come later or not at all (e.g., some species cannot fly or are reluctant to cross water gaps). In other words, some species might arrive on a new island within a year or so, but others might be such poor fliers that they never arrive.

Through knowledge of extinction and colonization curves, we can model the expected number of species at equilibrium on oceanic islands. We can also use the model to predict the effects on species when existing habitats are divided into isolated fragments. For example, if an island is carved out of a larger area (e.g., when rising sea levels separate two formerly connected landmasses or when woodland is surrounded by agricultural land), three effects operate to reduce the number of species present: (1) a smaller area only captures a subset of all species initially; (2) the extinction rate increases associated with a reduction in population sizes; and (3) the immigration rate declines because individuals now have to move across inhospitable habitat. An increase in sea level toward the end of the last ice age (from ~140 m below present levels 24,000 years ago, to ~30 m below present levels 10,000 years ago; see Fig. 25.2) resulted in the formation of many islands, which were previously part of a larger landmass (termed land bridge islands; see Fig. 27.1). Around New Guinea, land bridge islands averaging 1,000 km^2 have substantially more species (about 80 species present) than oceanic islands of the same size (about 50 species present). If one assumes that oceanic islands are in balance between extinction and colonization, this suggests that land bridge islands have still to lose some of their species. Hence, extinction of birds may take a long time on these fairly large islands. On the other hand, smaller land bridge islands have similar numbers to oceanic islands, implying they have lost species.

27.2 Loss of Species from Habitat Fragments in Southern California

To apply island biogeography theory to fragmented habitats in the context of conservation biology, we consider an example of birds in southern California. Since the early 1900s, housing estates in the San Diego area have been built around canyons. Pockets of natural habitat lie in the canyons, surrounded by buildings. City records of housing developments indicated the time each canyon was first cut off from other natural areas (varying from 2 to 86 years at the time of the study, 1997), and maps give canyon areas and distances. A survey of eight native chaparral bird species across a total of thirty-four canyons in San Diego demonstrated that smaller canyons have fewer species than larger ones.

Although small canyons contain fewer species than larger ones, this may be a consequence of the fact that the canyons were smaller in the first place, implying that when they were cut off, not all species in the area were present in them. Therefore, the first question we might ask is whether isolation of a canyon has led to any loss of species. To address this issue, a research team compared the number of species in canyons to those present in a similar-sized area embedded in relatively undisturbed habitat, that of Camp Pendleton marine base. Motorists on the interstate freeway traveling to Los Angeles pass a signboard announcing "Camp Pendleton protects California's natural resources." Although conservation is obviously not its main function, the military base does play a part, being the only large undeveloped area along the southern California coast. Smaller canyons in the natural matrix of Camp Pendleton have fewer species than larger canyons, as expected. However, small canyons surrounded by development have even fewer species than similar-sized canyons in Camp Pendleton (Fig. 27.3). The implication is that small canyons surrounded by development have indeed lost species.

In addition to small size, canyons that have been cut off for a long time also have especially few species. The effects of age and area are largely independent of each other and approximately the same. Small, long-isolated fragments held none of the eight species under study, whereas large, recently isolated, fragments contain up to seven of the eight species. This observation is in accord with the expectation from the theory developed above. Small isolated canyons should have few species, and they do.

Nestedness

Some species are predictably lost from the canyons first, whereas others take much longer. The wren-tit is present in many small and long-isolated canyons, but the California quail lives only in the large, recently isolated ones. Such a predictable ordering of species identity with species numbers is termed nestedness. Complete nestedness implies that all locations with three species contain the same three species, and that all locations with four species contain these three, plus an additional species that is always the same. Therefore, if a system shows complete

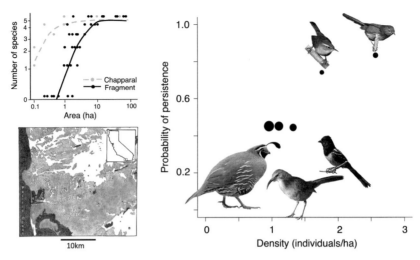

Figure 27.3 (see color plate) *Upper Left:* Number of bird species, out of the five common surveyed species in isolated canyons of different sizes in the continuous chaparral habitat of Camp Pendleton (gray) and the city of San Diego (three other species were rare in the area and so were not considered in this study). *Lower left:* The white shading indicates undeveloped areas in southern California. Camp Pendleton lies 40 km to the north. Note the presence of canyons. *Right:* Probability of persistence in 36 canyons, plotted against density in continuous chaparral, collected by a different researcher in a separate study. The size of the circle indicates the relative mass of each species. From left to right: California quail (170 grams), California thrasher (85 g), spotted towhee (39 g), Bewick's wren (10 g), and wren tit (15 g). Persistence is based on the data collected for the above left graph, which implied that the wren tit and spotted towhee were in all 36 canyons before they were cut-off by development, and each of the other three species in 31 canyons. (D. T. Bolger, A. C. Alberts and M. E. Soulé, [1991]. Occurrence patterns of birds in nested fragments: sampling, extinction, and nested species subsets. *American Naturalist* 137:155–166; Photos: Brad Imhoff (quail), thrasher (Mike Baird, CC 3.0). Others, wikipedia under CC 3.0 license)

nestedness, knowing the number of species in a location means knowing which species they are. The important consequence of nestedness is that, if only small fragments are left, some species will inevitably be lost and others will be retained.

27.3 Mechanisms of Species Loss from Small Areas

The canyon study is unusual in that the research team evaluated causes of species loss from canyons. Three factors have been identified: initial rarity, predators, and edge effects.

Rarity

Rare species are more likely to become extinct than common species (Fig. 27.3) because random fluctuations in population size are more likely to drive small populations to unsustainably low densities. Higher extinction rates of small populations appear to be a general phenomenon and are consistent with the theory of island biogeography developed above, which assumes that larger islands hold larger populations that are more resistant to extinction.

Predation

Predators destroy many bird nests in the canyons. Fifteen of nineteen predation events captured on video camera were by snakes. However, snakes and nest predation are more common in the larger canyons, implying snake presence cannot explain the decline in small birds in small canyons. In fact, this might have been predicted from the first principle given earlier: small fragments readily lose their predators because predators are generally rarer than their prey.

It turns out that predation of adult birds critically drives species loss in these canyons, and this predation risk arises from another of the big six COPHID threats: invasive species. Cats are common in the canyons. The 1997 survey indicated that half of the homeowners around a canyon owned on average 1.7 cats and that three-quarters of these owners let their cats go outside. A survey of 100 houses surrounding a midsized canyon revealed that cats brought back 840 small mammals, 525 birds, and 595 lizards to their homes in one year. These values must dramatically underestimate the total number of small animals killed in the canyon because cats presumably did not bring all their kills back to the house, and many cats living in the canyons have no owners.

Coyotes are like snakes in that they persist only in larger canyons. However, unlike snakes, coyotes eat cats. The coyote presence has a positive effect on the number of bird species present, which persists even after controlling for age and area (Fig. 27.4). Here we have yet another example of a trophic cascade: smaller canyons lose their coyotes, cat populations increase, and the species they eat decrease. Coyotes not only kill cats, but cat-owners keep their cats inside if coyotes are present, providing an example of a trait-mediated effect (the trait is cat confinement) on bird population sizes (Chapter 4).

The Argentine ant, so-named because it was originally from Argentina, is another invasive species in the canyons. This ant thrives in wetter areas, such as around people's watered lawns and in canyons that receive runoff from houses. The ant is aggressive, and both preys upon and competes with native ants. The Argentine ant appeared in one monitored canyon in San Diego in 1996. By

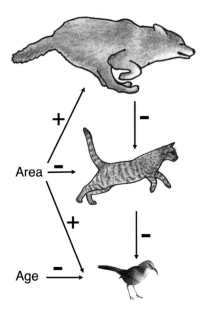

Figure 27.4 Paths indicating the effects of the density of coyotes and cats on the number of chaparral bird species (out of 8 total) in 28 San Diego canyons, as estimated from visual surveys. Area and age each have about twice the effect on number of bird species present than does cat or coyote presence, both of which are nevertheless significant. (K. R. Crooks and M. E. Soulé. [1999]. Mesopredator release and avifaunal extinctions in a fragmented system. *Nature* 400: 563–566). (coyote Yathin Krishnappa (CC BY-SA 3.0), cat (von Grzanka CC 3.0), thrasher (Mike Baird, CC 3.0))

2003, it had spread throughout the canyon, and in the process twenty-one of twenty-three native ant species were lost. The loss of native ants has been further connected to the decline of an intriguing reptile, the horned toad, which grows more quickly when eating native ants because they are relatively large (Fig. 27.5, at end of this chapter).

Edge Effects

Unlike true islands in the ocean, the edge of a protected area is subject to many incursions from humans, other species, and even abiotic factors such as wind, which disrupt the habitat. Daniel Janzen, a well-known tropical biologist, coined the term *eternal external threat* to summarize these edge effects. In the San Diego canyons, edge effects include owned cats, Argentine ants, and also invasive plants, many of which flourish in the human habitats around the canyons.

Figure 27.5 Argentine ants attacking a large native ant *(Pogonomyrmex subdentatus)*, which harvests seeds in the canyons of San Diego. (Photo by Alex Wild, www.alexanderwild.com)

A classic experiment in the tropics identified important roles for edge effects in affecting species persistence. In the early 1980s, near the Amazonian city of Manaus in Brazil, controlled forest clearing was used to generate twelve fragments ranging in size from 1 to 1000 ha. Within 10 years the smaller fragments lost many forest interior species, such as dung beetles and ground-dwelling birds. Edge effects were striking. Most notably, trees died. These effects were most apparent up to 100 m from the edge, but some wind damage extended as far as 300–400 m. However, they somewhat stabilized within a few years because of secondary growth.

27.4 Reserve Design

Data from multiple sources over timescales ranging from 10,000 years to 10 years (including the land-bridge islands in Oceania described earlier and the Manaus project) found that isolated tropical forest fragments the size of 10 km^2 typically lose a bird species within just 7 years, and even fragments 500 km^2 in area lose their first bird species within 40 years. For mammals, species loss has been documented from large parks. In Ghana, West Africa, six parks smaller than 5,000 km^2 lost at least one-quarter of their large mammals in 30 years. Eleven

292   ECOLOGY OF A CHANGED WORLD

of fourteen parks in western North America have lost at least one species in the past 80 years. In summary, large reserves are always better than small reserves because they preserve more species, but even quite large reserves lose species.

Suppose one could preserve a certain area of land, and do so by creating either two small reserves some distance apart or one large one of the same combined size. Separating populations may mean that threats do not affect both populations (e.g., from a disease outbreak, or a developer); given that a smaller patch should preserve nearly all the species of one twice the size, perhaps creating two reserves would be the better option (Fig. 27.1). This would be especially the case if species loss happened by chance, resulting in different species being maintained in different reserves. The major problem with this idea is that some species are lost more easily than others: nestedness is pervasive, meaning that all the smaller patches generally contain the same subset of species. The species that are predictably lost include larger predators that play important roles in the ecosystem (e.g., coyotes and snakes from the canyons of San Diego).

The debate about the merits of a Single Large Or Several Small reserves ("SLOSS") continues to occupy many pages of the conservation biology literature. In reality, it is not usual for a reserve to be designed solely with respect to the number of species it should preserve, for many political and logistical reasons. Conservation biologists instead emphasize not only the importance of large reserves but also the value of connections between reserves (i.e., corridors along which species can move).

27.5 Conclusions

Chapter 26 asked where to put reserves, and the present chapter asked how to design them. Recent analyses of studies along the lines of the San Diego canyons have found similar results, whereby it takes years to lose species from large patches (some have even gained in number from introduced species), but there is rapid loss from small patches. These studies also show that loss happens more quickly from tropical than temperate regions. Hence, threatened tropical habitats are particularly important places for reserves, and large reserves are disproportionately important to conservation. Two issues remain. The first is protection: hunting happens in many reserves (Fig. 21.4B), invasive species penetrate them (Fig. 24.1E), and even some of the legally most stringently protected areas (national parks) have agricultural land within them. Second, political and historical contingencies have meant that large reserves are often concentrated in regions that have small human populations that also tend to be species-poor. In the United States, the large protected areas are in the west, and small-ranged species in the southeast are particularly poorly protected. In

India, of the two largest protected areas one is in a desert and the other is above the treeline in the Himalaya, both of which have few species. In China, too, the large protected areas are in the drier west. However, both India and China also have some quite well-protected areas in their species-diverse tropical regions, which have been explicitly developed for flagship species (the tiger and giant panda, respectively).

28

The Value of Species

> What limits the annual fish catch—fishing boats (capital) or remaining fish in the sea (natural resources)? Clearly the latter. . . . Economic logic says to invest in and economize on the limiting factor. Economic logic has not changed; what has changed is the limiting factor. It is now natural resources, not capital, that we must economize on and invest in.
>
> —Herman Daly, former senior economist of
> the World Bank (2015)

In this chapter we ask why it is worth preserving species at all. The six major threats to nature, COPHID—Climate change, Overharvesting, Pollution, Habitat conversion, Invasive species, and infectious Disease—are to varying degrees detrimental to humans, despite the benefits some provide. The energy supply from fossil fuels drives much of the world's economy, but the consequent global warming is clearly threatening to humans, for it is causing large-scale disruption through hotter temperatures, redistribution of precipitation, and rising sea levels. Pollution is an unwanted side effect of human activity. Habitat conversion benefits humans by creating more agricultural land, but if this land were left in its original state, it would provide ecosystem services (contributions to human well-being, Chapter 14). Overharvesting is particularly egregious because with more restraint we could harvest at the maximum sustainable yield, raising food supplies. That reasoning is exemplified by the quote from Herman Daly, which encompasses the two reasons species may be harvested below maximum sustainable yield: depletion and conservation. Invasive species by definition cause economic and/or environmental harm. Diseases are not good in any way. Some are derived from habitat destruction bringing humans in closer contact with wild animals and others from exploitation of those animals.

Coupled with the adverse effects on humans, these threats affect other species and therefore, indirectly, they touch us. What are the benefits to humans of preserving species, thereby making the case for conservation? In this chapter, we first consider the case for preserving many species, whatever their identity. We then focus on the benefits, both present and into the future, of preserving specific species. We describe:

Ecology of a Changed World. Trevor Price, Oxford University Press. © Oxford University Press 2022.
DOI: 10.1093/oso/9780197564172.003.0028

(1) How ecosystems are affected by a reduction in plant species.
(2) How other ecosystem services have been disrupted by dependence on a few species, as illustrated by oysters and bees.
(3) Direct benefits to humans of individual species, such as food, medicine, and as tourist attractions.

Ecosystem Functions Affected by Loss of Biodiversity

The simplification of ecosystems—that is, a reduction in the number of species they contain—has negative effects on ecosystem functions, defined as ecological processes that control the fluxes of energy, nutrients, and organic matter through the environment.

Plant productivity is not simply a function of heat, water, and nutrients (Chapter 13) but also depends on number of species. Figure 28.1 shows the results of an experiment conducted at Minnesota's biological station on 147 experimental plots, each 3 m × 3 m in size. Plots differed in the number of species added, but all were seeded with the same total mass of seeds. The more species present, the higher the total plant productivity. Productivity gains are associated with a greater drawdown of nutrients in the soil (Fig. 28.1). Effects

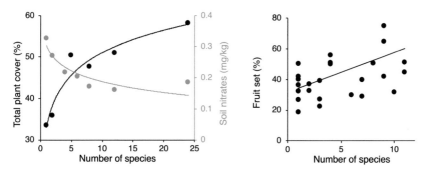

Figure 28.1 *Left:* In small experimental plots, the number of plant species correlates with plant cover and nutrient depletion of the soil (recorded after two seasons). *Right:* In a Californian almond grove, fruit set correlates with number of wild bee species. (D. Tilman, D. Wedin, and J. Knops. [1996]. Productivity and sustainability influenced by biodiversity in grassland ecosystems *Nature* 379: 718–720; D. Tilman, F. Isbell, and M. Cowles. [2014]. Biodiversity and ecosystem functioning, *Annual Review of Ecology, Evolution, and Systematics* 45:471–493); A-M. Klein, C. Brittain, S. D. Hendrix, et al. 2012. Wild pollination services to California almond rely on semi-natural habitat. *Journal of Applied Ecology* 49:723–732)

296   ECOLOGY OF A CHANGED WORLD

on productivity are magnified over time. On 9 m x 9 m plots established at the same place, after 10 to 20 years, the productivity of plots with many species was about twice that of plots with few species. Mechanisms include increased nutrient capture (e.g., from greater soil depths); an increase in the number of predatory insects driving a trophic cascade that reduced caterpillars and herbivorous insects and, consequently, leaf loss; feedbacks from plant matter accumulated in the soil adding extra nutrients.

Researchers have conducted more than 150 experiments along these lines. On average, the reduction of the number of species on a plot by about 30% decreased productivity by almost 10%. Productivity loss arises from two equally important contributions. First, the presence of more species means that, by chance, one is likely to grow especially well in the given conditions. Second, different plant species occupy different niches (e.g., one may grow rapidly and require abundant light, whereas another may grow more slowly and is shade tolerant) and therefore use resources more efficiently than a single species attempting multiple strategies. Other ecosystem functions besides productivity have been studied:

"Greater plant diversity is now known to lead to greater net primary productivity, greater root mass, greater use of limiting soil nutrients, greater rates of accumulation of soil organic carbon and organic nitrogen, greater ecosystem stability, greater resistance to invasion by exotic plant species, greater insect diversity, and lower incidences of species-specific plant diseases". Tilman, Hartley and Clark (2019).

In this quote, greater ecosystem stability means that the total number of individuals of all species combined varies less from one year to the next in more species-rich systems. Resistance to invasion by exotic species is attributed in part to more drawdown of nutrients (Fig. 28.1).

Bees

One the best examples of the dangers of community simplification comes from bees, which are important pollinators of crops (notably fruits, nuts, coffee, and many vegetables). A survey of 89 studies showed that the more bee species that are present locally, the greater the numbers of fruits that formed. For example, in a Brazil study of cashews, a location with five bee species visiting the cashews had a 10% fruit set (the fraction of flowers producing fruits), whereas one with ten visiting bee species had a 40% fruit set. In California, the presence of one bee species visiting almonds was associated with an average of a 35% fruit set, whereas the presence of 10 bee species resulted in a 55% fruit set (Fig. 28.1). In both cases, high species richness was associated with close proximity to natural habitats. The increase in fruit set results for two reasons that are of similar importance (and resemble those for plant productivity): First, the more species there are, the more bees there are in total, and second, different species visit flowers in different ways, thereby diversifying the chances of pollination.

These studies show how a diverse assemblage of wild bees is good for pollination. However, despite the presence of at least 17,000 species of bees in the world, pollination of crops is mostly conducted by just one, the honeybee (the honeybee contributed the greatest fruit set in both the examples given above). The honeybee is native to Eurasia and Africa. It has been domesticated for 6,000 years and was brought to North America in the 1600s, where it now persists both in the wild and in managed captive colonies. In the United States, across seven studied crops (apple, watermelon, blueberry, pumpkin, sweet cherry, tart cherry, and almond), the honeybee provides more than 50% of the pollination services for all except watermelon. Practically all of California's almonds are pollinated by bees coming out of 1 million hives (two-thirds of all hives in the United States). Here is a 2016 quotation from a bee keeper: "My bees are in California pollinating almonds" [in February]. "In the middle of March, they are going to be trucked all the way back to Florida to pollinate oranges, then they are trucked another thousand miles north to pollinate apples in Pennsylvania." In the summer many colonies are transported to the Great Plains, where the bees make honey. Dependence on this single species for so much pollination comes with risks. In North America, starting sometime before 2006, large numbers of hives were abandoned by the worker bees (a phenomenon known as "colony collapse disorder") particularly in the winter, for reasons that are mostly attributed to parasites (especially an invasive mite derived from an Asian bee), magnified by the effects of pesticides. The mite both eats bee parts and carries several debilitating viruses. In the years 2010–2020, annual colony losses ran at 30–40%. Presently, colony maintenance is so difficult that recent economic assessments have concluded that it is not profitable for beekeepers to rent out their hives to almond growers. The planting of hedgerows to encourage wild bees may help alleviate the problem.

The detrimental effects of simplification are often amplified by additional losses of species. In Chapter 21 we described how overfishing on coral reefs led to the rise of sea urchins because the large fish that ate them and competed with them for food were removed. These urchins then grew to high densities that promoted the spread of disease, which in turn caused urchin populations to decline in some places. The consequent algal overgrowth of coral reefs then led to barren reefs that are of little value to anyone.

298 ECOLOGY OF A CHANGED WORLD

28.2 Ecosystem Functions Affected by Declines
of Single Species

The concept of keystone species maintaining communities can be extended to that of keystone species maintaining ecosystem functions. The loss of a single species, such as killer whales and lions, lead to trophic cascades that have ramifying effects on ecosystems (Chapter 4). Single species contribute many other ecosystem functions beyond the prevention of trophic cascades:

Oysters

Consider the benefits of the American oyster in Chesapeake Bay. Oyster harvesting in the Chesapeake area is now less than 2% of what it was 150 years ago. The oysters are much smaller than they were in more pristine times. In ancient refuse dumps, some oyster shells are 1 ft (30 cm) across. Formerly, gray whales, dolphins, manatees, river otters, steelhead salmon, sea turtles, alligators, giant sturgeon, sharks, and rays were all present in the bay. The greatest decline in oysters happened toward the end of the 1800s, driven by overfishing and new technologies associated with fishing, notably dredging, which led to habitat destruction. In the 1950s, oyster declines were accelerated by diseases, at least one of which was a virus known as MSX that came from Asia along with an introduced oyster.

Of course, a direct cost to us is that we eat oysters, and overharvesting has led to a much lower harvest than would otherwise be possible. But a second cost is reduction in ecosystem function. Oysters are efficient grazers on plankton. They are estimated to have filtered the water of the bay every three days 100 years ago, compared with just one to two times a year now. Correspondingly, bottom-dwelling microorganisms in the bay have been largely replaced by planktonic microorganisms, generating algal blooms, eutrophication, anoxia, and fish deaths. In addition to the lack of filtering by oysters, the increase of algae in the bay also results from more nutrients in the water (i.e., pollution). Again we see an interaction of multiple threats: pollution, invasive species, and overharvesting have all contributed to the deterioration of the Bay's native species, both directly and indirectly, through the effects on other species, most notably the oyster.

28.3 Direct Value of Species

Species provide many direct benefits to humans beyond their role in ecosystems. According to India's greatest naturalist, Sálim Ali, wildlife conservation should

THE VALUE OF SPECIES   299

be for "scientific, cultural, aesthetic, recreational and economic reasons." The preamble to the U.S. Endangered Species Act (1973) states that "Species are of esthetic, ecological, educational, historical, recreational, and scientific value to the Nation and its people." Many animal and plant products are consumed (e.g., oysters and honey) and surely many more could be. The original DNA sequencing technology depended on an enzyme that comes from a hot spring bacterium. Nine of the top 10 prescription drugs come from organisms. Antibiotics are derived from fungi, but less than 10% of all fungi have been named, let alone evaluated for benefits. Indigenous people use many species of plants for medicinal purposes. Very many species have adaptations that we are only just beginning to understand. For example, octopi have fantastic camouflage that they rapidly alter to fit their background. How they do this has even interested the U.S. Department of Defense.

Consider the story of a poison dart frog. Poison dart frogs have long been of value to humans. Native South Americans use the frog skin excretion in their blow darts, with which they hunt birds and mammals. In 1974, John Daly from the U.S. National Institutes of Health collected samples of skin secretions from two populations of one species, *Epipedobates anthonyi,* in Ecuador (Fig. 28.3, end of the chapter). Small amounts of the toxin administered to mice appeared to block pain more effectively than morphine, but it was too toxic for use in humans. To produce a less toxic version, in 1976 the scientists returned to make more collections. One population had completely vanished: much of the surrounding forest had been felled and converted to banana plantations. At a second site they did find a few frogs. However, in 1984, passage of the Convention on International Trade in Endangered Species (CITES) Treaty (Chapter 20) made it very difficult to export what was then classified as a threatened species (it is now considered near-threatened). Instead, in 1990, technological breakthroughs meant that the chemistry of the sample collected earlier could be determined. Abbott Laboratories began to develop similar compounds and eventually created ABT-594, a nontoxic, nonaddictive painkiller potentially effective for treating several types of pain. Unlike morphine, ABT-594 apparently "promotes alertness instead of sleepiness and has no side effects on respiration or digestion." Its value seems clear, but the drug has not been marketed, primarily because the harmful side-effects of higher doses have yet to be resolved.

While many species do not provide obvious benefits, it is hard to know which ones will, and still harder to know how they will be valued in the future. For example, it may sometimes be profitable to harvest the last few individuals of a species of whale and invest the funds at a few percent interest. However, the value of a whale may increase much more quickly than interest, as shown by the whale-watching industry in California. The Florida grasshopper sparrow is an endangered subspecies that is declining due to habitat loss. The nature writer Ted

300 ECOLOGY OF A CHANGED WORLD

Williams famously said we should fight to preserve it "not because it is anything, only because it is" (i.e., we should preserve it for ethical reasons). Nevertheless, he noted that it is already a source of enrichment of at least some human lives. More tangibly, it likely provides ecosystem functions by consuming pests, and we might predict that the habitat associated with its preservation will become an economically profitable use of land in the future as recreational areas.

28.4 Prospects

This book has deliberately focused on summarizing the underpinnings to the biodiversity crisis. Many issues have not been touched upon that might be up for debate. New opportunities for conservation are opening up: management of the environment rather than simply leaving natural places to themselves is increasingly accepted. For example, cities are introducing green spaces and have attracted rare species such as peregrine falcons in Europe and the United States (but one reason is perhaps that cities are pesticide free, and another is the an abundance of pigeons). Ecosystem functioning has been altered by adding nonnative species to areas that used to have extinct species, as has happened to some extent already (e.g., horses in the United States), as well as to areas that were naturally species-poor (for example, Ascension Island's mountain peak was transformed from fern-dominated into luxuriant forest through introductions). Other ideas include moving species about to help them keep up with climate change, bringing mammoths back from extinction given their DNA has been preserved in ice, and introducing genes into invasive species that reduce their fertility. All these issues have been seriously raised in the literature. They may preserve some species, but not most, and they may also cause extinctions.

Graphs throughout this book show exponential—or sometimes greater than exponential—increases in the exploitation of natural resources, and we should return to this phenomenon to summarize the overarching threats. Figure 28.2 gives a final example. In 2017, imports of tropical logs into China increased four-fold over those of previous years to a level that was sustained in 2018 (China accounts for about 85% of the entire international trade in tropical logs). The imports are primarily from West Africa, New Guinea, and the Solomon Islands. Among the imports were rosewoods, which consist of about 300 tropical species, valued for making high-end furniture and for its fragrance. Combined across species, rosewoods are thought to be the world's most trafficked wild product by value, exceeding that of products from pangolin, elephant, and rhinoceros combined. In 2017, more than 4 million rosewood trees were exported from Nigeria to China, even though several species were on the CITES 2 list, requiring certification of legality by the exporting country, which was often obtained through

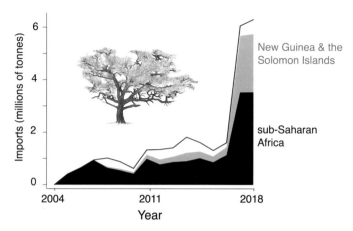

Figure 28.2 Tropical logs imported to China and two main sources. A tonne may correspond to 3–5 trees, implying a harvest of more than 18 million tropical trees for China in 2017. (Chatham House. 2018. resourcetrade.earth, http://resourcetrade.earth). Tropical log imports are not recorded in this source beyond 2018.

government corruption (the Ministry of the Environment can issue permits). The CITES Secretariat described this trade as "lawful but awful." Because of lobbying, import from Africa may have slowed after 2018 (tropical logs are now not separated out in the database), but this is unlikely for New Guinea and the Solomon Islands, and we anticipate increasing threats to the tropical parts of the Americas. It is impossible for increases such as these to be sustained for any length of time. Indeed, in some cases, exploitation is decreasing because of depletion (e.g., rosewoods from South Asia, fisheries as a whole). In other cases, technology may result in a switch from one depleted resource to another that satisfies the same demand (e.g., until recently, moving to a new fishery once one had been exploited). Such boom-and-bust cycles imply an ongoing decline in the earth's resources.

A feature of present exploitation is that much is driven by companies and governments, far removed from the threatened location. In tropical areas with extensive wilderness, such as the Amazon, West Africa, and New Guinea, this has placed Indigenous peoples under threats similar to those of the forests they depend upon, leading to substantial activism, albeit with modest success. In these areas, conservationists now emphasize the need to respect Indigenous cultures and to pay more attention to environmental justice and cultural issues. In other tropical areas that have been more heavily exploited, including Indonesia, India, east Africa, and the tropical part of China, the preservation of nature now relies on protected areas. While legislation and enforcement can help, again and again we see that successful conservation requires local support, as well as the

Figure 28.3 Poison dart frog, *Epipedobates anthonyi*. (Holger Krisp, CC 3.0)

incorporation of local knowledge, including an understanding of cultural and political forces.

Often, successful conservation efforts reflect the impact of dedicated individuals operating within either the government or nongovernmental organizations (NGOs). In the late 1990s, illegal logging, palm oil franchises, and hunting severely threatened Gunung Palung National Park in Indonesia (Chapter 13). Subsequently, local efforts have reduced pressures on the park. They have included negotiating with the government to acknowledge its importance, bringing cases against poachers, engaging schoolchildren, encouraging sustainable exploitation methods that were already in practice, and reviving ecotourism. Ultimately, it seems likely that the safest way to protect an area is through increased local incomes, ecotourism being the most obvious route. In Rwanda's Volcanoes National Park, the gorilla population has more than doubled over the past 20 years, and tourists are willing to pay thousands of dollars to spend a short time in the field with them. The government reported that $70 million were generated from gorilla trekking in 2019. Tourism is Rwanda's biggest source of foreign exchange, driven by its natural resources.

Over the past 50,000 years, our predecessors have removed some fantastic beasts. Unlike our predecessors, we are aware of the issues, and it is possible to stem the loss. While the future is uncertain, the possible and favorable outcome is that we will have met E. O. Wilson's challenge (Chapter 1) to move people out of poverty, while maintaining a substantial fraction of the world's current biodiversity, thereby bringing the world into balance.

Notes and References

With the exception of the Acknowledgments section, the first references to each chapter are often further general readings. See also the online material for appendices to the chapters and worked examples. All references to studies that come with associated figures or tables are in the figure and table legends and are not repeated here.

Preface

Barney Ronay quote: www.theguardian.com/football/audio/2020/jun/04/the-life-and-times-of-barney-ronay-football-weekly-extra. For references to bat disease, see Chapter 19.

Acknowledgments

Programming:

R Core Team (2020). *R: A language and environment for statistical computing.* R Foundation for Statistical Computing. Vienna, Austria.

E. Pebesma. (2018). Simple features for R: Standardized support for spatial vector data. *The R Journal* 10: 439–446.

R. J. Hijmans. (2020). Raster: Geographic data analysis and modeling. *R package version* 3.3–13.

Chapter 1: The Changed World

A concise summary of the issues facing us: S. Díaz, J. Settele, E. S. Brondízio, et al. (2019). Pervasive human-driven decline of life on earth points to the need for transformative change. *Science* 366:1327.

An early discussion of the six threats as we have categorized them: J. P. Collins and A. Storfer. (2003). Global amphibian declines: Sorting the hypotheses. *Diversity and Distributions* 9:89–98.

Four excellent books on topics covered in the book are as follows:

E. Kolbert. (2014). *The sixth extinction.* New York: Holt and Company. Account of threatened species.

P. Lymbery. (2017). *Dead zone.* London: Bloomsbury. Impact of modern farming on nature.

M. Lynas. (2020). *Our final warning: Six degrees of climate emergency.* New York: Harper Collins. Climate change.

C. Roberts. (2007). *Unnatural history of the sea.* Washington, DC: Island Press.

Income increase comes from the World Bank website (http://data.worldbank.org), based on per capita gross domestic product (GDP); note that this summarizes earnings and ignores investment returns. The meaning of GDP is considered further in Chapter 12.

304 NOTES AND REFERENCES

For references concerning climate change, see Chapters 10 and 15. Typically, only five mass extinctions, defined as >75% species lost in a reasonably short time interval (<1 million years; Chapter 25), are rated as having occurred over the past 500 million years, but some other times also experienced high extinction rates: R. K. Bambach. (2006). Phanerozoic biodiversity mass extinctions. *Annual Review of Earth and Planetary Science* 34:127–155.

Chapter 2: Population Growth

Status of HIV in the United States: https://www.hiv.gov/hiv-basics/overview/data-and-trends/statistics.
Malthus quotes: T. Malthus. (1993). An essay on the principle of population. Oxford: Oxford University Press, p. 61.

Chapter 3: Population Regulation

Swarm of desert locusts: M. Ullman. (2006). African desert locusts in Morocco in November 2004. *British Birds* 99:489–491.
Evidence for initial cod recovery: K. T. Frank, B. Petrie, J. A. D. Fisher, et al. (2011). Transient dynamics of an altered large marine ecosystem. *Nature* 477:86–89. (Considered further in Chapter 21)

Chapter 4: Interactions between Species: Mutualisms and Competition

R. H. MacArthur. (1972). *Geographical ecology*. New York: Harper and Row.
Number of species in Hawaii: A. Allison. (2003). Biological surveys? New perspectives in the Pacific. *Organisms, Diversity and Evolution* 3:103–110.
Fayatree story: P. M. Vitousek, L. R. Walker, L. D. Whiteaker, et al. (1987). Biological invasion by *Myrica faya* alters ecosystem development in Hawaii. *Science* 238:802–804; P. M. Vitousek and L. R. Walker. (1989). Biological invasion by *Myrica faya* in Hawai'i: Plant demography, nitrogen fixation, ecosystem effects. *Ecological Monographs* 59:247–265; L. R. Walker and P. M. Vitousek. (1991). An invader alters germination and growth of a native dominant tree in Hawai'i. *Ecology* 72:1449–1455.
Quotes from Grinnell: M. F. D. Udvardy. (1959). Notes on the ecological concepts of habitat, biotope and niche. *Ecology* 40:725–728.
Coexistence of Darwin's finches on the Galápagos: P. R. Grant and B. R. Grant. (2006). Evolution of character displacement in Darwin's finches. *Science* 313:224–226; P. R. Grant and B. R. Grant. (2011). *How and why species multiply: The radiation of Darwin's finches*. Princeton, NJ: Princeton University Press; T. Price. (1987). Diet variation in a population of Darwin's Finches. *Ecology* 68:1015–1028.
Alternative *Theories of Species Numbers:* D. M. Wilkinson. (1999). The disturbing history of intermediate disturbance. *Oikos* 84:145–147; S. P. Hubbell. (2001). *The unified neutral theory of biodiversity and biogeography*. Princeton, NJ: Princeton University Press.
Sink populations on mountains: A. L. Hargreaves, K. E Samis, and C. G. Eckert (2014). Are species' range limits simply niche limits writ large? A review of transplant experiments beyond the range. *American Naturalist* 183:157–173.

With reference to Figure 4.6, the basic ideas are in G. E. Hutchinson. (1959). Homage to Santa Rosalia or why are there so many kinds of animals? *American Naturalist* 93:145–159. A quantitative model is in: D. W. Schwilk and D. D. Ackerly. (2005). Limiting similarity and functional diversity along environmental gradients. *Ecology Letters* 8:272–281.

Chapter 5: Predation and Food Webs

Trophic cascades (resulting in the phrase "baboonification of Africa" in the popular press): J. A. Estes, J. Terborgh, J. S. Brashares, et al. (2011). Trophic downgrading of planet earth. *Science* 333:301–306.

The otter story (includes statements regarding otters eating lobsters): J. A. Estes, D. O. Duggins, and G. B. Rathbun. (1989). The ecology of extinctions in kelp forest communities. *Conservation Biology* 3:252–264. Otter distributions can be seen on the U.S. Geological Survey website (http://www.werc.usgs.gov).

Experiments on removal of starfish: R. T. Paine. (1974). Intertidal community structure: Experimental studies on the relationship between a dominant competitor and its principal predator. *Oecologia* 15:93–120.

Chapter 6: Parasites and Pathogens

C. Zimmer. (2000). *Parasite Rex*. New York: Free Press.

Parasites in Costa Rica: R. Carreno, L. A. Durden, D. L. Brooks, et al. (2001). *Parelaphostrongylus tenuis* and other parasites of white-tailed deer (*Odocoileus virginianus*) in Costa Rica. *Comparative Parasitology* 68:177–184; D. R. Brooks and D. A. McLennan. (2001). *The nature of diversity: An evolutionary voyage of discovery*. Chicago: University of Chicago Press.

Guinea worm infection numbers: World Health Organization (www.who.int).

Behavior of fish infected with fluke larvae: K. D. Lafferty and A. K. Morris. (1996). Altered behavior of parasitized killifish increases susceptibility to predation by bird final hosts. *Ecology* 77:1390–1397.

Years of life lost to COVID: H. Pifarré i Arolas, E. Acosta, G. López-Casasnovas, et al. (2021). Years of life lost to COVID-19 in 81 countries. *Scientific Reports* 11:3504.

Analysis of transmission dynamics: C. E. Mills, J. M. Robins, and M. Lipsitch. (2004). Transmissibility of 1918 pandemic influenza. *Nature* 432:904–906.

Measles vaccination in Romania: P. M. Strebel and S. L. Cochi. (2001). Waving goodbye to measles. *Nature* 414:695–696.

Chapter 7: Evolution and Disease

P. Ewald. (1994). *The evolution of infectious disease*. New York: Oxford University Press. Several examples are from this reference, such as the increase in virulence as a result of needle transfer of *Plasmodium knowlesi*.

S. M. Hedrick. (2004). The acquired immune system: A vantage from beneath. *Immunity* 21:607–615. This article asks the question: What would happen if we did not have an acquired immune system?

D. Quammen. (2015). *The chimp and the river*. New York: W. W. Norton. The origin of AIDS.

306 NOTES AND REFERENCES

Dobzhansky statement: T. Dobzhansky (1973). Nothing in biology makes sense except in the light of evolution. *American Biology Teacher* 35: 125–129

Malarial adaptations to the immune system: N. J. Spillman, J. R. Beck, and D. E. Goldberg. (2015). Protein export into malaria parasite-infected erythrocytes: Mechanisms and functional consequences. *Annual Review of Biochemistry* 84:813–841.

Chapter 8: The Human Food Supply: Competition, Predation and Parasitism

References to calorific intake are taken from the FAO website: www.fao.org/faostat/ Corn consumption: United States Department of Agriculture website (http://www.ers.usda.gov).

Crop loss to weeds from the FAO website: http://www.fao.org/3/a-y4751e/y4751e0l.htm.

100 rice plants and 10 weeds (barnyard grass): B. Graf and J. E. Hill. (1992). Modelling the competition for light and nitrogen between rice and *Echinochloa crusgalli*. *Agricultural Systems* 40:345–359.

Corn resistance to atrazine: R. H. Shimabukuro, D. S. Frear, H. R. Swanson, et al. (1971). Glutathione conjugation: An enzymatic basis for atrazine resistance in corn. *Plant Physiology* 47:10–14.

Herbicide resistance: C. N. Stewart (ed.). (2009). *Weedy and invasive plant genomics*. Ames, IA: Wiley-Blackwell, p. 4.

Biological control, cottony cushion scale insect: M. Begon, J. L. Harper, and C. R. Townsend. (1990). *Ecology: Individuals, populations and communities*. Malden, MA: Blackwell, pp. 571–572.

Chapter 9: Food Security

J. A. Foley, N. Ramankutty, K. A. Brauman, et al. (2011). Solutions for a cultivated planet. *Nature* 478:337–342.

H. C. J. Godfray, J. Pretty, S. M. Thomas, et al. (2011). Linking policy on climate and food. *Science* 331:1013–1014.

M. Springmann, M. Clark, D. Mason-D'Croz, et al. (2018). Options for keeping the food system within environmental limits. *Nature* 562:519–525.

Numbers of people severely undernourished are from the United Nations Food and Agriculture Organization website. The methodology to estimate these numbers has changed over time. The designation *Severely Undernourished* is considered to indicate >100 kilocalories below the mean daily energy requirement, which varies by age, sex, and activity, but for adults, averages about 1800 kcal (from http://www.fao.org/econo mic/ess/ess-fs/ess-fadata/en/#.VMVtmMZrxko). Income levels are from the World Bank website: www.worldbank.org/en/publication/global-monitoring-report.

Cropland and irrigation statistics are from the FAO website. https://www.fao.org/faostat.

Estimates of waste: M. V. Vilariño, C. Franco, and C. Quarrington. 2017. Food loss and waste reduction as an integral part of a circular economy. *Frontiers in Environmental Science* 5:21.

Calorific values after processing: E. S. Cassidy, P. C. West, J. S. Gerber, et al. (2013). Redefining agricultural yields: From tonnes to people nourished per hectare. *Environmental Research Letters* 8:UNSP 034015.

The assessment of aquaculture's contribution to our diet comes from the UN's 2020 Fish and Agriculture Report, plus a recent evaluation of wild-caught fish (Chapter 21).

Amount of ocean needed for aquaculture: R. R. Gentry, H. E. Froehlich, D. Grimm, et al. (2017). Mapping the global potential for marine aquaculture. *Nature Ecology and Evolution* 1:1317–1324.

Statistics on feed and conversion ratios for salmon and other animals, as well as proportions of fishmeal in salmon diets: Marine Harvest (OSE:MHG): 2018 Salmon Industry Handbook http://hugin.info/209/R/2200061/853178.pdf.

Discussion on alleviation of the impact of fishmeal: H. E. Froehlich, N. S. Jacobsen, T. E. Essington, et al. (2018). Avoiding the ecological limits of forage fish for fed aquaculture. *Nature Sustainability* 1:298–303. The figure of 2 billion that could be fed off the same land under pescatarian diets is my qualitative assessment from earlier statements, as well as consideration of the following article, which describes various scenarios for mixed land and aquatic diets: H. E. Froehlich, C. A. Runge, R. R. Gentry, et al. (2018). Comparative terrestrial feed and land use of an aquaculture-dominant world. *Proceedings of the National Academy of Sciences* 115:5295–5300. Scallops versus chicken eggs: B. S. Halpern, personal communication.

Farms in Zimbabwe: D. Tilman, C. Balzer, J. Hill, et al. (2011). Global food demand and the sustainable intensification of agriculture. *Proceedings of the National Academy of Sciences* 108:20260–20264.

Proportion of crops that are GM from the USDA: www.ers.usda.gov/data-products/adoption-of-genetically-engineered-crops-in-the-us/recent-trends-in-ge-adoption.aspx# GM cotton in China: F. Hossain, C. E. Pray, Y. Lu, et al. (2004). Genetically modified cotton and farmers' health in China. *International Journal of Occupational and Environmental Health* 10:296–303.

Purple tomatoes: Y. Zhang, E. Butelli, and C. Martin (2014). Engineering anthocyanin biosynthesis in plants. *Current Opinion in Plant Biology* 19:81–90.

Omega-3 fatty acids in plants: N. Ruiz-Lopez, R. P. Haslam, S. Usher, et al. (2015). An alternative pathway for the effective production of the omega-3 long-chain polyunsaturates EPA and ETA in transgenic oilseeds. *Plant Biotechnology Journal* 13:1264–1275. M. B. Betancor, M. Sprague, S. Usher, et al. (2015). A nutritionally-enhanced oil from transgenic *Camelina sativa* effectively replaces fish oil as a source of eicosapentaenoic acid for fish. *Nature Scientific Reports* 5:8104.

Chapter 10: Prediction

S. W. Pacala, E. Bulte, J. A. List, et al. (2003). False alarm over environmental false alarms. *Science* 301:1187–1188.

N. Silver. (2011). *The signal and the noise.* London: Penguin books. This work is by the person who accurately predicted the outcome of the 2012 presidential election and founded the 'prediction' website fivethirtyeight.com.

Quote on the similarity between ecology and economics: C. Roberts. (2007). *Unnatural history of the sea.* Washington, DC: Island Press, p. 340.

Predictions on climate change: Intergovernmental Panel on Climate Change's 5th report. IPCC (2013): Summary for Policymakers. In: T. F. Stocker, D. Qin, G.-K. Plattner, et al. (eds.) *Climate Change 2013: The Physical Science Basis. Contribution of Working Group I to the Fifth Assessment Report of the Intergovernmental Panel on Climate Change.* Cambridge: Cambridge University Press.

308 NOTES AND REFERENCES

Clouds and climate: O. Boucher, D. Randall, P. Artaxo, et al. (2013): Clouds and aerosols. In the same report as the previous citation.

Rise of syphilis in Baltimore: Centers of Disease Control: www.cdc.gov/mmwr/preview/mmwrhtml/00040527.htm. The example was described in M. Gladwell. (2002). *The tipping point: How little things can make a big difference.* New York: Little, Brown.

P. Ehrlich. (1968). *The population bomb.* New York: Ballantine Books

Ozone hole: https://ozonewatch.gsfc.nasa.gov/.

Chapter 11: Human Population Growth

J. Cohen. (1995). *How many people can the world support?* New York: W. W. Norton.

W. Lutz. (2009). Editorial: Towards a world of 2-6 billion well-educated and therefore healthy and wealthy people. *Journal of the Royal Statistical Society Series A-Statistics in Society* 172:701–705.

Chimpanzee population sizes: S. Strindberg, F. Maisels, E. A. Williamson, et al. (2018). Guns, germs, and trees determine density and distribution of gorillas and chimpanzees in western Equatorial Africa. *Science Advances* 4:eaar2964.

Effective population sizes of humans 200,000 years ago have been given as just ~10,000: H. Li and R. Durbin. (2011). Inference of human population history from individual whole-genome sequences. *Nature* 475:493–496. The effective population size is based on genetics and matches the true population size only under restrictive conditions. However, chimpanzees have greater genetic diversity than humans, despite a much smaller population at the present day, suggesting that in the past they had a much larger population than humans (J. Prado-Martinez, P. H. Sudmant, J. M. Kidd, et al. [2013]. Great ape genetic diversity and population history. *Nature* 499:471–475).

Human history: H. S. Groucutta, E. M. L. Scerria, L. Lewis, et al. (2015). Stone tool assemblages and models for the dispersal of *Homo sapiens* out of Africa. *Quaternary International* 38:8–30; C. Perreault and S. Mathew. (2012). Dating the origin of language using phonemic diversity. *PLoS ONE* 7:e35289; I. Hershkovitz, G. W. Weber, R. Quam, et al. (2018). The earliest modern humans outside Africa. *Science* 359:456–459; C. Barras. (2020). Controversial cave discoveries suggest humans reached Americas much earlier than thought. *Nature* 583:670–671.

Influenza kill numbers in 1918: P. Spreeuwenberg, M. Kroneman, and J. Paget (2018). Reassessing the global mortality burden of the 1918 influenza pandemic. *American Journal of Epidemiology* 187:2561–2567.

Famines: M. Davis (2001). *Late Victorian holocausts.* London: Verso.

Rwanda war: J. Diamond. (2004). *Collapse.* New York: Viking Press.

History of beer: D. J. Hanson. (2007). An overview of alcohol use in ancient history. In A. Manheimer (ed.), *Alcohol.* Detroit: Greenhaven Press, pp. 20–28.

Cave Paintings: A. Brumm, A. A. Oktaviana, B. Burhan, et al. 2021. Oldest cave art found in Sulawesi. *Science Advances* 7:eabd4648.

Fertility rates in Ethiopia: D. Canning, S, Raja, and A.Yazbeck, eds. *Africa's demographic transition: Dividend or disaster?* Washington, DC: The World Bank, 2015. https://www.loc.gov/item/2015007219, p. 14.

Malaria and Sri Lanka: C. M. Langford. (1996). Reasons for the decline in mortality in Sri Lanka immediately after the Second World War: A reexamination of the evidence. *Health Transition Review* 6:3–23.

Shanghai fertility estimate: M. G. Merli and S. P. Morgan. (2011). Below replacement fertility preferences in Shanghai. Population 66: 519–542.

Causes of declines in fertility in Kenya are from J. Cohen. (1995). How many people can the world support? New York: W. W. Norton, p. 66.

U.S. statistics: United States Census Bureau (http://www.census.gov/popclock). According to this source, in March 2019, there were about 7.5 births per minute, 2 immigrants (net—i.e., immigrants minus emigrants), and 6 deaths, implying population size in the United States is increasing by 3.5/minute.

Chapter 12: Growth of Wealth and Urbanization

J. B. Foster, B. Clark, and R. York (2010). *The ecological rift.* New York: Monthly Review Press. This book introduces Jevon's and Lauderdale's paradoxes and energy use in the last century.

C. Wheelan. (2010). *Naked economics.* New York: W. W. Norton. This book lists the price of chickens for different dates, as discussed in the text.

N. Klein. (2014). *This changes everything.* New York: Simon and Schuster. See p. 413 for a classic statement about imports of "useless stuff" from China.

Whale quotation: C. Roberts. (2007). *Unnatural history of the sea.* Washington, DC: Island Press, p. 112.

Technology: Support for the idea that technology can help the environment: M. R. Chertow. (2001).The IPAT equation and its variants: Changing views of technology and environmental impact. *Journal of Industrial Ecology* 4:13–29.

Stern report: N. H. Stern. (2007). *The Economics of Climate Change: The Stern Review.* Cambridge, UK: Cambridge University Press

Renewable costs: IRENA. (2021). *Renewable Power Generation Costs in 2020.* Abu Dhabi: International Renewable Energy Agency.

Energy supply by commodity. Statistical Review of World Energy 2020. www.bp.com.

"It is often stated that about half the impact on the environment over the past 40 years or so comes from population growth and half from increase in wealth" (P. Ehrlich, lectures).

The sulfur dioxide auction emissions program: the EPA website: https://www.epa.gov/airmarkets/so2-allowance-auctions.

Urbanization and urban centers: A. J. Florczyk, C. Corbane, M. Schiavina, et al. (2019). *GHS Urban Centre Database 2015, multitemporal and multidimensional attributes, R2019A.* European Commission, Joint Research Centre (JRC) http://data.europa.eu/89h/53473144-b88c-44bc-b4a3-4583ed1f547e.

Degree of urbanization: *Atlas of the human planet 2018: A world of cities.* (2018) European Commission, Joint Research Centre. Publications Office, LU.

Urbanization as a relief for wildlands: N. Myers. (1979). *The sinking ark.* Oxford: Pergamon Press.

Predictions and the urban vs. suburban comparison: United Nations, Department of Economic and Social Affairs, and Population Division. (2019). *World urbanization prospects: The 2018 revision.* https://population.un.org/wup.

Urbanization and the demographic transition: E. W. Sanderson, J. Walston, and J. G. Robinson. (2018). From bottleneck to breakthrough: Urbanization and the future of biodiversity conservation. *Bioscience* 68:412–426.

310 NOTES AND REFERENCES

Correlation of city size and impacts: L. Bettencourt. (2013). The origins of scaling in cities. *Science* 340:1438–1441.

Wildlands usurped: K. C. Seto, B. Güneralp, and L. R. Hutyra. (2012). Global forecasts of urban expansion to 2030 and direct impacts on biodiversity and carbon pools. *Proceedings of the National Academy of Sciences* 109:16083–16088.

Manaus and fish: D. J. Tregidgo, J. Barlow, P. S. Pompeu, et al. (2017). Rainforest metropolis casts 1,000-km defaunation shadow. *Proceedings of the National Academy of Sciences* 114:8655–8659.

Pollution: N. B. Grimm, S. H. Faeth, N. E. Golubiewski, et al. (2008). Global change and the ecology of cities. *Science* 319:756–760.

Education: R. Chepesiuk. (2007). Environmental literacy: Knowledge for a healthier public. *Environmental Health Perspectives* 115:A494–A499.

Biodiversity loss: E. Piano, C. Souffreau, T. Merckx, et al. (2020). Urbanization drives cross-taxon declines in abundance and diversity at multiple spatial scales. *Global Change Biology* 26:1196–1211.

Bee pollinators: D. M. Hall, G. R. Camilo, R. K. Tonietto, et al. (2017). The city as a refuge for insect pollinators. *Conservation Biology* 31:24–29.

Yellowstone visits: U.S. National Park Service: www.nps.gov. Corbett visits: M. Ghosh-Harihar, R. An, R. Athreya, et al. (2019). Protected areas and biodiversity conservation in India. *Biological Conservation* 237:114–124.

Chapter 13: Habitat Conversion

The UN Millennium ecosystem assessment rates habitat loss as the primary threat: http://www.millenniumassessment.org/en/index.html.

Living Planet Report: WWF. (2020). Living Planet Report 2020. Bending the curve of biodiversity loss. R. E. A. Almond, M. Grooten, and T. Petersen (eds)., Gland, Switzerland: WWF. https://livingplanet.panda.org/en-us. The Living Planet Report site presents data on land-use patterns (see Figs. 13.2, 13.3). The computations were made by the Global footprint network: footprintnetwork.org. The weighting factors and areas of land use falling into different categories (Table 13.1) are described in: B. Ewing, D. Moore, S. Goldfinger, et al. (2010.) *The ecological footprint atlas* 2010. Oakland: Global Footprint Network, http://www.footprintnetwork.org/en/index.php/GFN). This paper gives the technical details from which the salient points are presented in the Living Planet Report. It has been updated regularly (e.g., M. Borucke, D. Moore, G. Cranston, et al. [2013]. Accounting for demand and supply of the biosphere's regenerative capacity: The National Footprint Accounts' underlying methodology and framework. *Ecological Indicators* 24:518–533; D. Lin, L. Hanscom, A. Murthy, et al. (2018). Ecological footprint accounting for countries: Updates and results of the National Footprint Accounts, 2012–2018. *Resources* 7:58–22). The updates give areas (except for fishing grounds), but not the exact weightings. However, the data itself are available from the global footprint network: www.footprintnetwork.org/en/index.php/GFN/page/footprint_data_and_results. I used the data to confirm that weightings had not substantially changed from the 2010 report.

Loss of wetlands: U.S. Fish and Wildlife, http://www.fws.gov/wetlands/Status-and-Trends/index.html. Loss of tallgrass prairie is also discussed by the U.S. Fish and Wildlife Service, but the estimate of >99% loss in Iowa comes from: D. D. Smith. (1998). Iowa prairie: Original extent and loss, preservation and recovery attempts. *Journal of the Iowa Academy of Science* 105: 94–108.

NOTES AND REFERENCES 311

Most of the data on general trends in forest loss comes from the references in Figure 13.4. However, the 5-yearly FAO reports, especially the one from 2015, were consulted by referencing the paper R. J. Keenan, G. A. Reams, F. Achard, et al. (2015). Dynamics of global forest area: Results from the FAO Global Forest Resources Assessment 2015. *Forest ecology and management* 352:9–20, http://www.fao.org/forest-resources-assessment/current-assessment/en. The FAO report is referred to as (1) the authority on primary, regenerated, and planted forest and (2) as a statement that the rate of loss of forest is less than it was in the 1990s. The FAO study is mainly based on country-specific questionnaires, which generate heterogeneity and sometimes inflated estimates of forest gained. In particular, satellite data (Figure 13.5) measures both deforestation and gain, whereas the FAO data measures the net change in forest cover. Both give similar estimates of net decrease in Brazil and Indonesia over 2000–2010, about 10,000 km^2 per year in Indonesia, and 30,000 km^2 per year in Brazil.

For information on tropical species richness and for our uncertainty about the number of species, see Chapter 23.

For the crisis at the turn of the century in Gunung Palung National Park: L. M. Curran, S. N. Trigg, A. K. McDonald, et al. (2004). Lowland forest loss in protected areas of Indonesian Borneo. *Science* 303:1000–1003.

Amazonia has >15% world's locked-up carbon: T. R. Feldpausch, J. Lloyd, S. L. Lewis, et al. (2012). Tree height integrated into pantropical forest biomass estimates. *Biogeosciences* 9:3381–3403; S. Fauset, M. O. Johnson, M. Gloor, et al. (2015). Hyperdominance in Amazonian forest carbon cycling. *Nature Communications* 6:6857.

Forest loss: http://news.mongabay.com/2015/11/amazon-deforestation-jumps-in-brazil. Mongabay.com gives an accounting of Brazil, Indonesia, and the Congo forests, but for Indonesia I used https://nusantara-atlas.org.

Chapter 14: Economics of Habitat Conversion

The Economics of Ecosystems and Biodiversity website (www.teebweb.org).

C. Wheelan. (2010). *Naked economics*. New York: W. W. Norton. Accessible introduction to economics.

S. Polasky, C. L. Kling, S. A. Levin, et al. (2019). Role of economics in analyzing the environment and sustainable development. *Proceedings of the National Academy of Sciences* 116:5233–5238.

G. Chichilnisky and G. Heal. (1998). Economic returns from the biosphere. *Nature* 391:629–630.

New York City's water supply: An entry into the points raised here comes from the Economics of Ecosystems and Biodiversity website (www.teebweb.org); see especially the report on ecological and economic foundations (2010), Chapter 9 (www.teebweb.org/publication/the-economics-of-ecosystems-and-biodiversity-teeb-ecological-and-economic-foundations).

Travel costs and valuation: The figures on Indian national parks come from: R. Badola, S. A. Hussain, B. K. Mishra, et al. (2010). An assessment of ecosystem services of Corbett Tiger Reserve, India. *The Environmentalist* 30:320–329; A. Bharali and R. Mazumder. (2012). Application of travel cost method to assess the pricing policy of public parks: The case of Kaziranga National Park. *Journal of Regional Development and Planning* 1:41–50.

312 NOTES AND REFERENCES

No use costs of U.S. national parks: M. Haefele, J. B. Loomis, and L. Bilmes. (2016). Total economic valuation of the National Park Service Lands programs: Results of a survey of the American public. (June 28, 2016). HKS Working Paper No. 16-024, Available at SSRN: http://dx.doi.org/10.2139/ssrn.2821124.

Coral reefs and an entry into methods: L. Carr and R. Mendelsohn. (2003). Valuing coral reefs: A travel cost analysis of the Great Barrier Reef. *AMBIO* 32:353–357.

Mangrove cover: The 20% estimate comes from the FAO and the 50% estimate from the Thai Forestry Department: E. B. Barbier. (2007). Valuing ecosystem services as productive inputs. *Economic Policy* 49:178–229.

Two methods of evaluating storm protection: R. DasGupta and R. Shaw. (2013). Cumulative impacts of human interventions and climate change on mangrove ecosystems of South and Southeast Asia: An overview. *Journal of Ecosystems*:379429.

Mangrove loss (note that this is not for all Thailand, but for just the Tsunami affected region): C. Giri, Z. Zhu, L. L. Tieszen, et al. (2008). Mangrove forest distributions and dynamics (1975–2005) of the tsunami-affected region of Asia. *Journal of Biogeography* 35:519–528.

REDD and *REDD+* Information comes from the UN website, http://www.un-redd. org. One controversy over REDD was observed by myself in Nicaragua, where in the town of San Juan del Norte (also known as Greytown) in December 2014, posters opposing REDD+ were on prominent display. This issue of land appropriation to help with climate change amelioration was prominently raised at the COP26 meeting in 2021. Criticisms of the use of cost-benefit analyses in decision making: F. Ackerman. (2008).

Critique of cost-benefit analysis, and alternative approaches to decision-making: A Report to Friends of the Earth England, Wales and Northern Ireland.

Chapter 15: Climate Crisis: History

D. Archer. (2009). *The long thaw*. Princeton, NJ: Princeton University Press.

Apart from the citation to Figure 15.1, details on the experiments with *Foramanifera*, and a general description of what causes the ~140,000-year cycles are in the classic paper J. Zachos, M. Pagani, L. Sloan, et al. (2001). Trends, rhythms, and aberrations in global climate 65 Ma to present. *Science* 292:686–693.

Ice core records going back 800,000 years can be downloaded from http://ncdc.noaa.gov/ paleo/study/17975. Note that this site only has CO_2 records and that the oxygen isotope records span a shorter period.

Land warming faster than oceans: G. Jia, E. Shevliakova, P. Artaxo, et al. (2020). Land–climate interactions. In: P. R. Shukla, J. Skea, E. Calvo Buendia, et al. (eds.). *Climate change and land: An IPCC special report on climate change, desertification, land degradation, sustainable land management, food security, and greenhouse gas fluxes in terrestrial ecosystems*. https://www.ipcc.ch/srccl.

Energy budget: "a gallon of gas may be derived from 90 tonnes of plant matter": J. S. Dukes. (2003). Burning buried sunshine: Human consumption of ancient solar energy. *Climatic Change* 61:31–44.

Energy imbalance: J. Hansen, P. Kharecha, M. Sato, et al. (2013). Assessing "dangerous climate change": Required reduction of carbon emissions to protect young people, future generations and nature. *PLoS ONE* 8:e81648.

NOTES AND REFERENCES 313

CO_2: The compilation of CO_2 estimates used to buttress the statement that they are higher than they have been for as long as the past 20 million years: citation to Figure 15.1; Y. G. Zhang, Mark Pagani, Zhonghui Liu, et al. (2013). A 40-million-year history of atmospheric CO_2. *Philosophical Transactions of the Royal Society A* 371:20130096.

Discussions of chemical reactions and carbon cycles: D. Archer. (2010). *The global carbon cycle*. Princeton, NJ: Princeton University Press. Ocean currents as a driver of changing concentrations of atmospheric CO_2 are discussed in this book and in J. Yu, L. Menviel, Z. D. Jin, et al. (2020). Last glacial atmospheric CO_2 decline due to widespread Pacific deep-water expansion. *Nature Geoscience* 13:628–633.

Increase in CO_2 as a driver of increased uptake of CO_2 by increased plant growth: D. Schimel, B. B. Stephens, and J. B. Fisher (2014). Effect of increasing CO_2 on the terrestrial carbon cycle. *Proceedings of the National Academy of Sciences* 112:436–441.

The possibility that slowed plate movements resulted in reduced outgassing of CO_2 15–5 million years ago: P. Voosen. (2021). Slowdown in plate tectonics may have led to ice sheets. *Science* 371:14.

Ice lost from Greenland: A. Shepherd, E. Ivins, Eric Rignot, et al. (2019). Mass balance of the Greenland ice sheet from 1992 to 2018. *Nature* 579:233–239.

Amazon droughts: S. L. Lewis, P. M. Brando, O. L. Phillips, et al. (2011). The 2010 Amazon drought. *Science* 331:554; A. Erfanian, G. Wang, and L. Fomenko (2017). Unprecedented drought over tropical South America in 2016: Significantly underpredicted by tropical SST. *Scientific Reports* 7:5811.

Species responses: N. Waller, I. C. Gynther, A. B. Freeman, et al. (2017). The Bramble Cay melomys *Melomys rubicola* (Rodentia:Muridae): A first mammalian extinction caused by human-induced climate change? *Wildlife Research* 44:9–21; C. Parmesan and G. Yohe. (2003). A globally coherent fingerprint of climate change impacts across natural systems. *Nature* 421:37–42. Upslope movement of lower range limits: C. Román-Palacios and J. J. Wiens. (2020). Recent responses to climate change reveal the drivers of species extinction and survival. *Proceedings of the National Academy of Sciences* 117:4211–4217.

B. G. Freeman, Y. Song, K. J. Feeley, et al. (2021). Montane species track rising temperatures better in the tropics than in the temperate zone. *Ecology letters* 24:1697–1708.

Chapter 16: Predictions of Future Climate and Its Effects

M. Lynas. (2020). *Our final warning: Six degrees of climate emergency.* New York: Harper Collins.

Cumulative emissions scenarios: O. Edenhofer, R. Pichs-Madruga, Y. Sokona, et al. (eds.). (2014). IPCC: *Climate change 2014: Mitigation of climate change. Contribution of Working Group III to the Fifth Assessment Report of the Intergovernmental Panel on Climate Change.* Cambridge: Cambridge University Press

One degree: J. Hansen, P. Kharecha, M. Sato, et al. (2013). Assessing "dangerous climate change": Required reduction of carbon emissions to protect young people, future generations and nature. *PLoS ONE* 8:e81648.

Last interglacial temperatures: S. Bova, Y. Rosenthal, Z. Liu, et al. (2021). Seasonal origin of the thermal maxima at the Holocene and the last interglacial. *Nature* 589:548–553. This publication gives sea surface temperatures about 0.75°C warmer than the average of the past 1000 years, which was about 0.2°C more than during the 1880–1920 period.

314   NOTES AND REFERENCES

The estimate of sea surface temperatures is thus comparable to that of the present day (https://data.giss.nasa.gov/gistemp/graphs_v4).

Two and three degrees: T. F. Stocker, D. Qin, G.-K. Plattner, et al. (eds.). (2013). Technical Summary. In: *Climate Change 2013: The physical science basis. Contribution of Working Group I to the Fifth Assessment Report of the Intergovernmental Panel on Climate Change.* Cambridge Cambridge University Press.

Long term sea level rise: A. Levermann, P. U. Clark, B. Marzeion, et al. (2013). The multimillennial sea-level commitment of global warming. *Proceedings of the National Academy of Sciences* 110:13745–13750.

Amazon quote: M. Lynas. (2008). *Six degrees.* New York: Harpercollins, p. 143.

Climate velocity: S. R. Loarie, P. B. Duffy, H. Hamilton, et al. (2009). The velocity of climate change. *Nature* 462:1052–1055.

Issues associated with methods of prediction of species responses to climate change: T. P. Dawson, S. T. Jackson, J. I. House, et al. (2011). Beyond predictions: Biodiversity conservation in a changing climate. *Science* 332:53–58.

Few extinctions recorded during the ice ages: S. M. Kidwell. (2015). Biology in the Anthropocene: Challenges and insights from young fossil records. *Proceedings of the National Academy of Sciences* 112:4922–4929.

The episode of warming ~55.6 mya: F. A. McInerney and S. L. Wing. (2011). The Paleocene-Eocene thermal maximum: A perturbation of carbon cycle, climate, and biosphere with implications for the future. *Annual Review of Earth and Planetary Sciences* 39:489–516.

Coral reefs: O. Hoegh-Guldberg, P. J. Mumby, A. J. Hooten, et al. (2007). Coral reefs under rapid climate change and ocean acidification. *Science* 318:1737–1742.

Historical extinction of coral reefs: W. Kiessling and C. Simpson. (2011). On the potential for ocean acidification to be a general cause of ancient reef crises. *Global Change Biology* 17:56–67. At 500 ppm atmospheric CO_2, very few coral reefs may survive, although some species may prove particularly adaptable: It appears that reefs can physiologically change over time to deal with altered conditions, thereby helping them to persist. In an experiment associated with temperature, a slowing in the growth of reefs was compensated after 2 years of warm temperature: S. R. Palumbi, D. J. Barshis, N. Traylor-Knowles, et al. (2014). Mechanisms of reef coral resistance to future climate change. *Science* 344:895–898.

Reefs became rare in the tropics in the Pleistocene interglacial: W. Kiessling, C. Simpson, B. Beck, et al. (2012). Equatorial decline of reef corals during the last Pleistocene interglacial. *Proceedings of the National Academy of Sciences* 109:21378–21383.

Northward movements: H. Yamano. K. Sugihara, and K. Nomura. (2011). Rapid poleward range expansion of tropical reef corals in response to rising sea surface temperatures. *Geophysical Research Letters* 38:L04601.

Great Barrier Reef: A. Dietzel, M. Bode, S. R. Connolly, et al. (2020). Long-term shifts in the colony size structure of coral populations along the Great Barrier Reef. *Proceedings of the Royal Society of London B* 287:20201432.

Biofuels: "40% of corn grain in the US devoted to biofuel production": U.S. Department of Agriculture: www.ers.usda.gov/data-products/us-bioenergy-statistics and R. H. Mumm, P. D. Goldsmith, Kent D Rausch, et al. (2014). Land usage attributed to corn ethanol production in the United States: Sensitivity to technological advances in corn grain yield, ethanol conversion, and co-product utilization. *Biotechnology for Biofuels* 7:61. www.biotechnologyforbiofuels.com/content/7/1/61.

NOTES AND REFERENCES   315

The proportion of land devoted to corn and other crops: faostat3.fao.org/home/E.

Gasoline and biofuel consumption: the U.S. Energy Information Administration:http://www.eia.gov/forecasts/steo/data.cfm?type=tables.

Gas and coal compared for emissions are from the U.S. Energy Information website www.eia.gov/tools/faqs/faq.php?id=73&t=11. Coal emits 35% more CO_2 than oil and 85% more than gas for the same gain in energy.

New York becoming carbon neutral: Mark Z. Jacobsona, R. W. Howarth, M. A. Delucchi, et al. (2013). Examining the feasibility of converting New York State's all-purpose energy infrastructure to one using wind, water, and sunlight. *Energy Policy* 57:585–601.

Chapter 17: Pollution

Mining: T. T. Werner, A. Bebbington, G. Gregory, et al. (2019). Assessing impacts of mining: Recent contributions from GIS and remote sensing. *The Extractive Industries and Society* 6:993–1012.

Plastics in the sea: M. Eriksen, L. C. M. Lebreton, H. S. Carson, et al. (2014). Plastic pollution in the world's oceans: More than 5 trillion plastic pieces weighing over 250,000 tons afloat at sea. *PLoS ONE* 9:e111913.

Vulture declines: See Figure 24.1. Light pollution effects on insects: A. C. S. Owens, P. Cochard, J. Durrant, et al. (2020). Light pollution is a driver of insect declines. *Biological Conservation* 241:108259.

Pigs in USA: www.aphis.usda.gov/animal_health/emergingissues/downloads/1pigs.pdf.

Nitrogen inputs and consequences: R. W. Howarth. (2008). Coastal nitrogen pollution: A review of sources and trends globally and regionally. *Harmful Algae* 8:14–20. This paper states that about 80% of the nitrogen atoms in a typical human have been fixed industrially, but I have not been able to locate an original source for this number. See also D. Fowler, M. Coyle, U. Skiba, et al. (2013). The global nitrogen cycle in the twenty-first century. *Philosophical Transactions of the Royal Society B* 368:20130164.

J. W. Erisman, J. N. Galloway, S. Seitzinger, et al. (2013). Consequences of human modification of the global nitrogen cycle. *Philosophical Transactions of the Royal Society B* 368:20130116.

M. B. Soons, M. M. Hefting, E. Dorland, et al. (2017). Nitrogen effects on plant species richness in herbaceous communities are more widespread and stronger than those of phosphorus. *Biological Conservation* 212:390–397.

A. J. van Strien, C. A. M. van Swaay, W. T. F. H. van Strien-van Liempt, et al. (2019). Over a century of data reveal more than 80% decline in butterflies in the Netherlands. *Biological Conservation* 234:116–122. See especially Figure 5 of this article, which assesses declines within heathlands.

Direct effects of reactive nitrogen: J. A. Camargo and Á. Alonso. (2006). Ecological and toxicological effects of inorganic nitrogen pollution in aquatic ecosystems: A global assessment. *Environment International* 32:831–849.

Dead zones: R. J. Diaz and R. Rosenberg. (2008). Spreading dead zones and consequences for marine ecosystems. *Science* 321:926–929.

Doubling of reactive nitrogen inputs is beyond what the world can tolerate: W. Steffen, K. Richardson, J. Rockström, et al. (2015). Planetary boundaries: Guiding human development on a changing planet. *Science* 347:1259855. This refers to a group of

316 NOTES AND REFERENCES

researchers who have developed the concept of planetary boundaries. Planetary boundaries are values for such items as ocean acidification, freshwater cooption, and carbon in the atmosphere that we should not cross. Those considered to have crossed far beyond what the planet can tolerate were solely: (1) the rate of loss of species and populations (genetic diversity), (2) the annual inputs of phosphorus, and (3) annual inputs of nitrogen fertilizer. (Carbon was not included because a total concentration of 450 parts per million is considered the boundary, although many would argue this is beyond safety.)

Acid rain: EPA website. International air quality data is from the World Health Organization. Mortality from pollution: J. Lelieveld, K. Klingmüller, A. Pozzer, et al. (2019). Cardiovascular disease burden from ambient air pollution in Europe reassessed using novel hazard ratio functions. *European Heart Journal* 391:464–467. Air pollution and Covid: https://energyandcleanair.org/wp/wp-content/uploads/2020/04/CREA-Europe-COVID-impacts.pdf.

Ecotoxicology: H -R. Koehler and R. Triebskorn. (2013). Wildlife ecotoxicology of pesticides: Can we track effects to the population level and beyond? *Science* 341:759–765.

30 million kg of atrazine a year: U.S. Geological Service website: https://water.usgs.gov/nawqa/pnsp/usage/maps/show_map.php?year=2011&map=ATRAZINE&hilo=L&disp=Atrazine.

Effects of atrazine: L. N Vandenberg, T. Colborn, T. B. Hayes, et al. (2012). Hormones and endocrine-disrupting chemicals: Low-dose effects and nonmonotonic dose responses. *Endocrine Reviews* 33:378–455.

Atrazine effects in the food web: J. R. Rohr, A. M. Schotthoefer, T. R. Raffel, et al. (2008). Agrochemicals increase trematode infections in a declining amphibian species. *Nature* 455:1235–1240.

Pesticide side effects: K. A. Grogan. (2014). When ignorance is not bliss: Pest control decisions involving beneficial insects. *Ecological Economics* 107:104–113.

DDT story comes from the EPA and U.S. Fish and Wildlife websites:

"A wolf may eat about 100x its own weight in meat during its lifetime": This is conservative, based on the assumptions of 7-year life, 55 kg adult weight, and 2.5 kg meat per day.

Inuit data from the Government of Canada. Canadian Arctic Contaminants Assessment Report Series: Synopsis of Research Conducted under the 2012–2013 Northern Contaminants Program, p. 39 https://www.ic.gc.ca/eic/site/063.nsf/eng/h_97659.html.

Silver toxicity: http://www.inchem.org/documents/cicads/cicads/cicad44.htm#6.0.

Chapter 18: Invasive Species

J. Lockwood, M. F. Hoopes, and M. P. Marchetti (2007). *Invasion ecology*. New York: Blackwell.

Information on invasive species in the United States: the U.S. Department of Agriculture website: www.invasivespeciesinfo.gov.

North American starling current population size is estimated to be ~93 million, a decline from 175 million in 1970: K. V. Rosenberg, A. M. Dokter, P. J. Blancher, et al. (2019). Decline of the North American avifauna. *Science* 366:120–124.

NOTES AND REFERENCES   317

Framework for studying invasions: T. M. Blackburn, P. Pyšek, S. Bacher, et al. (2011). A proposed unified framework for biological invasions. *Trends in Ecology and Evolution* 26:333–339.

Cane toad: J. R. Jolly, R. Shine, and M. J. Greenlees. (2015). The impact of invasive cane toads on native wildlife in southern Australia. *Ecology and Evolution* 5:3879–3894.

Live plant trade and invasive species: A. M. Liebhold, E. G. Brockerhoff, L. J. Garrett, et al. (2012). Live plant imports: The major pathway for forest insect and pathogen invasions of the US. *Frontiers in Ecology and the Environment* 10:135–143.

Importance of aquaculture movements: A. Zhan, E. Briski, D. G. Bock, et al. (2015). Ascidians as models for studying invasion success. *Marine Biology* 162:2449–2470.

Importance of trade, plus reference to the Suez Canal and melting sea ice as ways of breaking down barriers: F. Essl, S. Bacher, T. M. Blackburn, et al. (2015). Crossing frontiers in tackling pathways of biological invasions. *BioScience* 65:769–782; D. Normile. (2004). Invasive species expanding trade with China creates ecological backlash. *Science* 306:968–969 (refers to the longhorn beetle, as well as invasive species that have arrived in China from the United States).

Descriptions of fungal infections on wheat, and reference to a possible role for Hurricane Ivan: A. Y. Rossman. (2009). Ecological impacts of non-native invertebrates and fungi on terrestrial ecosystems. *Biological Invasions* 11:97–107.

Success of bird invasions in urban rather than native habitats: D. Sol, I. Bartomeus, and A. S. Griffin. (2012). The paradox of invasion in birds: Competitive superiority or ecological opportunism? *Oecologia* 169:553–564.

Introduction effort and feeding generalism as an explanation of successful invasions: P. Cassey, T. M. Blackburn, D. Sol, et al. (2004). Global patterns of introduction effort and establishment success in birds. *Proceedings of the Royal Society of London B* 271:S405–S408; B. P. Beirne. (1975). Biological control attempts by introductions against pest insects in field in Canada. *Canadian Entomologist* 107:225–236.

The importance of local conditions (lack of competitors, high resources, no enemies): K. Shea and P. Chesson. (2002). Community ecology as a framework for biological invasions. *Trends in Ecology and Evolution* 17:170–176; D. W. Redding, A. L. Pigot, E. E. Dyer, et al. (2019). Location-level processes drive the establishment of alien bird populations worldwide. *Nature* 571:103–106.

Loss of parasites on introduced species: M. E. Torchin, K. D. Lafferty, A. P. Dobson, et al. (2003). Introduced species and their missing parasites: *Nature* 421:628–630; C. E. Mitchell and A. G. Power. (2003). Release of invasive plants from fungal and viral pathogens. *Nature* 421:625–627.

Burmese python: See Figure 24.1.

Brown tree snake on Guam: T. H. Fritts and G. H. Rodda. (1998). The role of introduced species in the degradation of island ecosystems: A case history of Guam. *Annual Review of Ecology and Systematics* 29:113–140. This article includes details on possible subsidies provided by introduced species. Plant species diversity also seems to have decreased as a consequence of ramifying effects through the food web. A colloquial account of spider encounters on Guam is in D. Quammen. (1997). *The song of the dodo: Island biogeography in an age of extinction*. New York: Scribner. Genetic studies on the origin of the brown tree snake: J. Q. Richmond, D. Wood, J. Stanford, et al. (2014). Testing for multiple invasion routes and source populations for the invasive brown treesnake *(Boiga irregularis)* on Guam: Implications for pest management. *Biological Invasions* 17:337–349.

318 NOTES AND REFERENCES

Threats of invasive species in causing extinctions: M. A. McGeoch, S. H. M. Butchart, D. Spear, et al. (2010). Global indicators of biological invasion: Species numbers, biodiversity impact and policy responses. *Diversity and Distributions* 16:95–108.

Economic costs in Great Britain, including discussion about Japanese knotweed: F. Williams, R. Eschen, A. Harris, et al. (2010). The economic cost of invasive non-native species on Great Britain. *Knowledge for Life*: www.cabi.org.

Obama quote: Personal observation in 2007.

Cannabis: M. P. Fleming and R. C. Clarke. (1998). Physical evidence for the antiquity of *Cannabis sativa* L. (*Cannabaceae*). *Journal of the International Hemp Association* 5:80–92.

Mammal replacements: E. J. Lundgren, D. Ramp, J. Rowan, et al. (2020). Introduced herbivores restore Late Pleistocene ecological functions. *Proceedings of the National Academy of Sciences* 117:7871–7878.

Hawaii food web: J. Vizentin-Bugoni, C. E. Tarwater, J. T. Foster, et al. (2019). Structure, spatial dynamics, and stability of novel seed dispersal mutualistic networks in Hawai'i. *Science* 364:78–82.

Chapter 19: Introduced Disease

C. C. Mann. (2005). *1491: New revelations of the Americas before Columbus.* New York: A. A. Knopf. This book describes the inferred impacts of disease on Native American populations.

Covid-19 overall summary: S. Cobey. (2020). Modeling infectious disease dynamics. *Science* 368:713–714.

Connection to colds: S. M. Kissler, C. Tedijanto, Edward Goldstein, et al. (2020). Projecting the transmission dynamics of SARS-CoV-2 through the post pandemic period. *Science* 368:860–868.

The failure to vaccinate in Africa may lead to 250,000 deaths from measles: D. Guha-Sapir, I. Keita, G. Greenough. et al. (2020). COVID-19 policies: Remember measles. *Science* 361:269.

White nose disease: E. Cornwell, D. Elzinga, S. Stowe, et al. (2019). Modeling vaccination strategies to control white-nose syndrome in little brown bat colonies. *Ecological Modelling* 407:108724.

Spread of *Batrachochytrium dendrobatidis,* and the associated disease chytridiomycosis: L. Berger, A. A. Roberts, J. Voyles, et al. (2016). History and recent progress on chytridiomycosis in amphibians. *Fungal Ecology* 19:89–99; S. J. O'Hanlon, A. Rieux, R. A. Farrer, et al. (2018). Recent Asian origin of Chytrid fungi causing global amphibian declines. *Science* 360:621–627; Costa Rica: H. Zumbado-Ulate, K. N. Nelson, A. García-Rodríguez, et al. (2019). Endemic infection of *Batrachochytrium dendrobatidis* in Costa Rica: Implications for amphibian conservation at regional and species level. *Diversity* 11:129.

Fungal diseases in crops: P. Corredor-Moreno and D. G. O. Saunders. (2020). Expecting the unexpected: Factors influencing the emergence of fungal and oomycete plant pathogens. *New Phytologist* 225:118–125.

General review of wheat blast's history: P. C. Ceresini, V. L. Castroagudín, F. Ávila, et al. (2018). Wheat blast: Past, present, and future. *Annual Review of Phytopathology* 56:427–456.

NOTES AND REFERENCES   319

Northward movement of plant pathogens: D. P. Bebber, M. A. T. Ramotowski, and S. J. Gurr. (2013). Crop pests and pathogens move polewards in a warming world. *Nature Climate Change* 3:985–988.

The story of rinderpest: P-P. Pastoret, K.Yamanouchi, U. Mueller-Doblies, et al. (2006). Rinderpest—An old and worldwide story: History to *c.* 1902. In: T. Barrett, P-P. Pastoret, and W. P. Taylor (eds.). *Rinderpest and peste des petits ruminants: Virus plagues of large and small ruminants.* Amsterdam: Elsevier; C. A. Spinage. (2003). *Cattle plague: A history.* Boston: Springer.

Pastoret quotes: G. Scott. (2000). The murrain now known as rinderpest. *Newsletter from the Tropical Agricultural Association, U.K.* 20:14–16.

Responses of mammals after the removal of rinderpest: A. R. E. Sinclair. (1979). The eruption of the ruminants. In: A. R. E. Sinclair and M. Norton-Griffiths (eds.). *Serengeti: Dynamics of an ecosystem.* Chicago: University of Chicago Press.

Bat removal effects on pesticides: E. Frank. (2021). *The economic impacts of ecosystem disruptions: private and social costs from substituting biological pest control.* Manuscript. https://www.eyalfrank.com/research

Chapter 20: Harvesting on Land

Ripple, W. J. (2015). Collapse of the world's largest herbivores. *Science Advances* 1:e1400103. This article not only documents the decline in large animals but gives some examples of cascading effects through the food chain.

Note that, economically, harvesting can be an important contributor to pushing people above the poverty line and increasing nutrition. In Nepal, harvesting of lichens and wild asparagus contributes more than half of a $200/year cash income for some villagers, which for 20% of the villagers is the only source of all income (the rest have very small smallholdings): T. N. Maraseni, G. P. Shivakoti, G. Cockfield, et al. [2006]. Nepalese non-timber forest products: An analysis of the equitability of profit distribution across a supply chain in India. *Small-scale Forest Economics, Management, Policy* 5:191–206. In Madagascar, child nutrition correlates with hunting: C. D. Golden, L. C. H. Fernald, J. S. Brashares, et al. (2011). Benefits of wildlife consumption to child nutrition in a biodiversity hotspot. *Proceedings of the National Academy of Sciences* 108:19653–19656.

J. E. Fa and D. Brown. (2009). Impacts of hunting on mammals in African tropical moist forests: A review and synthesis. *Mammal Review* 39:231–264.

A short summary with references to harvesting rates and sustainability: E. L. Bennett, E. J. Milner-Gulland, M. Bakarr. et al. (2002). Hunting the world's wildlife to extinction. *Oryx* 36:328–329.

Pangolin harvest in Africa: D. J. Ingram, L. Coad, L. K. Abernethy, et al. (2018). Assessing Africa-wide pangolin exploitation by scaling local data. *Conservation Letters* 11:e12389.

Ivory trade: E. L. Bennett. (2015). Legal ivory trade in a corrupt world and its impact on African elephant populations. *Conservation Biology* 29:54–60.

Meat eaten in Gabon: D. S. Wilkie, M. Starkey, K. Abernethy, et al. (2005). Role of prices and wealth in consumer demand for bushmeat in Gabon, Central Africa. *Conservation Biology* 19:268–274.

Proportion eaten: R. Nasi, A. Taber, and N. van Vliet (2011). Empty forests, empty stomachs? Bushmeat and livelihoods in the Congo and Amazon Basins. *International*

320   NOTES AND REFERENCES

Forestry Review 13:355–368. This reference also suggests that 250,000 km² of pasture would be equivalent to sustainable harvesting from the Congo Basin.

Cambodia religious release: M. Gilbert, C. Sokha, P. H. Joyner, et al. (2012). Characterizing the trade of wild birds for merit release in Phnom Penh, Cambodia and associated risks to health and ecology. *Biological Conservation* 153:10–16.

Harry Potter: V. Nijman and K. A.-I. Nekaris. (2017). The Harry Potter effect: The rise in trade of owls as pets in Java and Bali, Indonesia. *Global Ecology and Conservation* 11:84–94.

Review of Cites and IUCN: B. R. Scheffers, B. F. Oliveira, I. Lamb, et al. (2019). Global wildlife trade across the tree of life. *Science* 366:71–76, *Science* 369: eabd8164.

Pet trade: E. R. Bush, S. E. Baker, and D. W. Macdonald. (2014). Global trade in exotic pets 2006–2012. *Conservation Biology* 28:663–676.

170 studies of hunting depletion: A. Benítez-López, R. Alkemade, A. M. Schipper, et al. (2017). The impact of hunting on tropical mammal and bird populations. *Science* 356:180–183.

Cameroon/Nigeria border: J. E. Fa, S. Seymoura, J. Dupain, et al. (2006). Getting to grips with the magnitude of exploitation: Bushmeat in the Cross–Sanaga river region, Nigeria and Cameroon. *Biological Conservation* 29:497–510.

Issues of drivers of hunting pressures are in the Fa and Brown (2009) review cited above; see also Figure 20.1.

Forest elephant decline: F. Maisels, S. Strindberg, S. Blake, et al. (2013). Devastating decline of forest elephants in central Africa. *PLoS ONE* 8:e59469; G. Wittemyer, J. M. Northrup, J. Blanc, et al. (2014). Illegal killing for ivory drives global decline in African elephants. *Proceedings of the National Academy of Sciences* 111:13117–13121.

Use of the logistic model to estimate MSY: D. S. Wilkie, M. Wieland, and J. R. Poulsen. (2019). Unsustainable vs. sustainable hunting for food in Gabon: Modeling short- and long-term gains and losses. *Frontiers in Ecology and Evolution* 7:357.

Impacts of monkey harvesting: C. A. Peres, T. Emilio, J. Schietti, et al. (2016). Dispersal limitation induces long-term biomass collapse in overhunted Amazonian forests. *Proceedings of the National Academy of Sciences* 113:892–897.

Chapter 21: Harvesting in the Ocean

General history and excellent introduction: C. Roberts. (2007). *Unnatural history of the sea*. Washington, DC: Island Press.

J. B. C. Jackson. (2001). What was natural in the coastal oceans? *Proceedings of the National Academy of Sciences* 98:5411–5418.

Sharks and tuna in Indonesian archaeological sites: S. O'Connor, R. Ono, and Chris Clarkson (2011). Pelagic fishing at 42,000 years before the present and the maritime skills of modern humans. *Science* 334:1117–1121.

California Channel Islands and early cascades: T. C. Rick and J. M. Erlandson. (2009). Coastal exploitation. *Science* 325:952–953.

Shifting baselines: D. Pauly. (1995). Anecdotes and the shifting baseline syndrome of fisheries. *Trends in Ecology and Evolution* 10:430.

Cod: J. B. C. Jackson, M. X. Kirby, W. H. Berger, et al. Historical overfishing and the recent collapse of coastal ecosystems. (2001). *Science* 293:629–638; Evidence for the beginnings of a cod recovery: K. T. Frank, B. Petrie, J. A. D. Fisher, et al. (2011). Transient dynamics of an altered large marine ecosystem. *Nature* 477:86–89.

Worldwide decline in sharks: B. Wörm, B. Davis, L. Kettemer, et al. (2013). Global catches, exploitation rates, and rebuilding options for sharks. *Marine Policy* 40:194–204.

FAO reports entitled *"The State of World Fisheries and Aquaculture"* are published every 2 years; they give the status of reported stocks, as well as discussions on selected topics (information on minimal shark catch was taken from the 2014 report), http://www. fao.org/fishery/sofia/en. Using Google Earth: D. Al-Abdulrazzak and D. Pauly. (2014). Managing fisheries from space: Google Earth improves estimates of distant fish catches. *ICES Journal of Marine Science* 71:450–455.

Rays and scallops: R. A. Myers, J. K. Baum, T. D. Shepherd, et al. (2007). Cascading effects of the loss of apex predatory sharks from a coastal ocean. *Science* 315:1846–1850.

Coral and seagrass examples: J. B. C. Jackson, M. X. Kirby, W. H. Berger, et al. Historical overfishing and the recent collapse of coastal ecosystems. (2001). *Science* 293:629–638.

Steller's sea cow: J. A. Estes, A. Burdin, and Daniel F. Doak. (2016). Sea otters, kelp forests, and the extinction of Steller's sea cow. *Proceedings of the National Academy of Sciences* 113:880–885.

Chapter 22: Harvesting: Prospects

Tragedy of the commons: G. Hardin. (1968). The tragedy of the commons. *Science* 162:1243–1248.

Speculators benefiting from extinction: C. F. Mason, E. H. Bulte, and R. D. Horan. (2012). Banking on extinction: Endangered species and speculation. *Oxford Review of Economic Policy* 28:180–192.

Mediterranean tuna: International Commission for the Conservation of Atlantic Tunas (ICAAT) at https://www.iccat.int/en/; European Parliament summary: https://www. europarl.europa.eu.

What features make reserves work?: G. J. Edgar, R. D. Stuart-Smith, T. J. Willis, et al. (2014). Global conservation outcomes depend on marine protected areas with five key features. *Nature* 506:216–220.

Land overview of protected areas: W. F. Laurance, D. C. Useche, J. Rendeiro et al. (2012). Averting biodiversity collapse in tropical forest protected areas. *Nature* 489:290–294.

India: M. Ghosh-Harihar, R. An, R. Athreya, et al. (2019). Protected areas and biodiversity conservation in India. *Biological Conservation* 237:114–124.

Technology overview: ConservationXLabs. https://conservationxlabs.com/who-we-are.

Shark tracking: N. Queiroz, N. E. Humphries, A. Couto et al. (2019). Global spatial risk assessment of sharks under the footprint of fisheries. *Nature* 572:461–466.

Whale sharks: B. Norman, J. A. Holmberg, Z. Arzoumanian, et al. (2017). Undersea constellations: The global biology of an endangered marine megavertebrate further informed through citizen science. *BioScience* 67:1029–1043.

Whale genetics: C. S. Baker, M. Lento, F. Cipriano, et al. (2000). Predicted decline of protected whales based on molecular genetic monitoring of Japanese and Korean markets. *Proceedings of the Royal Society of London B* 267:1191–1199.

322 NOTES AND REFERENCES

Sushi restaurants: E. T. Spencer and J. F. Bruno. (2019). Fishy business: Red snapper mislabeling along the coastline of the southeastern United States. *Frontiers in Marine Science* 6:513.

Chapter 23: Species

General reference: S. L. Pimm, C. N. Jenkins, R. Abell, et al. (2014). The biodiversity of species and their rates of extinction, distribution, and protection. *Science* 344:1246752.

For species concepts: J. Coyne and H. A. Orr. (2004). *Speciation*. Sinauer. The definition of a phylogenetic species as used here is different from its original use by Joel Cracraft, which required that all individuals of one species are different from all individuals of another species in any trait, not necessarily DNA. For example, if one population contained red individuals and another blue, then these would be termed phylogenetic species, whereas here we call these morphological species. The original logic was that if one could determine the gene that affected morphology, then surely the gene would cluster in the same way as the morphology, implying that they are phylogenetic species for that gene. However, it is now clear that we need to study multiple genes, and the term *phylogenetic species* is used in the context of what the ensemble says.

Use of a 2-million-year yardstick: T. Price. (2008). *Speciation in birds*. Boulder, CO: Roberts and Co.

Assessment of phylogenetic/morphological bird species: G. F. Barrowclough, J. Cracraft, J. Klicka, et al. (2016). How many kinds of birds are there and why does it matter? *PLoS ONE* 11:e0166307.

Primate dates: A. Scally, J. Y. Dutheil, L. W. Hillier, et al. (2012). Insights into hominid evolution from the gorilla genome sequence. *Nature* 483:169–175. The base of the phylogeny in this article is stated to be 12–16 million years. We use 16 million to be consistent with the phylogeny as depicted in that paper. Note that the eastern and western gorilla are sometimes considered subspecies.

Slightly over 1.9 million animals and >450,000 plant species have been described: S. L. Pimm, C. N. Jenkins, R. Abell, et al. (2014). The biodiversity of species and their rates of extinction, distribution, and protection. *Science* 344:1246752.

Estimate of the number of species: C. Mora, D. P. Tittensor, S. Adl, et al. (2011). How many species are there on earth and in the ocean? *PLoS Biology* 9:e1001127. This study relies on the rate of description of new species. An alternative method controls for the number of people describing species, which has also increased, and the rate of discovery per person is going down. That led to a very low estimate of 2 million (M. J. Costello, S. Wilson, and B. Houlding [2012]. Predicting total global species richness using rates of species description and estimates of taxonomic effort. *Systematic Biology* 61:871–883), which, however, was revised upwards to 5 ± 3 (M. J. Costello, R. M. May, and N. E Stork. [2013]. Can we name earth's species before they go extinct? *Science* 339:413–416). Parasite numbers are very uncertain, with some researchers arguing that beyond vertebrates, there is less than one parasite per host (M. J. Costello. [2016]. Parasite rates of discovery, global species richness and host specificity. *Integrative and Comparative Biology* 56:588–599), but the possibility of cryptic parasite species remains to be assessed. The number of undescribed reptile species may be 50%: S. Meiri. (2016). Small, rare and trendy: Traits and biogeography of lizards described in the 21st century. *Journal of Zoology* 299:251–261.

NOTES AND REFERENCES   323

Cryptic species: "Five species of birds in the same museum drawer": P. Alström and U. Olsson. (1999). The golden-spectacled warbler: A complex of sibling species, including a previously undescribed species. *Ibis* 141:545–568.

Parasitoid wasps at Guanacaste: M. A. Smith, J. J. Rodriguez, J. B. Whitfield, et al. (2008). Extreme diversity of tropical parasitoid wasps exposed by iterative integration of natural history, DNA barcoding, morphology, and collections. *Proceedings of the National Academy of Sciences* 105:12359–12364.

Endangered Species Act: U.S. Fish and Wildlife Service: www.fws.gov/endangered.

Chapter 24: Population Declines

See, in general, Living Planet Index: https://livingplanetindex.org/home/index

"20% of 77 marine studies were in protected areas." The figure comes from a critique of a review of how species composition is changing over time: A. Gonzalez, B. J. Cardinale, G. R. H. Allington, et al. (2016). Estimating local biodiversity change: A critique of papers claiming no net loss of local diversity. *Ecology* 97:1949–1960. The original review, which found little evidence for change in species numbers, is: M. Dornelas, N. J Gotelli, B. McGill, et al. (2014). Assemblage time series reveal biodiversity change but not systematic loss. *Science* 344:296–299.

European studies on insect declines: C. A. Hallmann, M. Sorg, E. Jongejans, et al. (2017). More than 75% decline over 27 years in total flying insect biomass in protected areas. *PLoS ONE* 12:e0185809; S. Seibold, M. M. Gossner, N. K. Simons, et al. (2019). Arthropod decline in grasslands and forests is associated with landscape-level drivers. *Nature* 574:671–674. Debate on this paper: https://ecoevorxiv.org/cg3zs.

Summary of insect studies: R. van Klink, D. E. Bowler, K. B. Gongalsky, et al. (2020). Meta-analysis reveals declines in terrestrial but increases in freshwater insect abundances. *Science* 368:417–420.

Butterfly declines in the western United States: M. L. Forister, C. A. Halsch, C. Nice et al. (2021). Fewer butterflies seen by community scientists across the warming and drying landscapes of the American West. *Science* 371:1042–1045.

Monarch declines: S. B. Malcolm. (2018). Anthropogenic impacts on mortality and population viability of the monarch butterfly. *Annual Review of Entomology* 63:277–302.

Bumblebee declines: P. Soroye, T. Newbold, and J. Kerr. (2020). Climate change contributes to widespread declines among bumble bees across continents. *Science* 367:685–688.

Insect loss in Europe: F. Sánchez-Bayo and K. A. G. Wyckhuys. (2019). Worldwide decline of the entomofauna: A review of its drivers. *Biological Conservation* 232:8–27.

Costa Rican biologists: D. H. Janzen and W. Hallwachs. (2021). To us insectometers, it is clear that insect decline in our Costa Rican tropics is real, so let's be kind to the survivors. *Proceedings of the National Academy of Sciences* 118:e2002546117.

10% increase over the 16 years from 1995: O. Venter, E. W. Sanderson, A. Magrach, et al. (2016). Sixteen years of change in the global terrestrial human footprint and implications for biodiversity conservation. *Nature Communications* 7:12558.

An analysis of >150 studies of local harvesting has shown that even if harvesting is conducted sustainably, habitats are altered: D. Stanley, R. Voeks, and L. Short (2012). Is non-timber forest product harvest sustainable in the less developed world? A systematic review of the recent economic and ecological literature. *Ethnobiology and Conservation* 1:9.

324 NOTES AND REFERENCES

Hunting is "overwhelmingly the greatest conservation threat to Afrotropical forest wildlife": S. E. Koerner, J. R. Poulsen, E. J. Blanchard, et al. (2017). Vertebrate community composition and diversity declines along a defaunation gradient radiating from rural villages in Gabon. *Journal of Applied Ecology* 54:805–814, p. 811.

History of North American Breeding Birds Survey: J. R. Sauer, W. A. Link, J. E. Fallon, et al. (2013). The North American Breeding Bird Survey 1966–2011: Summary analysis and species accounts. *North American Fauna* 79:1–32; J. R. Sauer and W. A. Link. (2011). Analysis of the North American Breeding Bird Survey using hierarchical models. *The Auk* 128:87–98.

Sources of bird mortality: S. R. Loss, T. Will, and P. P. Marra. (2015). Direct mortality of birds from anthropogenic causes. *Annual Review of Ecology, Evolution and Systematics* 46:99–120. The estimate comes with large confidence limits. For cats, two-thirds of the birds and many more of the mammals are thought to be killed by cats with no owner (i.e., feral cats).

Neonicotinoids: Y. Li, R. Miao, and M. Khanna. (2020). Neonicotinoids and decline in bird biodiversity in the United States. *Nature Sustainability* 3:1027–1035; M. L. Eng, B. J. M. Stutchbury, and C. A. Morrissey.(2019). A neonicotinoid insecticide reduces fueling and delays migration in songbirds. *Science* 365:1177–1180.

Red List categories explained: S. H. M. Butchart, A. J. Stattersfield, L. A. Bennun, et al. (2004). Measuring global trends in the status of biodiversity: Red List indices for birds. *PLoS Biology* 2:e383. More detailed explanations are on the IUCN Red List website, http://www.iucnredlist.org/. The Red List INDEX computations: S. H. M. Butchart, H. R. Akçakaya, J. Chanson, et al. (2007). Improvements to the Red List Index. *PLoS ONE* 2:e140.

Chapter 25: Extinction

P. Brannen. (2017). *The ends of the world: Volcanic apocalypses, lethal oceans, and our quest to understand earth's past mass extinctions.* New York: HarperCollins. This book describes the history of mass extinctions in a very accessible manner.

G. Ceballos, P. R. Ehrlich, and P. H. Raven. (2020). Vertebrates on the brink as indicators of biological annihilation and the sixth mass extinction. *Proceedings of the National Academy of Sciences* 117:13596–13602.

Mass extinctions and recovery: D. Jablonski. (2005). Mass extinctions and macroevolution. *Paleobiology* 31:192–210; D. Jablonski and W. G. Chaloner. (1994). Extinctions in the fossil record. *Philosophical Transactions of the Royal Society, B* 344:11–16.

Differential fossilization of threatened and non-threatened species: R. E. Plotnick, F. A. Smith, and S. K. Lyons. (2016). The fossil record of the sixth extinction. *Ecology Letters* 19:546–555.

Other estimates for extinction rates are as high as 20%/100,000 years (implying a species persists for an average of 500,000 years) and as low as 1%/10 million years (implying a species persists for an average of 10 million years—respectively, these are from A. D. Barnosky, N. Matzke, S. Tomiya, et al. [2011]. Has the Earth's sixth mass extinction already arrived? *Nature* 471:51–57; S. L. Pimm, C. N. Jenkins, R. Abell, et al. [2014]. The biodiversity of species and their rates of extinction, distribution, and protection. *Science* 344:1246752). These seem less plausible than the intermediate rates presented here.

NOTES AND REFERENCES   325

Songbird dispersal and replacement: F. K. Barker, C. A. Schikler, P. Feinstein, et al. (2004). Phylogeny and diversification of the largest avian radiation. *Proceedings of the National Academy of Sciences* 101:11040–11045.

European fossil record: G. Mayr. (2005). The Paleogene fossil record of birds in Europe. *Biological Reviews* 80:515–542.

Nests as an advantage: S. L. Olson. (2001).Why so many kinds of passerine birds? *Bioscience* 51:268–269.

Mass extinctions. P. Brannen. (2017). *The ends of the world: Volcanic apocalypses, lethal oceans, and our quest to understand earth's past mass extinctions.* New York: HarperCollins; D. Irwin. (2015). *Extinction: How life on earth nearly ended 250 million years ago.* Updated Edition. Princeton, NJ: Princeton University Press.

Pleistocene effects: Chicago was under 1.5 km of ice 18,000 years ago: W. R. Peltier. (2004). Global glacial isostasy and the surface of the ice-age earth: The ice-5G (VM2) model and grace. *Annual Review of Earth and Planetary Sciences* 32:111–149.

Coral extinctions: A. O'Dea, J. B. C. Jackson, H. Fortunato, et al. (2007). Environmental change preceded Caribbean extinction by 2 million years. *Proceedings of the National Academy of Sciences* 104:5501–5506.

First human impacts and discussion of top down controls: Y. Malhi, C. E. Doughty, M. Galetti, et al. (2016). Megafauna and ecosystem function from the Pleistocene to the Anthropocene. *Proceedings of the National Academy of Sciences* 113:838–846.

Evidence for contribution of climate to extinctions at northerly latitudes. A. Cooper, C. Turney, K. A. Hughen, et al. (2015). Abrupt warming events drove Late Pleistocene Holarctic megafaunal turnover. *Science* 349:602–606.

Papers confirming the overwhelming role of humans: G. W. Prescott, D. R. Williams, A. Balmford, et al. (2012). Quantitative global analysis of the role of climate and people in explaining late Quaternary megafaunal extinctions. *Proceedings of the National Academy of Sciences* 109:4527–4531 (This reference provides a data set on extinction dates); C. N. Johnson, J. Alroy, N. J. Beeton, et al. (2016). What caused extinction of the Pleistocene megafauna of Sahul? *Proceedings of the Royal Society B* 283:20152399; C. Sandom, S. Faurby, B. Sandel, et al. (2014). Global late Quaternary megafauna extinctions linked to humans, not climate change. *Proceedings of the Royal Society B* 281:20133254.

Discussion of possible climate effects of the loss of large mammals: F. A. Smith, J. I. Hammond, M. A. Balk, et al. (2016). Exploring the influence of ancient and historic megaherbivore extirpations on the global methane budget. *Proceedings of the National Academy of Sciences* 113:874–879.

Pacific island extinctions: D. W. Steadman. (1995). Prehistoric extinctions of Pacific island birds—biodiversity meets zooarchaeology. *Science* 267:1123–1131; D. W. Steadman. (2006). *Extinction and biogeography of tropical Pacific birds.* Chicago: University of Chicago Press.

New Zealand moas: R. N. Holdaway and C. Jacomb. (2000). Rapid extinction of the moas (*Aves: Dinornithiformes*): Model, test, and implications. *Science* 287:2250–2254. These authors considered several scenarios beyond those outlined in the text. In one model, given a human rate of increase of 2.2% and the consumption of 1 moa/week/10 people (as in the text), it would take 70 years to extinction if 100 people arrived followed by habitat loss, and 50 years to extinction if 200 people arrived without habitat loss. Other alternatives included a human growth rate of 1% and 1 moa/week/ 10 people, which would take up to 100 years longer.

326 NOTES AND REFERENCES

Recent losses. Australian mammals: J. C. Woinarski, A. A. Burbidge, and P. L. Harrison (2015). Ongoing unraveling of a continental fauna: Decline and extinction of Australian mammals since European settlement. *Proceedings of the National Academy of Sciences* 112:4531–4540.

Comparison of extinction rates of recently described mammals: S. L. Pimm, C. N. Jenkins, R. Abell, et al. (2014). The biodiversity of species and their rates of extinction, distribution, and protection. *Science* 344:1246752.

Amphibian extinctions: J. Alroy. (2015). Current extinction rates of reptiles and amphibians. *Proceedings of the National Academy of Sciences* 112:13003–13008.

Mollusk extinctions: C. Régnier, G. Achaz, A. Lambert, et al. (2015). Mass extinction in poorly known taxa. *Proceedings of the National Academy of Sciences* 112:7761–7766.

Loss of bees and wasps from Britain: J. Ollerton, H. Erenler, M. Edwards, et al. (2014). Extinctions of aculeate pollinators in Britain and the role of large-scale agricultural changes. *Science* 346:1360–1362.

General review of insect declines: F. Sánchez-Bayo and K. A. G. Wyckhuys (2019). Worldwide decline of the entomofauna: A review of its drivers. *Biological Conservation* 232:8–27.

Snails on islands study: S. Chiba and K. Roy. (2011). Selectivity of terrestrial gastropod extinctions on an oceanic archipelago and insights into the anthropogenic extinction process. *Proceedings of the National Academy of Sciences* 108:9496–9501.

For a general discussion on threats to wide-ranging species: K. J. Gaston and R. A. Fuller. (2007). Biodiversity and extinction: Losing the common and the widespread. *Progress in Physical Geography* 31:213–225.

At least 16 bird species thought to be destined to go extinct have been saved through active intervention: S. Pimm, P. Raven, A. Peterson, et al. (2006). Human impacts on the rates of recent, present, and future bird extinctions. *Proceedings of the National Academy of Sciences* 103:10941–10946. More than 500 species of vertebrates are known to have population sizes of less than 1,000: G. Ceballos, P. R. Ehrlich, and P. H. Raven (2020). Vertebrates on the brink as indicators of biological annihilation and the sixth mass extinction. *Proceedings of the National Academy of Sciences* 117:13596–13602.

Chapter 26: Species across Space

General discussion of the latitudinal gradient: G. Mittelbach, D. W. Schemske, H. V. Cornell, et al. (2007). Evolution and the latitudinal diversity gradient: Speciation, extinction and biogeography. *Ecology Letters* 10:315–331.

One in 12 of all bird species in the northern Andes: C. Rahbek and G. R. Graves. (2001). Multiscale assessment of patterns of avian species richness. *Proceedings of the National Academy of Sciences* 98:4534–4539.

Panama vs. Illinois: T. W. Schoener. (1971). Large-billed insectivorous birds: A precipitous diversity gradient. *Condor* 73:154–161.

Wallace quote: Wallace, A. R. (1878). *Tropical nature and other essays.* New York: Macmillan, p.123.

Seasonality and species diversity: J. W. Valentine, D. Jablonski, A. Z. Krug, et al. (2008). Incumbency, diversity, and latitudinal gradients. *Paleobiology* 34:169–178.

Seasonality applied to altitudinal gradients: C. M. McCain. (2009). Vertebrate range sizes indicate that mountains may be "higher" in the tropics. *Ecology Letters* 12:550–560.

Number of trees in Peru and Missouri: A. H. Gentry. (1988). Changes in plant community diversity and floristic composition on environmental and geographical gradients. *Annals of the Missouri Botanical Gardens* 75:1. Data were downloaded from: https://ecologicaldata.org/wiki/alwyn-h-gentry-forest-transect-dataset-0.

The importance of considering phylogenetic diversity: W. Jetz, G. H. Thomas, J. B. Joy, et al. (2014). Global distribution and conservation of evolutionary distinctness in birds. *Current Biology* 24:919–930.

The approximate correspondence between Faith's PD and species richness across space: A. Voskamp, D. J. Baker, P. A. Stephens, et al. (2017). Global patterns in the divergence between phylogenetic diversity and species richness in terrestrial birds. *Journal of Biogeography* 44:709–721.

Reasons for exceptional threats to species in the tropics: (1) Small-range species often have low population sizes where they are found. J. H. Brown. (1984). On the relationship between abundance and distribution of species. *American Naturalist* 124:255–279. The pattern would be well worth a critical reexamination. (2) Tropical species in general are rarer than temperate ones. As in the first case, more data are needed, but see D. J. Currie, G. G. Mittelbach, H. V. Cornell, et al. (2004). Predictions and tests of climate-based hypotheses of broad-scale variation in taxonomic richness. *Ecology Letters* 7:1121–1134.

Atlantic rainforest: See Saving Nature. https://savingnature.com. Madagascar: from IUCN: https://www.iucn.org.

Chapter 27: Island Biogeography and Reserve Design

Original surveys in the San Diego canyons: M. E. Soulé, D. T. Bolger, A. C. Alberts, et al. (1988). Reconstructed dynamics of rapid extinctions of chaparral-requiring birds in urban habitat islands. *Conservation Biology* 2:75–92. (See also Figures 27.3 and 27.4.)

Nest failures: M. A. Patten and D. T. Bolger. (2003). Variation in top-down control of avian reproductive success across a fragmentation gradient. *Oikos* 101:479–488.

Cats in canyons: K. R. Crooks and M. E. Soulé. (1999). Mesopredator release and avifaunal extinctions in a fragmented system. *Nature* 400:563–566.

Argentine ant: C. V. Tillberg, D. A. Holway, E. G. LeBrun, et al. (2007). Trophic ecology of invasive Argentine ants in their native and introduced ranges. *Proceedings of the National Academy of Sciences* 104:20856–20861.

Tropical experiment: W. F. Laurance, T. E. Lovejoy, H. L. Vasconcelos, et al. (2002). Ecosystem decay of Amazonian forest fragments: A 22-year investigation. *Conservation Biology* 16:605–618.

Predicted loss of bird species from isolated fragments: W. D. Newmark, C. N. Jenkins, S. L. Pimm, et al. (2017). Targeted habitat restoration can reduce extinction rates in fragmented forests. *Proceedings of the National Academy of Sciences* 114:9635–9640.

Loss of mammals from western North America: W. D. Newmark. (1995). Extinction of mammal populations in western North American National Parks. *Conservation Biology* 9:512–526.

Loss of mammals from Ghana: J. S. Brashares, P. Arcese and M. K. Sam. (2001). Human demography and reserve size predict wildlife extinction in West Africa. *Proceedings of the Royal Society B* 268:2473–2478.

India reserves: M. Ghosh-Harihar, R. An, R. Athreya, et al. (2019). Protected areas and biodiversity conservation in India. *Biological Conservation* 237:114–124.

328　NOTES AND REFERENCES

Agricultural impacts on terrestrial reserves. V. Vijay and P. R. Armsworth. (2021). Pervasive cropland in protected areas highlight trade-offs between conservation and food security. *Proceedings of the National Academy of Sciences* 118:e2010121118.

Chapter 28: Value of Species

R. Dirzo, H. S. Young, M. Galetti et al. (2014). Defaunation in the Anthropocene. *Science* 345:401–406.

Daly quote: H. Daly. (2015). *From uneconomic growth to a steady-state economy.* Northampton, UK: Elgar, p. 52.

Ecosystem functions affected by biodiversity: B. J. Cardinale, J. E. Duffy, A. Gonzalez, et al. (2012). Biodiversity loss and its impact on humanity. *Nature* 486:59–67; D. Tilman, F. Isbell, and M. Cowles. (2014). Biodiversity and ecosystem functioning. *Annual Review of Ecology, Evolution and Systematics* 45:471–493.

Quote on plant diversity: D. Tilman, N. Hartline, and M. A. Clark. (2017). Saving biodiversity in the era of human-dominated ecosystems. In: T. E. Lovejoy and L. Hannah (eds.). *Biodiversity and climate change: Transforming the biosphere.* New Haven, CT: Yale University Press, p. 357.

Bees: M. Dainese, E. A. Martin, M. A. Aizen, et al. (2019). A global synthesis reveals biodiversity-mediated benefits for crop production. *Science Advances* 5:eaax0121.

Brazil: L. Flores, A. J. S. Pacheco Filho, C. Westerkamp, et al. (2012). A importância dos habitats naturais no entorno de plantações de cajueiro anão precoce (*Anacardium occidentale* L.) para o sucesso reprodutivo. *Inheringia* 67:189–197.

Basic bee/honeybee facts: R. Winfree, N. M. Williams, J. Dushoff, et al. (2007). Native bees provide insurance against ongoing honey bee losses. *Ecology Letters* 10:1105–1113.

Seven crops: J. R. Reilly, D. R. Artz, D. Biddinger, et al. (2020). Crop production in the USA is frequently limited by a lack of pollinators. *Proceedings of the Royal Society B* 287:20200922.

Bee trucker quotation is from *BBC*, news.bbc.co.uk/2/hi/science/nature/7925397.stm (accessed December 2016).

Colony collapse: F. Sánchez-Bayo, D. Goulson, F. Pennacchio, et al. (2016). Are bee diseases linked to pesticides? A brief review. *Environment International* 89–90:7–11; G. Degrandi-Hoffman, G. Graham, F. Ahumada, et al. (2019). The economics of honey bee (*Hymenoptera: Apidae*) management and overwintering strategies for colonies used to pollinate almonds. *Journal of Economic Entomology* 112:2524–2533.

Oysters: J. B. C. Jackson, M. X. Kirby, W. H. Berger, et al. (2001). Historical overfishing and the recent collapse of coastal ecosystems. *Science* 293:629–638.

Ali quote: Ali, S. 2007. *The Fall of a Sparrow.* Oxford: Oxford University Press.

China timber trade: See J. Saunders and M. Norman. (2020). *Conflict, Fragility, and Global Trade in High-Risk Timber.* https://www.forest-trends.org/publications/conflict-fragility-and-global-trade-in-high-risk-timber and *The Rosewood Racket.* https://eia-global.org/reports/the-rosewood-racket for discussion about Nigeria, and translation of trees into tonnes.

Poison dart frog, including ABT-594 quote: E. O. Wilson. (2002). *The future of life* New York: Alfred A. Knopf, pp. 121–122; Deriving a non-opiate painkiller [ABT-594] from Epipedobates tricolor. See also *Mongabay.* http://rainforests.mongabay.com/05epidatidine.htm.

Florida grasshopper sparrow: The Williams quotation comes from P. Marra and C. Santella. (2016). *Cat wars: The devastating consequences of a cuddly killer*. Princeton, NJ: Princeton University Press, p. 170.

Ascension Island: D. M. Wilkinson. (2004). The parable of Green Mountain: Ascension Island, ecosystem construction and ecological fitting. *Journal of Biogeography* 31:1–4.

Subject Index

For the benefit of digital users, indexed terms that span two pages (e.g., 52–53) may, on occasion, appear on only one of those pages.

Tables and figures are indicated by *t* and *f* following the page number

acid rain, 128–29, 180
acquired immune system, 67
Africa, 25–26, 48, 53–54, 57–59, 71, 110–11, 115, 117*f*, 162*f*, 204, 205, 207*f*, 208*f*, 211, 211*f*, 212*f*, 249*f*–50, 280*f*
age and area hypothesis, 277
agriculture, origins, 110–11
AIDS, 17–19, 18*f*, 58*t*, 64, 71, 114, 194
Alaska, 29, 46–47, 47*f*, 48, 229–30, 230*f*, 231
albedo, 268
Ali, S., 298–99
alien species, 185*f*, 186, 187–88, 191, 192*f*
Amazonia, 130, 138–39, 140, 147–50, 149*t*, 160–61, 162*f*, 163, 206, 207*f*, 210, 213–14, 273–74, 277, 279–80, 291, 301–2
America, North, 2, 46, 123*f*, 126*f*, 158, 163–64, 170, 186, 188, 189, 190*f*, 193, 196, 197–98, 249*f*–50, 251, 254–58, 255*f*, 256*f*, 257*f*, 270, 274*f*, 276*f*
America, South, 41–42, 139–40, 205, 281
anoxia (anoxic), 178
ant, 54, 289–90, 291*f*
Anthropocene, 2, 267*f*
antibiotics, 298–99
antibody, 67
antigen, 67
Archaea, 243–44
artificial selection, 89–90, 90*f*
Atlantic forest, 281
atrazine, 77–78, 180–81
Australia, 68–69, 110–11, 162–63, 164–65, 185*f*, 186, 202, 235, 263, 266, 269–70
AZT, 65, 66*f*, 67

B cells, 67
baboon, 48
baboonification, 48
barnacles, 36–37, 36*f*, 50–51, 53–54
bat, 59, 194–95, 201–2, 249*f*–50, 276

beer, 75, 110–11
Beijing, 182–83
bicarbonate, 159
biodiversity hotspots, 273, 281–82
biofuel, 75, 167*f*, 171–74, 172*t*, 174*f*
Bioko, 207–8, 208*f*
biological control, 80–81, 82, 186, 187
biological species, 239–43, 240*f*, 241*f*, 245–47, 246*f*
biomagnification, 176, 181–82, 182*f*
birth rate, defined, 10*t*, 11
Blue Economy Challenge, 235
Borneo, 140–41, 141*f*, 210–11, 214, 283
bottom up control, 45–46, 268
Brazil, 138–39, 140*f*, 174*f*, 201, 281, 291, 296–97
Bruegel, P., 90*f*
brussels sprouts, 89
Bt, 79–80, 91–92
bushmeat, 206, 208*f*, 213, 214
butterfly effect, 101
bycatch, 220, 228, 231–32

Cambodia, 208
canyon study, California, 286–88, 290*f*
carbon emissions, 162*f*
carbonic acid, 161–62
Carson, R., 107, 182–83
cash crops, 92, 143
cattle ranching, 145–47, 146*f*, 281
cement, 160–61, 168*f*
Centers for Disease Control, 102, 256
cercaria larva, 57, 58*f*
chaos, 101
Chesapeake Bay, 298
China, 26, 86, 91, 116–17, 116*f*, 118*f*, 119, 124*f*, 126*f*, 137*f*, 168*f*, 182–83, 186–87, 194–95, 232–34, 279–80, 292–93, 300–1, 301*f*
chiton, 50–51
chlorofluorocarbons (CFCs), 107

332   SUBJECT INDEX

cholera, 57, 64, 72–73
cholera, el tor, 73
citizen science, 235
Clean Air Act (USA), 128–29, 173–74, 182–83
Clean Water Act, 128–29, 173–74, 182–83
climate velocity, 169–70
clouds, 103–4, 140, 158, 200
coal, 130, 160–61, 171–72
Cocoliztli, 196f
coexistence, intermediate disturbance, 40–41
coexistence, neutral theory, 41
coffee, shade, 152, 228
Cohen, J., 112–13, 114
Collapse, 114
common cold (cold), 64, 72, 194–95
competition, exploitative, 23, 33
competition, interference, 23, 33
competition, interspecific, 32f, 33
competition, intraspecific, 22–23, 23f, 26f
confidence limit, 106–7, 255f
Congo, 138–39, 142, 206–7, 213, 280f
Conservation International, 281
consumer surplus, 145–47
Convention on International Trade in
    Endangered Species (CITES), 209–10, 299,
    300–1
COPHID, 4, 5–6, 5f, 132, 248, 289, 294
coral reef, 170–71, 221, 264–65, 266f
coronavirus, vii, 19, 194–95
corridors, 292
COVID-19, 57–59, 83, 113, 114, 168f, 194–95
critically endangered species, 199f, 249f–50,
    258, 281–82
cropland, 135t
cryptic species, 245–46, 246f

DALY, 57–59
Darwin, C., 50–51
Darwin's finches, 34, 35f, 39–40, 41–42, 69,
    189–91
DDT, 79–80, 107, 181–82
death rate, defined, 10t, 11
delta variant (δ), 194–95
demand function, 146–47, 146f
demographic transition, 113, 113t, 114–19,
    118f, 120f, 129–30
density dependence, 22–24, 23f, 30
density independence, 22–24
density, delayed dependence, 27
desertification, 84–85
developmental abnormalities, 179f, 180
diagnostic traits, 239, 240–42
Diamond, J., 114

diarrheal disease, 58t, 72–73
discount rate, 147–50, 148f, 149t, 151t, 153–54
disease, density-dependent transmission, 55–56
disease, frequency-dependent transmission,
    55–56
DNA sequencing, conservation tool, 234–35
DNA sequencing, phylogeny construction, 243
DNA sequencing, species delimitation, 240–43
Dobzhansky, T., 64
domestic cats, 289, 290f
doubling time, 16, 18, 61–62, 112–13, 114,
    123–24, 130–31
drought, 57, 158, 163
dust bowls, 158

ectoparasite, 52, 188
Ehrlich, P., 107
el Niño, 104–6, 163
emergent infectious disease (EID), 194, 195
Endangered Species Act, 247, 298–99
endoparasite, 52
Environmental Protection Agency (EPA), 128–29
epidemic, 49f, 55f, 62f, 68–69, 68f, 71, 102, 194,
    196, 200, 201f
eternal external threat, 290
Europe, 79, 113, 163–64, 178–79, 180, 185f,
    251, 270
Exclusive Economic Zone, 229
exponential growth, 10t, 13–16, 13f, 21, 24, 101,
    112–13
exponential, negative, 258, 261–62, 262f
externality, 150–51, 152–53, 227

Faith's Phylogenetic Diversity, 279–80, 280f
feedback, negative, 5f, 101, 103–4
feedback, positive, 32–33, 101, 103–4, 163
fertility rate, 114–17, 116f, 117f, 118f, 119
fertilizer, 84, 87–90, 87f, 88f, 92, 134, 136, 144,
    176, 177–78, 182–83
fire, 133–35
fisheries, maximum sustainable yield in, 25f, 27
flightless rails, extinction of, 268–69
fluke (flatworm), 54, 57, 58f, 180–81, 270–71
Food and Agriculture Organization (FAO), 75,
    83, 84–85, 84f, 125f, 134, 135, 136, 138–39,
    229
food chain, 43–48, 176, 181, 182f, 189–91,
    216–17
food security, defined, 83
food web, 48–50, 50f, 51, 57, 58f, 180–81, 182f,
    195, 204, 213–14, 224–25
fossil, 155–56, 217, 221, 224–25, 261, 262f, 266f,
    267f, 269, 283

fracking, 124–25, 173–74
fungal diseases, 79, 163–64, 194, 195, 197–98, 200–2, 203*f*, 204
fungicide, 78–79, 201, 204

Gabon, 206, 210, 211*f*, 249*f*–50
Galápagos, 34, 35*f*, 38–39, 41–42, 189–91, 284*f*
Gates foundation, 89–90
Gause, W., 33–34
genetically modified food (GM), 88*f*, 91–92
geometric growth, 11, 13–14, 15, 16, 19–20
geometric series, 148–49
George's Bank, 28, 29, 29*f*, 30, 218*f*, 219–20
Ghana, 252–53, 291–92
ghost fishing, 220
Global Footprint Network, 135
globalization, 126*f*
goatscapes, 227
Google Earth, 215
grazing land, 133–34, 135, 136
Great Barrier Reef, 147, 171
great elephant survey, 211
green mountain (Ascension Island), 300
green revolution, 89–90, 107
greenhouse gas, 2, 3*t*, 103–4, 106–7, 158–62, 172*t*, 173–74, 268
Grinnell, J., 33–34, 38–39
Gross Domestic Product, 121–22
Guam, extinction on, 189–91

harlequin toad, 198, 200, 201*f*
Harry Potter, 208
Hawaii, 37–38, 160, 161*f*, 187–88, 193, 216–17, 269
heat waves, vii, 4, 162–63, 167–69
herbicide, 76–78, 82, 84, 88–89, 91–92, 176, 180–81
herd immunity, 59–61
human immunodeficiency virus (HIV), 17–19, 18*f*, 59, 65, 66, 66*f*, 67, 71, 114, 194
How many people can the earth support?, 112–13
Human Appropriated Net Primary Productivity (HANPP), 132–35, 133*f*
hybrid, 39, 83, 90–91, 110–11, 202
hydrothermal vents, 245–46
hypoxia, 178

ice sheet, 155–56, 156*f*, 162–63, 278–79
immune system, 64, 65, 66–67, 70, 194–95
India, 37, 89–90, 92, 107, 124*f*, 126*f*, 131, 134–35, 137*f*, 143, 147, 150, 168*f*, 176, 183, 185*f*, 193, 194–95, 249*f*–50, 252–53, 253*f*, 292–93

Indonesia, 110–11, 134, 141*f*, 174*f*, 208, 215, 249*f*–50, 279–80, 282, 301–2
Inflation, 117*f*, 121–22, 123*f*, 148
influenza, 59, 61–62, 62*f*, 66–67, 113–14, 196, 197–98
innate immune system, 66–67
intercropping, 84–85
interglacials, 2, 156–57, 167–69, 170–71, 264
Intergovernmental Panel on Climate Change (IPCC), 97–98, 100, 103, 104, 106–7, 166, 167*f*, 169
International Commission for the Conservation of Atlantic Tunas (ICCAT), 231–32
International Union for the Conservation of Nature (IUCN), 199*f*, 209, 241*f*, 248, 258, 260, 262–63, 270, 275*f*, 281–82
intrinsic rate of increase, *r*, 10*t*, 24, 28
invasive species act, 192
invasive species, defined, 184
Irish potato famine, 75, 78
irrigation, 84–85, 133–35
isotope, 2, 155–56, 156*f*, 159, 160
ivory trade, 205, 206

Janzen, D., 246*f*, 290
Jevons, W., 123–24, 127–28, 171
Jevons' paradox, 127–28, 171

kelp forest, 46–47, 47*f*, 48, 49*f*, 50–51, 220–21

la Niña, 104–6
Lambir Hills National Park, 214, 249*f*–50
land sharing, 253–54, 253*f*
land sparing, 253–54, 253*f*
leishmaniasis, 194
limpet, 50–51
Line Islands, 222*f*, 232–34
Living Planet, 134–35, 248–51
locust, 26, 27, 28
logistic growth, 24–27, 25*f*, 28–29, 33–34, 34*f*, 39–40, 212–13, 218–19
Lynas, M., 169

MacArthur, R. H., 285*f*
MacDonald's, 126*f*
Madagascar, 207*f*, 266–67
malaria, 57, 58*t*, 64, 71–72, 113, 167–69, 181–82
Malthus, T., 19–20, 113–14
mangrove, 150–52, 151*t*
Marine Stewardship Council (MSC), 228
Mass extinction, 1, 2, 224–25, 264
maximum sustainable yield (MSY), 27–29, 29*f*, 212–13, 218*f*, 226, 227–28, 229, 294

334 SUBJECT INDEX

Mayr, E., 240f
measles, 25–26, 57–59, 58t, 60–61, 60f, 66–67, 194, 196, 197
medieval warm period, 158
merit release, 208
methane, 158–59, 173–74, 176, 268
Mexico, 196, 196f
moas, extinction of, 269
model, defined, 9
monoculture, 78, 79, 84, 200–1
morphological species, 241f, 243
Mount Pinatubo, 104
mussel, 50–51, 185f, 191, 192
mutualism, 1, 32–33, 32f
Myxomatosis, 68–69, 68f

National Park, Corbett, 131
National Park, Gunung Palung, 140–41, 141f, 173, 302f
National Park, Yellowstone, 131
natural selection, 64–65, 67, 69, 89, 115–16, 194–95
negative discounting, 144
nematode, 52, 53, 55
neonicotinoids, 257
nestedness, 287–88, 292
Net Primary Productivity (NPP), 132–35, 133f, 136, 273–74, 296
New Guinea, 141, 189–91, 281, 284f, 286, 300–1
New York's water supply, 144
niche, 31, 38–40, 41, 50, 269, 274–76, 277
nitrogen, as a limit on plant growth, 87, 160–61
nitrogen, as a pollutant, 177–78, 179
nitrogen, reactive, 32–33, 37–38
no-use value, 145
North American Breeding Bird Survey, 235, 254–58

Obama, B., 192
omicron variant, 194–95
opportunity cost, 150, 151t, 154f
organic farming, 79–80, 152, 228
ozone, 107, 176, 178–79

Paleocene-Eocene Thermal Maximum (PETM), 170
Paraguay, 153, 154f
Paramecium, 33–34, 40
path diagram, 5, 43, 44f, 52
Pauly, D., 215
Polychlorinated biphenyls (PCBs), 181, 182f
per capita growth rate, r, 10t, 12, 13f, 14, 24, 27, 28–29, 61–62, 111f, 187–88, 216–17
pescatarian, 86

pet trade, 186, 189, 199f, 208, 209–10, 232
phylogenetic species, 241f, 242–43
plantation, 136, 140–41, 173, 251, 252f, 281, 299
Pleistocene, 2, 262f, 265f, 266f, 267f, 270–71, 277, 278f
poison dart frog, 299
pollination services, 295f
Population bomb 107
population projections, 119
population pyramid, 118f
population size, humans, 12t, 155
precipitation, predicted changes, 168f
probability, 97–98
productivity gradient, 276
productivity, net primary, 132, 295–96
Purchasing Power Parity (PPP), 121–22

quotas, 229–32, 234–35

range limits, 38f, 163–64, 254
red list, 199f, 258
Reducing Emissions from Deforestation and Forest Degradation in Developing Countries (REDD), 152–53
renewables, 128, 166, 174–75
reproductive isolation, 239–40
reproductive value, 59, 61–62, 194–95
reservoirs, disease, resistance, drug, 64–65
restoration, 143, 226
rice blast, 79
risk aversion, 109
risk prone, 109
Roberts, C., 100, 124
rosewoods, 300–1
roundup, 77–78, 78f, 91–92
Rwanda, 114, 301–2

salmon, 86, 218f, 229, 247, 298
savingnature.org, 281
sea level, 156f, 162–63, 166–69, 264–65, 265f, 286
sea urchins, 46, 47f, 48, 49f, 221
susceptible-exposed-infectious-recovered (SEIR) model, 62, 197–98, 203f
Serengeti, 25–26, 26f, 200
shark, 44, 215, 216–17, 218f, 220, 222f, 234–35
shrimp farming, 86, 150–52
Silent Spring, 107, 182–83
single large or several small reserves (SLOSS)?, 292
sixth mass extinction, 261, 271, 272
snakes, predation by, 189–91, 289
solar, 128, 171–72, 174–75
species durations, 261–62, 263

species-area relationships, 283, 284*f*
spillovers, 194
statistical noise, 100, 103, 104–6
stem rust, 79
Stern Report, 128
structural uncertainty, 98, 99*f*, 101–3, 104, 166
subsistence, 143, 206–8, 209*f*, 253*f*
subsistence, hunting, 206, 209*f*
subspecies, 240–42, 241*f*, 243, 299–300
syphilis, 72, 102

T cells, 67
technology transfer, 87–88
technology, agriculture, 88–90
technology, conservation, 127–29, 234–35
technology, harvesting, 207–8, 215
Thailand, 86, 150–52, 151*t*
Tilman, D., 88*f*, 174*f*, 295*f*, 296
time series, 210, 211, 216, 217, 229
tolerance vs. resistance, 197, 199
tongue worm, 53*f*
top-down control, 45–46, 268
tragedy of the commons, 227–28, 230*f*
tragedy of the tragedy of the commons, 231
travel costs, 145–47, 146*f*
tree diversity, 274*f*, 277–78
trip-generating function, 145–47, 146*f*
trophic cascade, 48, 220–21, 268, 289, 295–96, 298
tropical rainforest, 139–40, 139*f*, 273–74, 275–76
tuberculosis, 17, 58*t*, 66

urbanization, 59, 114, 129–30
utility curve, 109

vaccines, 37, 59, 60–61, 60*f*, 67, 194–95, 197–98
value uncertainty, 98–101, 99*f*, 103–6, 166
vegetarian, 85–86, 85*f*
virulence, evolution of, 69, 71–73
volcanic eruptions, 37–38, 98, 104, 264
vultures, 176, 249*f*–50

Wallace, A. R., 277
war, 19–20, 97, 113–14
water flea, 53–54
weed, 75, 76–78, 76*f*, 78*f*, 88–89, 91–92, 180, 252–53
West Nile virus, 256, 257*f*
wetlands, 138, 180–81
whales, 46–47, 123–24, 217–18, 218*f*, 219*f*
whaling, 47, 123–24, 217, 218–19, 219*f*
wheat blast, 79, 201
white blood cell, 65, 66–67
white nose syndrome, 197–98, 198*f*, 204, 256
Williams, T., 299–300
Wilson, E. O., 4, 5–6, 179*f*, 302
wind, 128, 171–72, 174–75
World Health Organization (WHO), 53–54, 57–59, 181–82

yield gap, 87–88, 87*f*
younger dryas, 268

zoonose, 194

Species Index

For the benefit of digital users, indexed terms that span two pages (e.g., 52–53) may, on occasion, appear on only one of those pages.

Tables and figures are indicated by *t* and *f* following the page number

Acropora coral, 221
African leopard, *Panthera pardus,* 206–7
agouti, *Dasyprocta* spp., 209*f*
Agrobacterium tumefacians, 91
American coot, *Fulica americana,* 58*f*
American lobster, *Homarus americanus,* 235
Argentine ant, *Linepithema humile,* 289–90, 291*f*
Asian longhorned beetle, *Anoplophora glabripennis,* 186–87

Bacillus thurigiensis, 79–80, 91
Balanus balanoides, 36*f*
bald eagle, *Haliaeetus leucocephalus,* 79–80, 107, 181–82
Batrachochytrium dendrobatidis, 198
Bewick's wren, *Thryomanes bewickii,* 288*f*
blue duiker, *Philantomba monticola,* 211*f*, 212–13
bonobo, *Pan paniscus,* 243–44, 244*f*, 263, 279–80, 280*f*
Bougainville monarch, *Monarcha erythrostictus,* 241*f*
brown tree snake, *Boiga irregularis,* 189–91, 269
Burmese python, *Python bivittatus,* 189, 249*f*–50

cabbage, *Brassica oleracea,* 89
cactus finch, *Geospiza scandens,* 39
California killifish, *Fundulus parvipinnis,* 58*f*
California quail, *Callipepla californica,* 288*f*
California thrasher, *oxostoma redivivum,* 288*f*
calthrop, *Tribulus cistoides,* 35*f*
cane toad, *Rhinella marina,* 186
cannabis, *Cannabis sativa,* 193
Caribbean monk seal, *Neomonachus tropicalis,* 224–25
Carolina parakeet, *Conuropsis carolinensis,* 269–70
chestnut-bellied monarch, *Monarcha castaneiventris,* 241*f*

chicken, *Gallus domesticus,* ii, 31, 85–87, 122, 124–25, 144, 206
chili pepper, *Capsicum* spp., 79
chimpanzee, *Pan troglodytes,* 17, 59, 110, 243–44, 244*f*, 245–46, 280*f*
Chthamalus stellatus, 36*f*
cod, *Gadus morhua,* 28, 29, 217, 218*f*, 219–20
coffee, *Coffea* spp., 143, 152, 228, 281, 296–97
common barberry, *Berberis vulgaris,* 79
common vetch, *Vicia sativa,* 77*f*
corn rootworm, *Diabrotica* spp., 79–80, 80*f*, 81*f*
corn, *Zea mays,* 75, 77, 79, 88–89, 89*f*, 91–92, 172*t*, 173
cotton, *Gossypium* spp., 79–80, 91–92
cottony cushion scale, *Icerya purchasi,* 80–81, 188
cownose ray, *Rhinoptera bonasus,* 220–21
coyote, *Canis latrans,* 144, 289, 290*f*, 292
crested mona monkey, *Cercopithecus pogonias,* 211*f*
curious skink, *Carlia ailanpalai,* 191

Diprotodon octatum, 3*f*
dune fescue, *Vulpia fasciculate,* 23*f*

eastern Atlantic bluefin tuna, *Thunnus thynnus,* 231–32
emerald ash borer, *Agrilus planipennis,* 185*f*, 191, 192*f*
Epipedobates anthonyi, 299, 302*f*
Eurasian cuckoo, *Cuculus canorus,* 53*f*
European rabbit, *Oryctolagus cuniculus,* 31–32, 50–51, 68–69, 68*f*, 189

fayatree, *Myrica faya,* 37–38, 187–88
Florida grasshopper sparrow, *Ammodramus savannarum,* 299–300
forest elephant, *Loxodonta cyclotis,* 206–7

giant trevally, *Caranx ignobilis,* 216*f*
ginger, *Zingiber officinale,* 79
golden spectacled warbler, *Seicercus* spp., 246*f*

338 SPECIES INDEX

gorilla, *Gorilla* spp., 206–7, 243–44, 244*f*, 279–80, 280*f*, 301–2
gray fox, *Urocyon cinereoargenteus,* 189
gray heron, *Ardea cinerea,* 21–22, 22*f*, 23–24, 28, 31
gray-cheeked mangabey, *Lophocebus albigena,* 211*f*
great blue heron, *Ardea herodias,* 57, 58*f*
greater spot-nosed monkey, *Cercopithecus nictitans,* 211*f*
greenish warbler, *Phylloscopus trochiloides,* 37, 38*f*
grey steppe cattle, 200
ground ivy, *Glechoma hederacea,* 32*f*, 33
Guam swiftlet, *Aerodramus bartschi,* 189–91
guinea worm, *Dracunculus medinensis,* 53–54

hairy nightshade, *Solanum sarrachoides,* 78–79
harlequin poison frog, *Oophaga histrionica,* 199*f*
herring, *Clupea harengus,* 28, 219–20, 229–31, 230*f*
honey bee, *Apis mellifera,* 297
horned lark, *Eremophila alpestris,* 255*f*
horned toad, *Phrynosoma,* 286

Japanese knotweed, *Reynoutria japonica,* 191

killer whale, *Orcinus orca,* 46–47, 47*f*, 220–21, 298

lady beetle, *Rodolia cardinalis,* 188
large ground finch, *Geospiza magnirostris,* 34, 35*f*, 38–39, 41–42
lentil, *Lens culinaris,* 77*f*
lion, *Panthera leo,* 48, 266–67
little brown bat, *Myotis lucifugus,* 249*f*–50
long-billed vulture, *Gyps indicus,* 249*f*–50

macaque, *Macaca* spp., 72
medium ground finch, *Geospiza fortis,* 34, 35*f*, 38–39, 41–42
monarch butterfly, *Danaus plexippus,* ix, 251

North Atlantic right whale, *Eubalaena glacialis,* 235
northern long-eared bat, *Myotis septentrionalis,* oil palm, 92, 136, 138–39, 140–41, 143, 173, 249*f*–50, 301–2

olinguito, *Bassaricyon neblina,* 244–45, 279*f*
orangutan, *Pongo* spp., 140–41, 243, 244*f*, 263
oyster, 86, 295, 298–99

Palaeolama mirifica, 267*f*
Panamanian golden toad, *Atelopus zeteki,* 199*f*, 201*f*
pangolin, 205, 206, 209, 300–1
Paramecium aurelia, 33–34, 34*f*, 40
Paramecium caudatum, 33–34, 34*f*, 40
passenger pigeon, *Ectopistes migratorius,* 269–71
peregrine falcon, *Falco peregrinus,* 79–80, 300
pickleweed, *Salicornia virginica,* 58*f*
Plasmodium falciparum, 71–72
Plasmodium knowlesi, 72
Plasmodium malariae, 71–72
Plasmodium ovale, 71–72
Plasmodium vivax, 71–72
Pseudogymnoascus destructans, 197–98

racoon, *Procyon lotor,* 249*f*–50, 279*f*
rapeseed, *Brassica napus,* 92
red grouse, *Lagopus lagopus,* 55–56, 56*f*, 59–60
red-brocket deer, *Mazama americana,* 209*f*
red-eyed vireo, *Vireo olivaceus,* 254–55, 256, 257*f*
Rhinoceros, *Rhinoceros* spp., 210–11, 300–1
rice, *Oryza sativa,* 75, 76–77, 76*f*, 79, 89–90
rinderpest, *Rinderpest morbillivirus,* 25–26, 194, 199–200, 204
rose-ringed parakeet, *Psittacula krameri,* 185*f*

sabre toothed cat, 267*f*
saola, *Pseudoryx nghetinhensis,* 244–45
savannah elephant, *Loxodonta africana,* 211, 212*f*
scallops, 86–87, 220–21, 264–65, 266*f*
sea otter, *Enhydra lutris,* 46–47, 47*f*
seagrass, 221
sheep, *Ovis aries,* 50–51, 54, 227–28, 227*t*
Shigella, 73*f*
skipper butterflies, 246*f*
southern right whale, *Eubalaena australis,* 218–19, 219*f*, 235
soybean, *Glycine max,* 32–33, 75, 79, 201–2
spider monkey, *Ateles* spp., 209*f*, 213–14
spiny lobster, *Panulirus interruptus,* 48–50
spotted towhee, *Pipilo maculatus,* 288*f*
starling, *Sturnus vulgaris,* 186, 188, 263
Steller's sea cow, *Hydrodamalis gigas,* 224–25
striped hyena, *Hyaena hyaena,* 204
striped shore crab, *Pachygrapsus crassipes,* 58*f*
Swainson's thrush, *Catharus ustulatus,* 256, 257*f*

tapir, *Tapirus terrestris,* 209*f*, 213–14
tea viburnum, *Viburnum setigerum,* 190*f*

SPECIES INDEX 339

tiger, *Panthera tigris,* 107, 210–11, 243–44, 248–51, 292–93

tricolored bat, *Perimyotis subflavus,* turtle grass, *Thalassia testudinum,* 221

variable antshrike, *Thamnophilus caerulescens,* 164f

viburnum leaf beetle, *Pyrrhalta viburni,* 189, 190f

viburnum, *Viburnum* spp., 189, 190f

whale shark, *Rhincodon typus,* 235

wheat, *Triticum* spp., 75, 79, 89–90, 90f, 201

white-backed vulture, *Gyps africanus,* 249f–50

white-capped monarch, *Monarcha richardsii,* 241f

white-tailed deer, *Odocoileus virginianus,* 52, 189

wild dog, *Lycaon pictus,* 204

wild strawberry, *Fragaria vesca,* 32f, 33

wildebeest, *Connochaetes taurinus,* 25–26, 26f, 28, 200, 204

woodpecker finch, *Camarhynchus pallidus,* 41–42

woolly monkey, *Lagothrix* spp., 213–14

wren tit, *Chamaea fasciata,* 287–88, 288f

yellow-browed warbler, *Phylloscopus humei,* 37, 38f

zebra mussel, *Dreissena polymorpha,* 185f, 191, 192